BACTERIAL DISEASES OF PLANTS AND THEIR MANAGEMENT

— *A Textbook* —

THE AUTHORS

Dr. M.R. Ravikumar is working as Professor of Plant Pathology (Specialization in Plant Bacteriology) at College of Agriculture, Hanumanamatti (UAS, Dharwad) for the last 25 years. He has published 25 International and 75 National research papers in reputed International and National journals. He has presented more than 75 research papers in International/National symposiums/Conferences, One book in English and 10 Books in kannada and more than 50 Popular articals in both Kannada and English.

Prof. Shamarao Jahagirdar, serving as Professor(Plant Pathology) and Chief Editor, Publication Center, UAS, Dharwad and also served as Former Dean and Head of Institute, PAJANCOA and RI, Karaikal (Puducherry UT) and also National Principal Investigator (Soybean Pathology), IISR, Indore. He was a post doctoral fellow at National Institute of Agricultural Science and Technology, South Korea (2010). Prof. Jahagirdar has published 150 papers in journals,100 papers in conferences, deposited 30 sequences of Colletrotrichum trancatum causing anthracnose of soybean in Blast sequence. 50 articles and 5 books (both in English and Kannada), 5 chapters in books and 12 reports.

Dr. Dinesh Singh (Principal Scientist), ICAR- IARI, New Delhi born in Mirzapur, Uttar Pradesh on 23 February 1968. He joined ARS services in January 20, 1997 at IIVR, Varanasi as Scientist. He has guided 7 Ph. D. and 2 M. Sc. students in the field of plant bacteriology. He has published 162 research papers in pear reviewed reputed journals, 44 popular articles, 32 books, 100 book chapters, 5 technical bulletins, 9 training manuals and 2 practical manuals and 1 extension bulletin.

Dr. K. Dinesh, Ph.D. (Plant Pathology) is working as Assistant Professor in department of Plant Pathology, College of Horticulture, Anantharajupeta. He has completed his B.Sc. from Dr. YSR Horticultural University, Venkataramannagudem, A.P. in 2016 and M.Sc. (Plant Pathology) in 2018 from Central Agricultural University, Imphal (Manipur) and Secured 4th (AIR) in ICAR-JRF/SRF(PGS)-2018 and completed his Doctoral degree from University of Agricultural Sciences, Dharwad, Karnataka in 2021. He is the author/co-author of 12 research articles and 6 book chapters in reputed journals like Indian Phytopathology, Folia Microbiologica and other esteemed publishers like Springer, Elsevier, CRC press and IOP publishers.

BACTERIAL DISEASES OF PLANTS AND THEIR MANAGEMENT

— *A Textbook* —

— Authors —

M.R. Ravikumar

Shamarao Jahagirdar

Dinesh Singh

K. Dinesh

2023

Daya Publishing House®

A Division of

Astral International Pvt. Ltd.

New Delhi – 110 002

ISBN: 9789359190945 (Int. Edition)

Published by : **Daya Publishing House®**
A Division of
Astral International Pvt. Ltd.
– ISO 9001:2015 Certified Company –
4736/23, Ansari Road, Darya Ganj
New Delhi-110 002
Ph. 011-43549197, 23278134
E-mail: info@astralint.com
Website: www.astralint.com

Printed at : **Replika Press Pvt. Ltd.**

University of Agricultural Sciences, Dharwad-580 005, Karnataka, India

Dr. R. Basavarajappa
Vice Chancellor, UAS, Dharwad

Foreword

Plant pathogens are ubiquitous in nature, causing crop diseases, affecting yield of the crops, through-out the globe under changing climate. The nature of infestation varies from mild symptoms to catastrophes destroying large area both locally and globally. The plant pathogens are difficult to manage with a single window of application method because of diversity of the pathogen in terms of races pathovars *etc.*, either single or group of pathogens involve in causing disease complex. Diversity of races, evolution of new population of pathogens are the major stumbling blocks in IDM. Major food crops are subject to several diseases starting from the nursery of field stage to postharvest apart from protected cultivation, which affect both the quantity and quality of the produce, incurring loss to the growers. These biotic constraints along with several biotic factors can, at times, seriously compromise food security. In order to address these issues in a holistic way the author felt the need for technology and resource compilation from various sources, to present under basic and emerging facts of plant bacteriology. The book entitled "Bacterial Diseases of Crop Plants and their Management" authored by Dr. M.R. Ravikumar, Dr. Shamarao Jahagirdar, Dr. Dinesh Singh and Dr. K. Dinesh is an attempt towards achieving this objective. This book covers the basic and applied strategies such as diseases of economically important crops such as fruit crops, field crops, medicinal crops, aromatic crops and ornamental crops, morphology, structure and composition of bacterium, isolation of bacterium, classification, nutrition and genetics of bacteria and integrated management of bacterial disease. I congratulate the authors for providing the fundamentals and applied details of plant bacterial diseases. I am sure this book serves as a ready reckoner's for students, teachers, extension workers, researchers and all other stakeholders in discipline of plant pathology in particular and agriculture in general.

Dr. R. Basavarajappa
Vice Chancellor, UAS, Dharwad

Preface

This would be a very long preface if we were to set down in detail the reasons why we attempted to write this book in present form. Suffice it to say that text is based on courses which we have been teaching to undergraduate and postgraduate students. The purpose is to provide the basic and emerging facts whereby the students may be introduced to the scientific foundations of this noble subject called Plant Bacteriology.

Bacterial pathogens are serious threat to the economically important, crop plants causing many diseases and more than 10 per cent of global food production is lost due to these bacterial diseases. The losses caused by diseases in plants can be minimised by correct diagnosis and by adopting integrated management approaches.

The diseases of economically important crops such field crops, fruits crops, medicinal crops, aromatic crops and ornamental crops have been covered, describing their losses, symptoms, etiology, disease cycle, epidemiology and Integrated management practices.

Up-to-date and recent classification bacterial pathogens, morphology, structure and composition of bacterium,nutrition and genetics of bacterium and integrated diseases management are included in the book. A special feature of this book is as it contains many high quality coloured photographs of disease symptoms that illustrate most of the diseases which will help in the identification and understanding them.

M.R. Ravikumar

Shamarao Jahagirdar

Dinesh Singh

K. Dinesh

Contents

	Foreword	*v*
	Preface	*vii*
1.	**Introduction**	**1**
2.	**Morphology, Structure and Composition**	**7**
3.	**Classification of Bacteria**	**15**
4.	**Isolation and Characterization**	**33**
5.	**Nutrition and Growth**	**51**
6.	**Survival and Dispersal of Bacteria**	**61**
7.	**Diagnosis of Bacterial Diseases**	**65**
8.	**Management of Bacterial Diseases**	**97**
9.	**Bacterial Diseases of Fruit Crops**	**115**
10.	**Bacterial Diseases of Vegetable Crops**	**149**
11.	**Bacterial Diseases of Field Crops**	**211**
12.	**Bacterial Diseases of Spices, Medicinal and Aromatic Crops**	**231**
	References	**255**
	Index	**279**
	Colour Plates	**I-VII**

Chapter 1

Introduction

Declining crop productivity due to unsuitable agricultural practices over the years and an alarming rate of population growth have both put up a severe strain on the food supply situation in the world. Pertaining to massive population pressure, increase in food grain production is an uphill task in today's world. The need of the day is sustainable agriculture without harming the delicate balance of soil ecology. For achieving this, a clear understanding about the microbes which are the cause of losses that is plant pathogens and proper mitigation of them by various approaches is necessary. Integrated disease management is becoming the most viable option for management of disease. Among these plant pathogens bacterial pathogens are gaining upper hand in recent scenario, this might be due to changing climatic conditions and other agricultural practices like monoculture of crops. In natural environments there is typically a great diversity of species within the plant community and a corresponding diversity of plant pathogens, all at relatively low level. The destruction of such communities and the development by man of large-scale cultivation of single crops has led to the selection of particular crop-related pathogens and major outbreaks of bacterial diseases being followed. With many bacterial diseases, the region and date of origin (or initial observation) can be historically defined and the geographical spread of the pathogen within a particular crop documented. A good example of this is provided by bacterial leaf blight of rice, caused by *Xanthomonas oryzae* pv. *oryzae* and one of the most destructive diseases of rice in Asia. This disease was initially observed in Japan in 1884. By 1908, it was common in the south-west part of the country and in 1926 the disease was recorded in the northwest. After 1950, the disease increased markedly and by 1960 was known to occur throughout Japan, with the exception of the island of Hokkaido. In India, bacterial leaf blight was initially recorded in the

Bombay state in 1951 and has since become widespread throughout the country with the introduction and cultivation of new high yielding but susceptible rice cultivars. Now it became a serious threat for rice production in almost all major rice cultivated states of the country.

Another important reason for this increase in bacterial diseases is the fitness of a bacterial pathogen which can be quantified by measuring its reproductive rate, rate of multiplication, efficiency of infection and the amount of disease caused by its aggressiveness. The presence of excess genes for virulence, however, imposes a fitness penalty upon the pathogen. Therefore, a mutation from a virulence to virulence occurs only if needed to overcome an R gene for resistance; that is, only if the mutation is absolutely necessary for the pathogen's survival. New biotypes of bacteria seem to arise with varying frequency by means of at least three sex-like processes: (i) conjugation, which occurs when two compatible bacteria come into contact and a small portion of the chromosome or plasmid from one is transferred to the other through a conjugation bridge or pilus; (ii) genetic transformation, which occurs when bacterial cells absorb and incorporate genetic material that is secreted by other bacteria or released during their rupture and (iii) transduction, which occurs when a bacterial virus (bacteriophage) transfers genetic material from the bacterium in which the phage was produced to the bacterium, it will infect next (Agrios, 2005).

Bacterial plant diseases are the most frequent and severe in tropical and subtropical places, where warm and humid conditions are ideal for bacterial growth. Indeed, consistent annual crop losses are recorded in all countries. A number of bacterial diseases are of major economic importance with direct financial loss due to decreased production and indirect loss due to the implementation of expensive management strategies. The financial loss in agricultural production typically arises both due to a direct effect on the quality (ex. citrus canker) and quantity (ex. Bacterial leaf blight of rice) of the agricultural product as well as an overall deleterious effect on the plant itself (ex. Wilt of solanaceous crops). Post-harvest damage to stored plant products by bacterial pathogens (particularly soft rot bacteria) is also an area of major economic importance. In general, plant pathogenic bacterial species belonging *to Xanthomonadaceae, Pseudomonadaceae* and *Enterobacteriaceae* families target all types of plants that can supply them with appropriate food and shelter. The most devastating plant pathogens belong to genera such as *Erwinia, Pectobacterium, Pantoea, Dickeya, Agrobacterium, Pseudomonas, Ralstonia, Burkholderia, Acidovorax, Xanthomonas, Clavibacter, Streptomyces, Xylella, Spiroplasma and Phytoplasma* (Kannan and Bastas, 2015).

1. How Prokaryotes can be Distinguished from Eukaryotes

Bacteria are small, microscopic, simple, unicellular microorganisms. The biological properties and predominant reproduction by binary fission relate them in prokaryotes. Majority of bacteria are strictly saprophytes and they are highly beneficial to man. However, several bacterial species are known to cause many dreaded diseases both in human beings and animals like Cholera, Leprosy, Tuberculosis, Plague, Syphilis, Anthrax. Besides these, some bacterial species which generally live in/around the crop plants in which they incite various diseases of economic significance are known as phytopathogenic bacteria.

Eukaryotic Cells

"**Eukaryotic**" means "**true nucleus**". Eukaryotic cells will be characterized by following properties as mentioned below:

- ☆ **Eukaryotic cells possess extensive internal membrane systems:** All these cells will be surrounded by a plasma membrane. In these eukaryotic cells the interior of the cell is filled with additional membranes which carryout specialized tasks.
- ☆ **Eukaryotic cells exhibit internal compartmentalization:** The interior of eukaryotic cell will be divided into different regions or compartments by various membranes.
- ☆ **Eukaryotic cells possess membrane-enclosed cell organelles:** Various compartments within the eukaryotic cell forms structures called cell organelles. Each organelle performs various specific functions such as respiration by mitochondria, photosynthesis by chloroplasts and other functions by other organells.
- ☆ **Eukaryotic DNA is confined within a membrane-bound structure called nucleus:** This allows the eukaryotic cells to have a greater amount of DNA than that of prokaryotic cells, permitting great revolutionary diversity.

Prokaryotic Cells

"**Prokaryotic**" means "**pre-nucleus**". This name reflects the idea that prokaryotes are believed to be the first cell evolved. It is believed that eukaryotic cells are evolved from prokaryotic ancestors in the process of evolution. Prokaryotic cells are much simpler and primitive than eukaryotic cells. Microorganisms having prokaryotic cell forms include the eubacteria ("true bacteria") and the archaebacteria. Prokaryotic cells have been characterized by the following properties:

- ☆ **Prokaryotic cells lack extensive internal membrane systems:** Like eukaryotic cells, prokaryotic cells are also surrounded by a plasma membrane. However, there is very little internal membrane structure. The plasma membrane in some prokaryotes possess internal folds of that are believed to carry out specific functions (respiration, ATP synthesis *etc.*).
- ☆ **Prokaryotic cells lack extensive internal compartmentalization.**

- **Prokaryotic cells lack membrane enclosed organelles:** Specialized regions *i.e.* the internal folds of the plasma membrane in prokaryote performs various functions which are generally carried out by membrane bound organelles in eukaryotes (such as respiration, protein secretion and photosynthesis).
- **Prokaryotic DNA do not have membrane-bound nucleus:** It is confined within a specialized region of the cell termed as "nucleoid". These cells lack nuclear membrane. Prokaryotic cells have much lesser DNA than that of eukaryotic cells.

Eubacteria have Single-Celled or Colonial Arrangement

In almost all the bacterial species, each cell is structurally identical but metabolically independent. For this reason, bacteria generally regarded as "single-celled" organisms. In general, most of the bacteria grow in specific groupings or arrangements (Different arrangements and shapes of bacterial cells are discussed more thoroughly in the following section of book on bacterial classification).

Eubacteria have a Distinct Ribosome Structure (70S)

Ribosome structure have played an important role in establishing the phylogenetic relationships among various organisms. Biologists in general examine two parameters when they compare the ribosomes of different species.

- **Size**: Prokaryotic organisms (eubacteria and archaebacteria) have ribosomes (70S) that are significantly smaller with fewer protein subunits than eukaryotic ribosomes (80S). Ribosomes are measured in units called Svedberg (S), a unit related to the velocity at which the ribosomes sediments in a high-speed centrifuge. Prokaryotic ribosomes are 70S type and eukaryotic ribosomes are 80S type.
- **Ribosomal RNA nucleotide sequence**: Most of the taxonomists believe that ribosomal RNA (rRNA) nucleotide sequences are good indicators of evolutionary relationships among organisms. As evolution proceeds, changes in rRNA sequences of more closely related two species are noticed. Nucleotide sequence analysis of rRNA has indicated surprisingly that the archaebacteria and eukaryotes shared a more recent common ancestor with each other than with the eubacteria. This was surprising as earlier opinion was that the archaebacteria represented the most primitive prokaryotic organisms.

Eubacteria have a single chromosome (circular) per cell, some bacteria will have plasmid DNA (extra chromosomal). There are several differences between the DNA of prokaryotes and eukaryotes regarding structure and composition.

- **Bacteria typically will have a single DNA molecule (chromosome) in each cell:** In contrast, eukaryotic cells usually have several chromosomes (the exact number is a characteristic to each species).

- ☆ **Prokaryotic DNA is not contained within a membrane-bound nucleus:** Instead, it is found in the form of "nucleoid" in the cell's cytoplasm.
- ☆ **Prokaryotes have less total DNA per cell (about 10 times less) than eukaryotes:** In eukaryotes, there are many highly repetitive DNA sequences; such repetitiveness is not observed in prokaryotic DNA.
- ☆ **Prokaryotic chromosomes are circular:** The DNA double helix forms a covalently closed circle, whereas Eukaryotic chromosomes are believed to be linear in nature, having two ends.
- ☆ **Eukaryotic DNA is organized into nucleosomes which are short segments of DNA wound around protein cores all along the chromosome:** Therefore, the DNA of a eukaryotic chromosome resembles "beads-on-a-string", whereas DNA of Prokaryotes is not packed into nucleosomes.
- ☆ **A prokaryotic cell may contain plasmids:** A plasmid is a short, self-replicating piece of circular DNA (extra chromosomal). Plasmids usually encode special functions like antibiotic resistance, conjugation(the ability to transfer DNA between cells) *etc.*

History of Plant Bacteriology

Table 1: Historical Events in Bacterial Plant Pathology

Name of the Scientist	*Year*	*Event*
W. B. Leismaan	1858	Demonstrated that root nodules of legumes were formed by a specific group of bacterial cells.
M. S. Woronin	1876	Isolated root nodule bacteria from plants.
Prilleaux	1879	Reported bacterial decay of wheat kernel.
T. J.Burrill	1878	University of Illionis, USA discovered bacteria as etiological agents of plants.
T. J. Burrill	1882	Published fire blight pathogen *Micrococcus amylovorus*
J. H. Wakker	1883	Established yellow slime disease of water hyacinth incited by a bacterium
Luigi Savastano	1887	Described bacterial knot of olives and isolated casual organism *Bacillus oleaetuberculosis* in 1889.
S. N. Winogradsk	1889	Discovered and described autotrophic bacteria that play important role in iron and sulphur cycles on Earth in. Later, he described free living nitrogen fixing bacteria such as *Azotobacter* and *Nitrobacter*.
J. Omori	1896	Soft rot pathogen *Bacillus* spp.
E. F. Smith	1901	Validation of bacteria as plant pathogens. He worked on bacterial wilt (*Erwinia tracheiphilum*) of cucurbits which was first reported 1893. Later he also worked on wilt of solanaceous plants and black rot of crucifers. He was regarded as "Father of phytobacteriology".

Name of the Scientist	*Year*	*Event*
C. O. Jensen	1910	Demonstration of evidence between crown galls of plants and cancer of animals.
L. R. Jones	1910	Isolated *Erwinia carotovora* (*Pectobacterium*) from soft rotting vegetables. The rot was due to an enzyme pectinase. He described the disease and life cycle not the pathogen.
S. Hori	1912	Publication of soft rot pathogen.
G.H. Koons and J.E. Kotila	1925	Isolation of bacteriophages of *Bacillus carotovorus*
F. J. Griffith	1928	Discovered transformation in bacteria (*Streptococcus pneumoniae*)
Selman Abraham Waksman	1944	Discovered antibiotics Streptomycin produced by actinomycetes, *Streptomyces griseus*
Oswald T. Aver and co-workers	1944	Demonstrated that DNA is the genetic material of bacteria.
Joshua Lederberg and Edward Tatum	1946	Discovered bacterial conjugation in *Escherichia coli.*
Joshua Lederberg	1952	Coined the term plasmid.
Norton Zinder and Joshua Lederberg	1952	Discovered transduction in *Salmonella typhi.*
R. M. Klein	1954	Basic process of transformation in crown gall.
A. C. Braun	1955	Wild fire toxin and its action.
A. K. Chatterjee and M. P. Starr	1972	Conjugative gene transfer in bacteria.
P. B. New and A. Kerr	1972	Bio control of crown gall (*Agrobacterium radiobacter* strain 84)
I. Zaenen *et al.*	1974	Demonstration of Ti plasmids in *Agrobacterium tumefaciens*
G.P. Fox and C.R. Woese	1977	Suggested that prokaryotes are two types that true bacteria and archaebacteria.
D. W. Dye *et al.*	1980	Introduced pathover system in bacterial taxonomy
Staskawicz *et al.*	1984	Cloned first avirulence gene from *Pseudomonas syringae* pv. *phaseolicola.*
Martin *et al.*	1993	Cloned of *Pto* resistance gene from tomato. It was map based cloning of protein kinase gene conferring disease resistance in tomato bacterial speck disease caused by *Pseudomonas syringae* pv. tomato.
Simpson *et al.*	2000	Published the full genome sequence of plant pathogenic bacteria *Xylella fastidiosa* casuing pierce disease of grapes.

Chapter 2

Morphology, Structure and Composition

A. Morphology of Bacterial Cell

The structure of the bacterial cell envelop is important for pathogenicity because host recognition is determined by the interactions between the host cell wall and extracellular polysaccharides or lipopolysaccharides of bacteria.

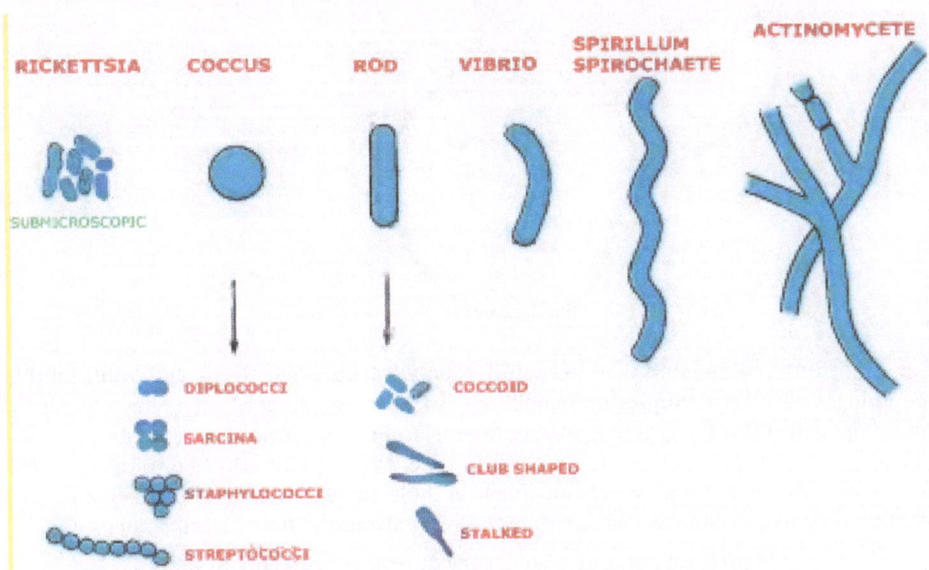

(*Source*: Phytobacteriology Principles and Practice by Janse, 2006).

Eubacteria have three basic phases. They are SPHERICAL, ROD and SPIRAL, also referred as COCCUS (plural. cocci), BACILLUS (pl. bacilli) and SPIRILLUM (pl. spirilla) respectively. Most plant pathogenic bacteria fall within the range of 1.0-5.0 x 0.5-1.0μm. *Streptomyces* also called ray fungi and possibly a link between bacteria and fungi, it is characterized by the nature of forming of highly branched mycelium, 0.5-2.0 μm in width and form spores as chains at tips of aerial hyphae.

The gross morphology and structure of representative Gram-negative bacteria is shown in Figure 1.

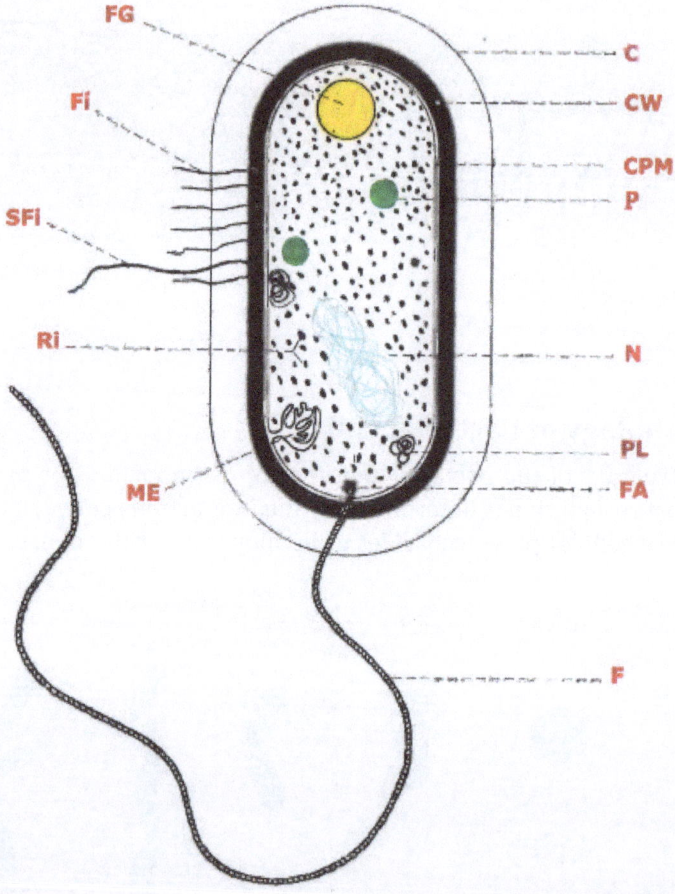

Explanation: C = capsule of extracellular polysaccharides; CW = cell wall, mainly peptidoglycan; CPM = cytoplasmic membrane; F = polar flagellum; FA = flagellar attachment in cytoplasm; FG = fat granule, storage material; Fi = fimbriae; ME = mesosome; N = nucleosome (nucleic acid strand or 'chromosome' at one point attached to the cell wall, more than 1000 μm long); P = polyphosphate granule, storage material; PL = plasmid (short strand of closed, circular, free nucleic acid); Ri = ribosomes; Sfi = sex fimbrium or pilus.

Figure 1. Schematic Structure of Gram–ve Bacterial Cell.

(*Source*: Phytobacteriology Principles and Practice by Janse, 2006).

B. Cell Envelope – Structure

The cell envelope of Gram negative bacteria consists of the cell wall, the cell membrane and the outer membrane. The outer membrane is absent in the Gram + ve bacteria (Table 2).

The Cell Wall

It is the rigid frame work responsible for cell shape maintenance and it's integrity by protecting from external stresses. Cell wall of the Gram negative bacteria ranges from 10 to 15 nm in thickness and from 5 to 10 per cent in dry weight of the total cell envelope. These values in Gram positive ranges from 20 to 80 nm and 40 to 90 per cent. The basic component is a macromolecule called peptidoglycan. Chemically, it is a hetropolymer having straight chains of alternating units of N-acetylglucosamine and N-acetylmuramic acid and short peptide for cross linking. The cell wall of Gram + ve bacteria contains polysaccharides, teichoic acid and sometimes proteins.

Outer Cell Membrane

The cell envelope of Gram – ve bacteria has the outer membrane outside of the cell wall. It has a complex chemical structure consisting of phospholipids, lipopolysaccharides and proteins. This structure functions as follows:

1. Provides channels for passive diffusion of nutrients of hydrophilic solutes.
2. Establishes permeability barrier to toxic substances particularly antibiotics.
3. Provides receptor sites.
4. Facilitates pairing during conjugation.
5. Endows hydrophilicity.

Cell Membrane

It is a unit membrane located inside the cell wall. It consists of a lipid bilayer of 50 to 75 per cent protein and 20 to 35 per cent lipid. The cell membrane contains various enzymes for energy yielding metabolism. In some bacteria, such as *Clavibacter*, *Streptomyces* and *Bacillus*, a part of cell membrane envaginates into the cytoplasm to form complex membrane infoldings called MESOSOMES. Mesosomes are associated with respiratory activity, nuclear division, septum formation, spore formation and secretion of hydrolytic enzymes.

Periplasm

It is a matrix of polypeptide and saccharides. It contains various enzymes such as cellulases and pectinases.

Gram-reaction

Gram staining was discovered empirically in 1884 by Christian Gram. The steps of Gram staining are as follows:

Heat fix the bacterial smear on slide

↓

Stain with crystal violet

↓

Wash with water

↓

Add diluted iodine solution

Briefly wash with alcohol (decolourizing agent)

Deep purple colour retained ⟷ Decoloured or counter stain with safranin

↓ ↓

Purple — Red

Gram +ve — Gram –ve

KOH Test-It is another simple test for routine study of gram +ve and gram –ve bacteria

Suspend 24-48 h old bacterial culture in 3 per cent of KOH solution

↓

No change ⟷ Vicidity develops

Gram + ve — Gram – ve

Table 2. Difference between Gram Positive and Gram Negative Bacteria

Gram Positive Bacteria	*Gram Negative Bacteria*
Cell wall is thicker (20-80 nm) and contains traces of lipids.	Cell wall is thin (10-15 nm) and may contain upto 20 per cent lipids.
Striking simplicity of amino acids but glutamic acid occurs in large amounts.	All the amino acids are present.
Diaminopimelic acid is observed in some bacteria.	Diaminopimelic acid is observed in some bacteria.
More muramic acid is present.	Less muramic acid is present.
Teichoic acid is present.	Teichoic acid is absent.
The outer membrane is not present.	The outer membrane is mostly present.

External Structures

Flagellum (pl. Flagella)

Most plant pathogenic bacteria are motile by means of flagella except for *E. stewartii* and are coryneform. The flagellum has a helical shape. The base of the flagellum forms a slightly wide shaped hook penetrating the cell envelope that connects with basal body and are 30-40nm in diameter. Flagella helps in movement of bacteria, act as sensory organs to detect temperature, pH and chemotaxis. The flagell structure is divided into three parts *i.e.* basal body, hook and filament. Basal body is attached to cell membrance and cytolasmic membrane. It consists of rings surrounding a pair of proteins called MotB. The rings include: L-ring (outer ring anchored in the lipopolysaccharide layer and in gram the bacteria); P-ring (anchored in the peptidoglycan layer); C-ring (anchored in the cytoplasm); M-S ring (anchored in the cytoplasmic membrane). Hook is a broader area present at the base of the filament and connects filament to the motor protein in the base. The hook length is grater in the gram bacteria. Filament is a thin hair like structure arising from the hook.

The arrangement as well as number of flagella are of taxonomic importance. Flagellar arrangements (Figure 2) are grouped as

1. MONOTRICHOUS (a single flagellum at one end of the cell). *e.g., Xanthomonas* spp.
2. CEPHALOTRICHOUS (several (tuft) of flagella at one polar end only). *e.g., Pseudomonas fluorescens*
3. AMPHITRICHOUS (one flagellum at each polar end of the cell). *e.g., Pseudomonas* spp.
4. SUBPOLAR (lateral flagellum at one side or both side). *e.g., Agrobacterium* spp.

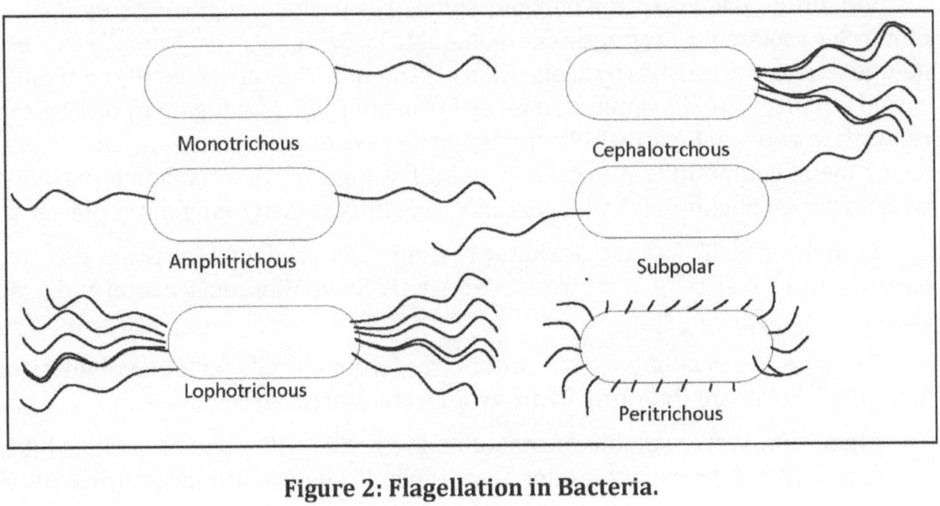

Figure 2: Flagellation in Bacteria.

5. LOPHOTRICHOUS (two or more flagella at one or both ends of the bacterial cell). *e.g., Spirillum* spp.
6. PERITRICHOUS (huge number of flagella all over the bacterial cell). *e.g., Erwinia, Dickeya, Brenneria, Pantora.*

Pilus or Fibriae

Many filamentous hair like appendages are formed large numbers on the surface of majority of Gram negative bacteria. They are called pilli or fimbrae. It is 3 to 14nm in diameter and 0.2 to 20μm in length. According to function and morphology fimbrae fall in 6 categories *e.g.,* SEX PILLI act as specific bridge for gene transfer in conjugation. Another act as adhesive organelles.

Extracellular Polysaccharides (EPS)

Bacteria often produce EPS outside the cell envelop. The small and dense layer with defined boundry is the CAPSULE. Water soluble and less adherent layer without definite boundry is SLIME LAYER.

Internal Structure

Cytoplasm: The cytoplasm contains ribosome granules, enzymes, coenzymes, intermediate metabolic products and inorganic substrates. It is the basin for protein synthesis and active metabolism.

Chromosome: In prokaryotes the terms NUCLEUS, GENOME and CHROMOSOME can be used interchangeably. The nuclear material within the cell is not separated by a membrane as in eukaryotic cells. There is basically one chromosome per cell but two or more can be seen in the exponential growth phase. The chromosome is highly folded and forms a compact mass and remain attached to cell membrane at replication site.

Plasmids: Bacteria often harbour small, extracellular, replicons (replication units). Such self replicating entities are PLASMIDS. Bacterial cells generally express new genetic characteristics by acquisition of plasmids. They are generally covalently closed circular DNA. Plasmids can be self transmissible (conjugative) or non self replicating (nonconjugative). Plasmids borne resistance against antibiotics and heavy metal compounds is known in plant pathogenic bacteria. Some plasmids with unknown phenotypic traits are called cryptic/resident/indigenous plasmids.

Endospores: These are dormant resting cells produced by Gram positive bacteria which resists against stress like heat, UV irradiation, desiccation and toxic chemical. *e.g., Bacillus* spp.

Reserve Materials: Bacteria contain reserve materials include polymers of inorganic phosphate (bolutin), hydroxybutyrate, starch, glycogen *etc.*

Pigments: Water soluble or insoluble pigments, with colours such as blue, green, yellow, red, brown *etc.* under aerobic conditions are produced by bacteria.

Examples

1. ***Pseudomonas fluorescens***: Fluorescent pigments, pyoverdin, siderophores(Fe III chelators)
2. ***Erwinia chrysanthemi***: Water soluble, blue pigment (indigoidine); *E. rhapontici*: Iron chelators; *E. herbicola*: yellow carotenoid pigments, *E. rubrifaciens*- pink colour, rubrifacine pigment.
3. ***Xanthomonas* spp.**: Yellow pigment, (Xanthomonodin), water insoluble
4. ***Streptomyces scabiei:***Melanin, brown water- soluble on tyrosine containing medium.
5. ***Ralstonia solanacearum***: Melanin, brown water- soluble on tyrosine containing medium.

Chapter 3

Classification of Bacteria

Content of Taxonomy

Taxonomy encompasses the study of classification, identification and nomenclature of bacteria. Classification is the grouping of organisms in orderly fashion. Identifications is a comparative process of determining taxonomic position of unknown bacteria with those of known bacteria. Nomenclature requires following the rules in the "International code of nomenclature of bacteria" in naming any bacterial taxa.

Concept of Bacterial Species

Bacterial species are the taxonomic groups of strains defined on the basis of phenotypic and genotypic characteristics in common.

Taxospecies

It refers to a group of strains that have a number of phenotypic characteristics in common and can be grouped into distinct phenotypic groups. *e.g., Erwinia amylovora* or *Ralstonia solanacearum* are the taxospecies.

Nomenspecies

It is the species with binomial names given in accordance with the rules. The names may, therefore, be claimed for another taxa on the basis of some criteria. *e.g., Xanthomonas campestris* is a nomenspecies superseded by *Xanthomonas campestris* pv. *campestris.*

Molecular Species

This term is applied to a group of strains with a high degree of nucleic acid

(DNA or RNA) homology. The sequence of NA establishes and preserves the identity of species.

Genospecies

It is a group of strains that can accomplish genetic exchange that occurs at the species level, genus level or at family level.

A. Nomenclature and Classification of Bacteria

i. Binomial Nomenclature

It is a system of naming organisms in a scientific manner using TWO NAMES FOR EVERY ORGANISM: The GENUS and the SPECIES.

The system follows certain rules:

1. The scientific name must be in **Greek** or **Latin**.

   This helps for accurate communication of information to other biologists around the world who use various languages for naming. This can be done by assigning a unique two-word scientific name to each organism. (BINOMIAL NOMENCLATURE)
2. The first part of the name is referred as **Genus** and the second part of the name is as **species.**
3. The Genus name refers to the relatively small group of organisms to which a particular type of organism belongs to.
4. The SECOND part of the name is the SPECIES.
5. The Species name is usually a Latin description of some important characteristic of the organism (SPECIES means **IDENTIFIER**).

Binomial nomenclature is the system of giving a scientific name to an animal or a plant or a microorganism. The name consists of two words, the first one referring to the name of the genus and the second to the name of the species. Binomial nomenclature helps to avoid the confusion created by naming using common names. Certain guidelines issued by the International Council of Binomial Nomenclature (ICBN) had to be followed in the use of binomial nomenclature. The most important is that name of the genus should start with a capital letter and name of the species should start with a small letter.

ii. Classification

The process of keeping similar things into groups.

History of Classification

- ☆ 4000 B.C. Aristotle (Greek Philosopher) had created the first written classification scheme. 1500's - 1700's.
  - ❖ Many different classification systems created.

- ✫ Most of them really complicated.
- ✫ Names based on common names – Lead to confusion.
- ✫ Names also based on lengthy scientific definitions.
- ✫ 1700's - Carols Linnaeus – A Swedish Biologist.
    - ❖ Established a very simple system for classifying and naming organisms.
    - ❖ Based on structural similarities of the organisms.
    - ❖ Binomial Nomenclature - two name naming system - still in use today.
    - ❖ Created a system of groups called as TAXA or TAXON.
    - ❖ Each Taxon is a category into which related organisms are placed together. Approximately 2.5 million kinds of organisms identified
- ✫ Linnaeus (1758) proposed a classification in the natural system that is the two-kingdom, which placed living organisms into two kingdoms - Plantae and Animalia.
- ✫ The plant kingdom proposed by Linnaeus and subsequent works of that period comprised of bacteria, fungi, algae, liverworts, masses, ferns, conifers and flowering plants.
- ✫ Animal group - anything that lives on land, in the water or in the air.
- ✫ Plant group - based this on their difference in stems.
- ✫ The animal kingdom proposed by Linnaeus and others includes the unicellular and multicellular animals.
- ✫ The two-kingdom classification has certain demerits, particularly regarding to lower forms of life which share common features with both plants as well as animals.
- ✫ In 1858 Earnt Haeckel proposed the three-kingdom classification with Plantae, Animalia and a new kingdom named Protista (includes all the unicellular organisms, both from plants and animals).
- ✫ Copeland (1966) suggested the addition of kingdom-Monera a fourth kingdom to include the prokaryotic organisms like bacteria and blue-green alga. This creation led to of four kingdom classification.
- ✫ The advent of electron microscope helped scientists, biologists in discovering the existence of two forms of cells in living organisms - prokaryotic and eukaryotic.
- ✫ In 1969, Robert H. Whittaker proposed the creation of a new kingdom called Mycota to include the fungi exclusively. This led to creation of five kingdom classification based on mode of nutrition like ingestion, absorption and photosynthesis. The five kingdoms are
    1. Monera,
    2. Protista,

3. Fungi,
4. Plantae
5. Animalia.

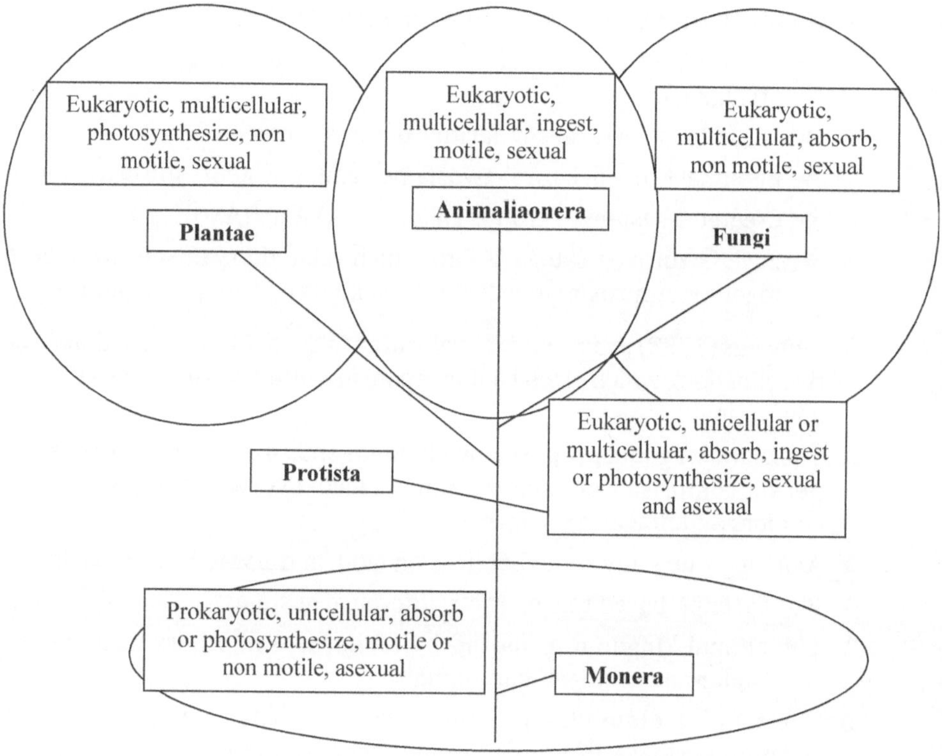

Five Kingdom Classification Proposed by Robert H. Whittakar in 1969.

Merits and Demerits of Five Kingdom Classification

- It is the most accepted system of modern classification because of the phylogenetic placing of different groups of living organisms. This system of classification looks more natural and scientific because of the following considerations:
- Separation of prokaryotes from others and making independent kingdom is justifiable because they differ from all other living organisms in their general organization.
- Grouping of all unicellular eukaryotes under the kingdom Protista has solved many problems related to placing of many groups, particularly related to the position of organisms like *Euglena*.
- Elevation of the group fungi to the status of a kingdom was justifiable because fungi totally differ from other primitive eukaryotes like algae and protozoans.

- ☆ The kingdoms Metaphyta and Metazoa are now more homogenous groups than they were in the two separate kingdom classification as it shows the phylogeny of different life styles.
- ☆ This five-kingdom classification gives a clear indication of cellular organization and modes of nutrition, the characters which appeared very early in the evolution process of life.

However, it has above merits, the five-kingdom classification has certain drawbacks also, particularly with reference to the lower forms of life as described below:

- ☆ The kingdoms Monera and Protista include heterogeneous and diverse forms of life. Both of these kingdoms comprises of photosynthetic (autotrophic) as well as non-photosynthetic (heterotrophic) organisms.
- ☆ Both the kingdoms include organisms with cell wall as well as without cell wall.
- ☆ None of the three higher kingdoms include a single ancestor of all its forms. Multicellular lines have originated from protistans several times (polyphyletic).
- ☆ Unicellular green algae like Volvox and Chlamydomonas have not been included under Protista because of their resemblance to other forms like green algae.
- ☆ Slime moulds differ totally from other members of Protista in their structural and functional organization.
- ☆ Viruses have not been given proper position in this system of classification.

Nevertheless, the five-kingdom classification has received a wide acceptance from the biologists all over the world.

Three Domain System of Life

Carl Woese (1977) proposed six kingdom classification which is an evolutionary model of classification based on differences in the sequences of nucleotides in the cell's ribosomal RNAs (16S rRNA), as well as the cell membranes lipid structure and its sensitivity to antibiotics. Comparing rRNA structure is very useful, because rRNA molecules throughout nature in all organisms carry out the same function, their structure changes very little over time. Therefore, similarities and dissimilarities in rRNA nucleotide sequence are a good indication of related or unrelated different cells and organisms. In 1990, he proposed three domain system and this system proposes that a common ancestor cell gave rise to three different cell types, each representing an individual domain, namely these three domains are the *Archaea* (archaebacteria), the *Bacteria* (eubacteria) and the *Eukarya* (eukaryotes). The *Eukarya* are further divided into 4 kingdoms: Protists, Fungi, Animalia and Plantae.

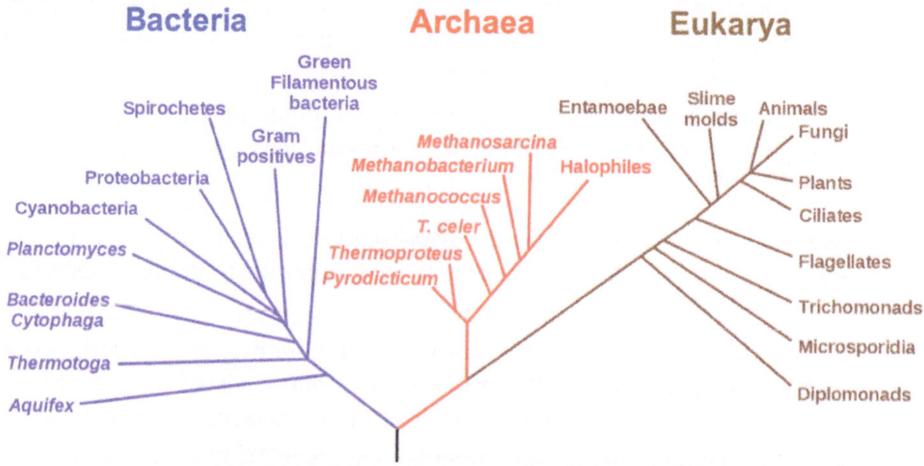

A phylogenetic tree based on rRNA data, presenting the separation of bacteria, archara and eukaryota domains. (*Source*: Public Domain, https://commons.wikimedia.org/w/index.php?curid=1201601)

Common Ancestor

Characterization of Domain

i. *The Archaea (archaebacteria)*

The *Archaea* possess the following characteristics:

a. *Archaea* are prokaryotic cells.

b. The Archaea have membranes consisting of branched hydrocarbon chains (many also containing rings within the hydrocarbon chains) bound to glycerol by ether linkages, unlike the Bacteria and the Eukarya.

c. Archaea's cell walls contain no peptidoglycan.

d. Archaea is not susceptible to some bacteria-affecting antibiotics, but is susceptible to some Eukarya-affecting antibiotics.

e. In extreme conditions, archaea frequently live and include methanogens, extreme halophiles and hyperthermophiles. One reason for this is that in the Archaea membranes, the ether-containing linkages are more stable than in the Bacteria and Eukarya ester containing linkages and are better able to withstand higher temperatures and higher acid concentrations.

ii. *The Bacteria (Eubacteria)*

Bacteria (also known as eubacteria or "true bacteria") are prokaryotic cells that are common encounter many more times than the archaebacteria in human daily life. Eubacteria are found almost everywhere and kill thousands and thousands of people every year, but also serves as antibiotic producers and food digesters. The *Bacteria* possess the following characteristics:

- ✰ *Bacteria* are having **prokaryotic cells.**
- ✰ Like the *Eukarya*, bacteria have membranes composed of **unbranched fatty acid chains attached to glycerol with ester linkages.**
- ✰ The cell walls of *Bacteria*, unlike the *Archaea* and the Eukarya, **contain peptidoglycan.**
- ✰ *Bacteria* are sensitive to general antibacterial antibiotics but are resistant to most antibiotics that can affect *Eukarya.*
- ✰ *Bacteria* possess **rRNA that is unique to the *Bacteria*** as indicated by the presence molecular regions distinctly differentiated from the rRNA of *Archaea* and *Eukarya.*
- ✰ *Bacteria* includes mycoplasmas, cyanobacteria, Gram-positive and Gram-negative bacteria.

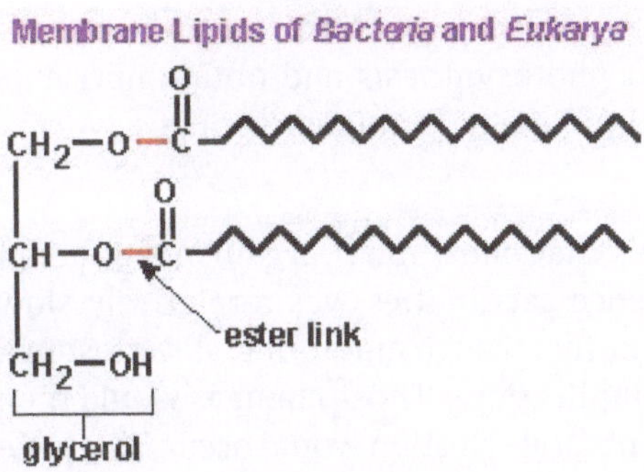

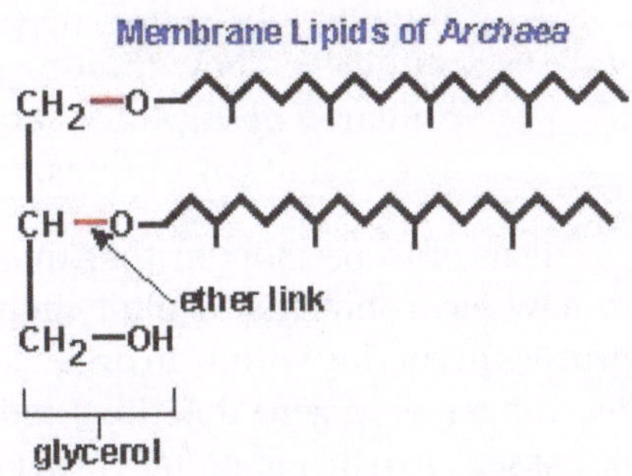

Membrane Lipids of Archaea, Bacteria and Eukarya. The Bacteria and the Eukarya have membranes composed of unbranched fatty acid chains attached to glycerol by ester linkages. The Archaea have membranes composed of branched hydrocarbon chains attached to glycerol by ether linkages.

iii. The Eukarya (eukaryotes)

The *Eukarya* (also spelled *Eucarya*) contains the following characteristics:

a. *Eukarya* have eukaryotic cells.

b. Like the *Bacteria*, they have membranes composed of unbranched fatty acid chains attached to glycerol with ester linkages.

c. Not all *Eukarya* possess cells with a cell wall, but for those *Eukarya* having a cell wall, peptidogly can is absent.

d. *Eukarya* are resistant to antibacterial antibiotics but are sensitive to other class of antibiotics that can affect eukaryotic cells.

e. *Eukarya* contain rRNA regions that is unique to the *Eukarya* as indicated by the presence molecular regions distinctly differentiates from the rRNA of *Archaea* and *Bacteria*.

The *Eukarya* are subdivided into the following four kingdoms:

1. **Protista Kingdom**: Protista are simple, predominately unicellular eukaryotic organisms. Examples: slime molds, euglenoids, algae and protozoans.
2. **Fungi Kingdom**: Fungi are unicellular or multicellular organisms with eukaryotic cells. The cells have cell walls but are not organized into true tissues. They do not carry out photosynthesis and mode of nutrient obtain is through absorption. Examples: sac fungi, club fungi, yeasts and molds.
3. **Plantae Kingdom**: Plants are multicellular organisms with eukaryotic cells. The cells are organized into tissues and possess cell walls. They obtain nutrients by photosynthesis and absorption. Examples: mosses, ferns, conifers and flowering plants.
4. **Animalia Kingdom**: Animals are multicellular organisms made up of eukaryotic cells. The cells are organized into tissues and lack these cell walls. They do not carry out photosynthesis and obtain nutrients primarily by ingestion mode of obtainment. Examples: sponges, worms, insects and vertebrates.

It used to be thought that the changes that allow microorganisms to adapt to new environments or alter their virulence capabilities was a relatively slow process occurring within an organism primarily through mutations, chromosomal rearrangements, gene deletions and gene duplications. Those changes would then be passed on to that microbe's progeny and natural selection would occur. This gene transfer from a parent organism to its offspring is called vertical gene transmission.

It is now known that microbial genes are transferred not only vertically from a parent organism to its progeny but also horizontally to relatives that are only distantly related, *e.g.*, other species and other genera. This latter process is known as horizontal gene transfer. Through mechanisms such as transformation, transduction and conjugation, genetic elements such as plasmids, transposons, integrons and even chromosomal DNA can readily be spread from one microorganism to another. As a result, the old three-branched "tree of life" in regard to microorganisms now appears to be more of a "net of life."

Prokaryotic Domain

The two prokaryotic domains (or empires), "Bacteria" (or "*Eubacteria*") and "Archaea" (or "*Archaeobacteria*") are subdivided into 35 phyla (or divisions): 30 phyla in the domain "Bacteria" and 5 phyla in the domain "Archaea".

Domain "Bacteria"

Domain bacteria has 30 Phyla as Acidobacteria, Actinobacteria, Aquificae,

Armatimonadetes, Bacteroidetes,Caldiserica, Chlamydiae, Chlorobi, Chloroflexi, Chrysiogenetes, Cyanobacteria, Deferribacteres, Deinococcus, Thermus, Dictyoglomi, Elusimicrobia, Fibrobacteres, Firmicutes, Fusobacteria, Gemmatimonadetes, Lentisphaerae, Nitrospira, Planctomycetes, Proteobacteria, Spirochaetes, Synergistetes, Tenericutes,Thermodesulfobacteria, Thermomicrobia, "Thermotogae, Verrucomicrobia.

Out of 30 phyla of Domain Bacteria only four phyla such as Actinobacteria, Firmicutes, Proteobacteria and Tenericutes are causing diseases to the plants are described here.

Phylum: Proteobacteria Stackebrandt *et al.,* 1988

- ☆ The Proteobacteria are a major and diversified group (phylum) of bacteria. They include a wide variety of pathogens, such as *Escherichia, Vibrio, Helicobacter* and many other notable genera. Includes free-living and many of the bacteria responsible for nitrogen fixation.
- ☆ Because of the great diversity of forms (shapes) found in this group, the Proteobacteria are named after Proteus (Greek god of the sea) capable of assuming many different shapes.
- ☆ All proteobacteria are negative for Gram's reaction with an outer membrane mainly composed of lipopolysaccharides. Many move by means of flagella, but some are non-motile or rely on bacterial gliding. The last includes the myxobacteria, a unique group among bacteria that can aggregate to form multicellular fruiting bodies.
- ☆ Most members are facultative or obligatory anaerobic, chemoautotrophs and heterotrophic, but there are several exceptions. A variety of genera, which are not closely related to each other, convert energy from light through photosynthetic process. These are called purple bacteria, referring to their mostly reddish pigmentation.

Phylum: Firmicutes Corrig. Gibbons and Murray, 1978

- ☆ The Firmicutes (Latin: *firmus*, strong and *cutis*, skin, referring to the cell wall) are a phylum of bacteria, most of them are Gram-positive in nature.
- ☆ A few however, such as Megasphaera, Pectinatus, Selenomonas and Zymophilus have a porous pseudo-outer-membrane that makes them to stain Gram-negative.
- ☆ In Firmicutes include all Gram-positive bacteria, but have recently defined them to be of a core group of related forms called the low-G+C group, in contrast to the Actinobacteria. They have round cells, called cocci (singular coccus), or rod-shaped forms.
- ☆ Many *Firmicutes* produce endospores which are resistant fordesiccation and survives extreme conditions.
- ☆ They are found in various environments and the group includes some notable pathogens also.

- Those in one family, the heliobacteria, produces energy through photosynthetic process. Firmicutes also play an important role in beer, wine and cider spoilage.

Phylum: Tenericutes Murray, 1984

Class Mollicutes (In the next edition of the *Bergey's Manual of Systematic Bacteriology*, the class Mollicutes will be excluded from the phylum "*Firmicutes*" and classified in the phylum "*Tenericutes*". A collection of bacterium in a biological division within the Monera. The Tenericutes are prokaryotic cells that are pliable and they lack cell wall. They are instead bound by a plasma membrane. The bacteria are single-celled. Some of which may cause infections and disease in animals, humans as well as plants.

Phylum: *Actinobacteria* (ex Margulis, 1974) Cavaliur-Smith, 2020

- The phylum *Actinobacteria* is made up of gram-positive organisms possessing a high mole per cent G+C composition with (> 55 per cent G+C).
- This group comprises 39 families and 130 genera making it as one of the largest phyla within *Bacteria*.
- They encompass a wide range of morphological forms like coccoid (*e.g. Micrococcus*), rod-coccoid (*e.g. Arthrobacter*), fragmenting hyphal forms like *Nocardia* to those with permanent and highly differentiated branched mycelium (*e.g. Streptomyces*).
- Many of the Actinobacteria are spore formers which range from motile zoospores to specialized propagules.
- They share very diverged physiology which is evident by their production of numerous extracellular enzymes and by the thousands of other metabolic products (including antibiotics) they synthesize and excrete.
- *Actinobacteria* especially members of *Streptomycetaceae* are the major antibiotic producers in the pharmaceutical and agricultural sectors. However, a few *Actinobacteria* are important human, animal as well as plant pathogens.

Domain: "Bacteria"

Phylum: Proteobacteria

Class I: Alphaproteobacteria

Order: Rhizobiales

Family: Beijerinckiaceae, Genera: *Beijerinckia, Chelatococcus, Methylocapsa*

Family: Bradyrhizobiaceae, Genera: *Agromonas, Bradyrhizobium, Nitrobacter*

Family: Rhizobiaceae : ***Agrobacterium***, *Allorhizobium*, ***Rhizobium***, *Sinorhizobium*

Family: Xanthobacteraceae, Genera: *Azorhizobium, Xanthobacter*

Family: Phyllobacteriaceae, Genus: *Candidatus* Liberibacter

Order: Rickettsiales

Family: Anaplasmataceae, Genera: *Aegyptianella, Anaplasma, Cowdria, Ehrlichia*

Family : Rickettsiaceae Genera: *Orientia, Rickettsia*

Order: Sphingomonadales

Family: Sphingomonadaceae, Genera: ***Rhizomonas*** (rejected name), ***Sphingomonas***

Class II: Betaproteobacteria

Order: Burkholderiales

Family: Burkholderiaceae, Genera: ***Burkholderia***, *Robbsia*

Family: Ralstoniaceae,Genus: ***Ralstonia***

Family: Comamonadaceae, Genus: ***Acidovorax, Xylophilus***

Family: Oxalobacteraceae, Genus: ***Janthinobacterium***

Order Nitrosomonadales

Family: Nitrosomonadaceae, Genera: *Nitrosolobus, Nitrosomonas, Nitrosospira*

Class III: Gammaproteobacteria

Order: Xanthomonadales

Family: Xanthomonadaceae, Genera: ***Xanthomonas, Xylella***

Order: Pseudomonadales

Family: Moraxellaceae, Genus: Acinetobacter

Family: Pseudomonadaceae; Genera:*Azotobacter*, ***Pseudomonas***, *Rhizobacter*,

Order: Enterobacteriales

Family: Enterobacteriaceae, Genera: ***Enterobacter, Gibbssiella***, *Salmonella*

Family: Erwiniaceae, Genera: ***Erwinia, Pantoea, Tatumella***, *Phaseolibacter*

Family: Pectobacteriaceae, Genera: ***Pectobacterium, Brenneria, Dickeya***

Family: Yersiniaceae, Genera: ***Ewingella, Samsonia, Serratia***

Class IV: Deltaproteobacteria or Deltabacteria

Order: Bdellovibrionales

Family: Bdellovibrionaceae, Genus: *Bdellovibrio, Micavibrio, Vampirovibrio*

Phylum: Firmicutes

Class: Bacilli (or Firmibacteria)

Order: Bacillales

Family: Alicyclobacillaceae, Genus: *Alicyclobacillus*

Family: Bacillaceae

Genera: *Aeribacillus, **Bacillus**, Filobacillus, Geobacillus, Gracilibacillus, Halalkalibacillus, Halobacillus, Halolactibacillus, Lentibacillus, Lysinibacillus, Marinococcus, Natronobacillus, Oceanobacillus, Tenuibacillus, Thalassobacillus, Tumebacillus,Virgibacillus, Viridibacillus, Vulcanibacillus*

Family: Listeriaceae, Genera: *Brochothrix - Listeria*

Family: Paenibacillaceae, Genera: *Brevibacillus, Fontibacillus, Thermobacillus*

Order: Lactobacillales

Family: Lactobacillaceae, Genera: *Lactobacillus, Paralactobacillus, Pediococcus,*

Family: Leuconostocaceae, Genera: *Fructobacillus, Leuconostoc,Oenococcus,*

Family: Streptococcaceae, Genera:*Lactococcus, Lactovum,Streptococcus*

Class: Clostridia

Order Clostridiales

Family Clostridiaceae, Genus: ***Clostridium***

Family Eubacteriaceae, Genera: *Acetobacterium, Eubacterium*

Phylum: Tenericutes

Class: Mollicutes

Order: Acholeplasmatales

Family: Acholeplasmataceae, Genus: Acholeplasma

Order :Anaeroplasmatales

Family: Anaeroplasmataceae, Genera: *Anaeroplasma, Asteroleplasma*

Order: Entomoplasmatales

Family: Entomoplasmataceae, Genera: *Entomoplasma, Mesoplasma*

Family: Spiroplasmataceae, Genera: ***Spiroplasma***

Order :Haloplasmatales

Family: Haloplasmataceae, Genera: *Haloplasma*

Order Mycoplasmatales

Family: Mycoplasmataceae, Genera: *Eperythrozoon, Haemobartonella, Mycoplasma,* Phyoplasma: ***Candidatus* Phytoplasma**

Phylum: *Actinobacteria*

Class: Actinobacteria

Subclass: Actinobacteridae

Order: Actinomycetales

Suborder: Actinomycineae

Family: Actinomycetaceae

Genera: *Actinobaculum, Actinomyces, Arcanobacterium*

Suborder: Corynebacterineae

Family: Corynebacteriaceae, Genera: *Bacterionema, Caseobacter, Corynebacterium*

Family: Mycobacteriaceae, Genera: *Amycolicicoccus, Mycobacterium*

Family: Microbacteriaceae, Genera: ***Arthrobacter, Clavibacter, Curtobac-terium, Rathayibacter, Leifsonia***

Family : Nocardiaceae, Genera: ***Nocardia, Rhodococcus***

Suborder: Streptomycineae

Family : Streptomycetaceae, Genus: ***Streptomyces***

System of Classification

i. **Artificial system:** A system of classification based on arbitrarily chosen criteria is termed as an artificial system.

ii. **Natural system:** The Linnaen classification expressed the biological nature of the objects.

Numerical Taxonomy

In the era of Koch and Pasteur bacteria were mainly classified according to morphological, physiological and biochemical criteria. Michel Adanson (a contemporary of Linnaeus) suggested Adansonian (numerical taxonomy). In this case, all phenotypic characters are given equal weightage and very large numbers of characters are examined. Numerical taxonomy does not have the evolutionary connotations of phylogenetic taxonomy, but it proves a more objective and table basis for the construction of taxonomic grouping. The number of common characters among them was considered as a quantitative measure for the taxonomic relationship. Similarities are quantitatively expressed by the matching coefficient of Sokal and Michener

- ☆ The similarity coefficient (Sj) does not take into account the characters negative for both the organisms (being based on only on positive matches).

☆ The matching coefficient (Ss) includes both positive as well as negative matches and expressed in per cent.

$$Sj = \frac{a}{a + b + c}$$

$$Ss = \frac{a + d}{a + b + c + d}$$

where,

a = number of characters positive in both the strains

b = number of characters positive in strain-1 and negative in strain -2

c = number of characters negative in strain-1 and positive in strain -2

d = number of characters negative in both the strains

Similarity coefficients are more reliable. The percentage of characteristics that two organisms or groups have in common is calculated. Ten or more characteristics of both the bacteria are used for calculation and then match the organisms. Example: A and B share 8 characters out of 10: similarity coefficient (Sj) is 8/10 (80 per cent). It can use many of such values to establish similarity matrix. Dendrograms help in displaying this information clearly. Grater similarity coefficient, closer relatedness between organisms is inferred. If similarity coefficient is greater than 70 per cent and inference is that two compared bacteria are of the same species.

Numerical taxonomy does not have the evolutionary connotation of phylogenetic taxonomy but it proves a more objectives and stable basis for the construction of taxonomic grouping. A great advantages is that a large number of characters must be analysed which result in a thorough examination of phenotypes.

Methods of Classification

Several methods of classifications have been developed from time to time. They include conventional taxonomy, numerical taxonomy, molecular taxonomy (DNA base composition, RFLP, DNA-DNA homology, DNA-rRNA homology, protein profiles, isozymes, gene transfer), chemotaxonomy, pyrotaxonomy and serological taxonomy.

Conventional Taxonomy

In conventional taxonomy strains are classified on the basis of morphological, cultural, biochemical, physiological and pathological characteristics. This has been the most common approach of bacterial taxonomy and many of the currently recognized taxa have been established by this method earlier. In conventional taxonomy the investigators used to attempt to select certain definitive properties like morphological or physiological or biochemical or pathological to take precedence over others. The taxa thus established doesn't have any phylogenetic relationship. The taxonomic significance of each definitive property being used in the conventional taxonomy varies depending on the type of bacteria. Therefore, minimal standards of phenotypic properties are anticipated with each taxonomic category to which plant pathogenic bacteria belongs to.

Molecular Taxonomy

In molecular taxonomy, bacterial strains are clustered into taxa on the basis of their DNA parameters such as base composition expressed as guanine and cytosine content (G + C per cent), homology, restriction fragment-length polymorphism (RFLP) and gene transfers. The molecular approach attempts to determine the relatedness of chromosomal DNA of two different strains in order to obtain a more direct relationship than conventional and numerical taxonomy which are based on the relatedness of phenotypic properties that can reveal only a part of their bacterial genome. Molecular taxonomy is therefore said to be a more rational taxonomic approach towards a better understanding of the ill-defined concept of species and genus.

DNA Base Composition

The G + C content of bacteria vary from 25 to 75 per cent depending on the taxon. When species are well described, the intraspecific variation of G+C content is about ± 1 per cent and for the genera, the intrageneric variation is about ±5 per cent. When variations are larger than these levels, the taxa are assumed to be heterogeneous in nature. Thus, the GC content has become one of the basic features in the description of new bacterial taxa.

DNA-DNA Homology

This, the method elucidates the complementarity of sequences of the nucleotides (A, T, G, C) between two DNA molecules of two different strains. A single strand of DNA derived from one bacteria is radioactively labeled with 14C or 32P and mixed with an unlabeled single strand DNA of different bacteria. The relatedness of base sequences of the two DNA molecules is analyzed based on the ratio of hybrid DNA developed in the mixture relative to homologous DNA. The DNA-DNA homology is well suited to classification of taxa below the species level. There is the possibility that the method of gene diagnosis *i.e.*, hybridization between chomosomal DNA and certain probe DNA encoding structural or any enzyme proteins, can be effectively applied to the classification of bacteria in the future. This belief is encouraged by findings that some monoclonal antibodies have revealed the presence of proteins that reflect the relatedness of plant pathogenic bacteria at different taxonomic ranks and that some genes associated with pathogenicity, *e.g.*, hrp genes are well conserved in certain groups of bacteria.

DNA-rRNA Homology

In this method, the 14C-labeled ribosomal RNA from the reference bacterial is hybridized with DNA of the test bacteria. It is good for detecting similarities between distantly related organism at genus or family levels. This is because the genetic loci encoding rRNA will be conserved without significant variation during the process of evolution.

Restriction Fragment-Length Polymorphism (RFLP)

Isolated DNA from different strains are cleaved with restriction endonuclease enzymes and the fragments so generated are separated on agarose gels according to size. These fragments are hybridized with the independently isolated and labeled DNA probes. The distribution patterns of the fragments which have the sequence homology with the probes can be compared among DNA molecules of different origin. The usefulness of RFLP in bacterial taxonomy is yet to be confirmed but it is widely used in the identification of strains at the species level or lower.

Protein Profiles

Bacteria produce a particular set of cellular and enzyme proteins as gene products. Taxonomy based on the comparison of such protein profiles can be referred to as indirect molecular taxonomy. The proteins of two or more strains are simultaneously compared by polysaccharide gel electrophoresis either in one or two dimensions with isoelectric focusing in the first dimension. When a more number of strains are used, the similarity of the stained gels or zymograms of the proteins can be analyzed by the computer.

Isozymes

Enzyme proteins with the same functions may be different in chemical nature. These enzymes are called isozymes and separated by starch electrophoresis. In *Escherichia coli, Rhizobium* spp., *Frankia,* spp. and *Aeromonas* spp. a high correlation was noticed between isozyme patterns and DNA homology and RFLP indicating usefulness in the classification and identification of intrasubspecific subdivisions such as biovars and pathovars. In plant pathogenic bacteria, the pathological types of citrus canker organism, *X. citri* pv. *citri* can be differentiated by the isozyme patterns, indicating its usefulness as a differential marker of intra sub specific grouping of bacteria.

Gene Transfer

Gene transfer in bacteria occurs through various mechanisms such as conjugation, transformation, transduction, lysogenization, or phage infection. Successful gene transfers between two taxa have some taxonomic significance but taxonomic evaluation of gene transfer depends mainly on stability of transferred genes in the recipient cell, *i.e.*, integration and/or replication. The relative importance of these phenomena in bacterial taxonomy can be considered in decreasing order of gene integration (*e.g.*, transformation, transduction and conjugation associated with chromosome recombination and phage conversions), gene transmission and expression (*e.g.*, plasmid transfer, phage infection, phage lysis or transfection) and associated genetic phenomena (*e.g.*, phage adsorption).

Chemotaxonomy

The formation of stable cellular components such as metabolic profiles, or pigments is all under genetic control and usually is a stable feature of individual strains. For example, the chemical composition of peptidoglycans, in particular to sugars and amino acid components has generally been accepted as an important criterion in the taxonomy of actinomycetes group. Various components such as isoprenoidquinons, mycolic acids, polar lipids, polyamines, fatty acids and cytochromes have also proved to be useful as taxonomic important criterion. With the development of relatively fast and reliable analytical devices attention has been increased significantly.

Pyrotaxonomy

Pyrolysis depends upon the general chemical composition of organisms which is an indirect expression of genes (DNA). A small amount (10100 ^g) of whole cells from colonies grown on a culture plate is directly pyrolysed at high temperatures between 20 and 120 °C for a few seconds and the high molecular weight ions generated are being analyzed by mass spectrometry or gas-liquid chromatography. Pyrolysis may be used for relatively quick identification of microorganisms and is popularized as pyrotaxonomy.

Serological taxonomy

The most common use of serology in bacterial taxonomy is the infra-sub-specific grouping of bacteria *i.e.* serovars according to the antigenic structures of bacterial cells (somatic and flagellar antigens). In serological works with polyclonal antibodies the results often vary depending on the strains, investigators and laboratories. Such inconsistency is mainly due to heterogeneous nature of polyclonal antibodies. Monoclonal antibodies prepared by an entirely different process react with certain specific antigen compounds. When the proper antibodies are obtained, they will be quite promising in identification of bacteria at the different taxonomic ranks.

Chapter 4

Isolation and Characterization

Isolation of Plant Pathogenic Bacteria from Infected Plant Parts

Small infected tissue of fruits/leaves were cut aseptically from typical symptom spots along with a little portion of healthy tissue. The infected bits were surface sterilized in 70 per cent alcohol/1.0 per cent sodium hypochlorite and washed in three series of sterile distilled water to remove traces of alcohol/sodium hypochlorite. The infected bits were then suspended in a small test tube (sterilized) containing 3.0 ml of sterilized distilled water for 10 min till water become slightly turbid due to oozing of bacterial cells from the cut ends of the diseased tissue. Later, bacterial suspension was serially diluted in 9 ml sterile distilled water. The inoculation is through pour plate method. One ml of the diluted bacterial cell suspension was added into each sterilized Petri plates and melted Nutrient Agar (NA)/Yeast Dextrose Calcium Carbonate Agar (YDCA) was poured. The plates were rotated gently in clock wise and anti-clock wise directions, so as to distribute the bacterial cell suspension uniformly in the plates and to obtain separated bacterial colonies. The inoculated plates were incubated at room temperature for 72 hour. Observations were made for development of well separated bacterial colonies on NA/YDCA media.

Common Media (Per liter) Used for Isolation of Plant Pathogenic Bacteria

1. Nutrient Agar (NA)

Chemicals	*Gram/Litre*
Beef extract	3.0 g
Peptone	5.0 g
Agar	15.0 g
Water	1.0 litre

2. Nutrient Glucose Agar (NGA)

Beef extract	3.0 g
Peptone	5.0 g
Glucose	2.5 g
Agar	15.0 g
Water	1.0 litre

3. Yeast Dextrose Calcium Carbonate Ahar (YDCA)

Yeast extract	10.0 g
Dextrose (Glucose)	20.0 g
Calcium carbonate (USP Light powder)	20.0 g
Agar	15.0 g
Water	1.0 litre

4. Nutrient Broth Yeast Extract Agar (NBYA)

Nutrient broth	8.0 g
Yeast extract	2.0 g
$K_2H\,PO_4$	2.0 g
KH_2PO_4	0.5 g
Glucose	2.5 g
Agar	15.0 g
Water	1.0 litre

5. King's B Medium

Chemicals	*Gram/Litre*
Proteose peptone (Difco)	20.0 g
K_2HPO_4	1.5 g
$MgSo_4$. 7 H_2O	1.5 g
Agar	15.0 g
Glycerol	15.0 ml
Water	1.0 litre

6. Casein Agar Plus Glucose

Casein acid hydrolysate	10.0 g
Yeast extract	5.0 g
Glucose	5.0 g
K_2HPO_4	4.0 g
Agar	17.0 g
Water	1.0 litre

7. Tyrosine Agar

Sodium caseinate	25.0 g
Sodium nitrate	10.0 g
L-Tyrosine	1.0 g
Agar	15.0 g
Water	1.0 litre

8. SX Agar

Starch (Soluble- potato)	10.0 g
Beef extract	1.0 g
Ammonium chloride	5.0 g
Dipotassium phosphate	2.0 g
Methyl violet 2 B	1.0 ml *
Methyl green	2.0 ml **
Agar	15.0 g
After autoclaving add cycloheximide	5.0 ml ***

* 1 per cent solution in 20 per cent ethanol -Be sure to use methyl violet 2 B and and not methyl violet B as the latter is more toxic to *X. campestris* pv. *campestris.* ** 1 per cent aqueous solution. *** 5.0 g to 10.0 ml methanol, make up the volume to 100 ml with water and filter sterilize (0.22 mm membrane).

SX medium can be used in isolation of *X. campestri.* pv. *campestris* and several other starch positive species from plant tissues and soil. It is also useful in direct plating of contaminated seeds.

9. Tween Medium for *X. vesicatoria*

Chemicals	*Gram/Litre*
Peptone	10.0 g
Potassium bromide	10.0 g
Calcium chloride	0.25g
Agar	15.0 g
After autoclaving add	
Tween 80	10.0 ml
Cephalexin	0.025 g
5 - fluorouracil	0.012g
Tobramycin	0.004g
Cycloheximide	0.05g
Water	1.0 litre

Characterization of Phytopathogenic Bacteria

Bacteria are prokaryotic microorganisms that cannot or can hardly been seen with the naked eye. Bacteria are characterized by the absence of a true nucleus with a membrane. The nuclear material (naked DNA) occurs free in the cell. Mitochondria are lacking and ribosome are small (70S). Generally chloroplasts do not occur in bacteria but photosynthetic bacteria contain chlorophylls in the cytoplasm or in so called thyllakoid membranes. Cell wall of bacteria contains peptidoglycan or peptidogly can – like polymers but do not contain chitin or cellulose.

Bacteria can be found almost everywhere. In soil, they are found the most abundantly in superficial layers of soil and 1 g of soil may contain about 10^8cfu and per hectare live weight of bacteria is about 100 – 4000 kg. In air, many bacteria or bacterial spores are found. They are often attached to particles, so that dusty rooms contain many bacteria, presenting problems when sterile work is performed. In polluted water a large number of bacteria (10^7 cells/ml) are found. Their number is dependent on the source of water, O_2 tension and degree of pollution. The pathogen may also present in the water. On or in the living organism bacteria are always present *e.g.* on the skin and in the gut of man and animals and on or near the roots of the plants.

Like man and animals, plants are affected by a number of diseases caused by bacteria. Bacterial diseases in plants can be as disastrous, especially because direct control is impossible in many cases. Bacteria that cause plant diseases are gram positive or negative rods with the exception of *Streptomyces* spp. that are filamentous. The phytopathogenic bacteria belong to 31 genera such as *Acidovorax, Agrobacterium, Arthrobacter, Brenneria, Burkholderia, Clavibacter, Curtobacter, Dickeya, Enterobacter, Erwinia, Herbaspirillum, Janthinobacterium, Leifsonia, Candidatus Liberobacter, Nocardia, Pantoea, Pectobacterium, Pseudomonas, Ralstonia, Rathayibacter, Rhizobacter, Rhodococcus, Samsonia, Serratia, Sphingomonas, Spiroplasma, Streptomyces, Xanthomonas, Xylella* and *Xylophilus.* The most plant pathogenic bacteria are intercellular (living between plant cells) and/ or necrotrophic pathogens. The group of the fastidious xylem- or phloem limited bacteria formerly called as Rickettsia- like bacteria and now Phytoplasma and fastidious xylem – and phloem limited bacteria (FXLB and FPLB) are non culturable. The brief description of important phytopathogenic genera is given below-

Acidovorax

The genes *Acidovorax* was originally proposed in 1990 by Williams *et al. (*1990). In 1992, Williams *et al.* proposed-transferring those plant pathogens previous ally classified as *Pseudomonas avenae, P. psecudoalcaligenes* subsp *citrulli* and *P. psudoallaligenes* subsp. *konjaci* into newly proposed genus as sub sp. of *Acidovarax avenae.* Bacterium within this genus belongs to the *Acidovorax* rRNA complex in rRNA super family III which are included in the β subclass of the proteobacteria. In *Acidovorax avenae* three subspecies were recognized: subsp. *avenae, citrulli* and *cattleyae.* The bacterium is a strictly-aerobic, Gram negative, straight to slightly curved rod shaped, 0.2-0.8x1.0-5.0 μm, motile by a single polar flagellum (rarely by 2 or 3 polar flagella). All strains are oxidase positive. No pigment is produced on nutrient agar. The strains of *A. konjaci* turn certain media (KB medium, YP agar and potato glucose agar) deep brown. Plant pathogenic strains within this genus cause seedling blights and fruit blotches. The bacterium causes diseases in plants as bacterial leaf spot and bud rot or brown spot of orchids (*A. avenae* subsp. *cattleyae*), bacterial blight disease of oat (*A. avenae* subsp. *avenae*) and blight disease in watermelon (*A. avenae* subsp. *avenae).*

Rhizobium (Previously known as *Agrobacterium*)

The genus *Agrobacterium* belongs to the family Rhizobiaceae. The bacterium is aerobic, Gram negative, rod shaped, 0.6 - 1.0 x 1.5-0 μm, occur singly or in pairs, non endospore forming, motile by one to 6 peritrichous flagella. On PDA + $CaCO_3$, *Agrobacterium* strains grow well and their colonies are distinguishable from other species. They are convex, glistering, circular with an entire edge and white to tannish cream in colour. *Agrobacterium* causes crown gall disease in rose, dahlia, chrysanthemum *(A. tumefaciens),* hairy root disease in tomato *(A. rhizogenes),* cane gall in *Rubus (A. rubi)* and gall on branches of grapevine *(A. vitis).*

Arthrobacter

The bacterium is Gram positive, irregular coccoid rods, measuring 0.6 - 0.7 x 0.9 - 1.0 μm, V-shaped, with clubbed ends without any filaments or motile by one sub polar or few lateral flagella. Colonies appear on B medium yellow, 3 - 5mm in diameter, convex in centre and embedded in mucous externally. They are obligate aerobes. The cell wall peptidogly can contain lysine as di-amino acid. Optimum temperature for growth is 25.3°C and twist *Arthrobactor* species grow in the range of about 10-35°C. DNase is produced and gelatin is usually liquefied. The G+C content of the DNA most species is in the range of 59 - 66 mol per cent. *A. lilicis* causes vascular wilting in *Ilex opaca* and blight in American holly.

Burkholderia

Burkholderia was proposed as a genus in 1992 by Yabuuchi *et al.* (1992) contained plant pathogens of the rRNA –DNA homology group II pseudomonads of Palleroni *et al. (*1973). Current taxonomy recognizes 18 species in the genus *Burkholederia* of which 8 are plant pathogenic. They are Gram negative, straight curved rods (0.5 - 1.0 x 1.5 - 4.0 μm), polar or multitrichous flagella, catalase positive and accumulates poly-β-hydroxybutyrate (PHB). Most are strict aerobes and do not produce fluorescent pigment under iron limiting conditions. *B. cepacia* produces antibiotics which act as biocontrol agent against many fungal, bacterial and nematode pathogens of plants. Plant disease symptoms produced by *Burkholderia* include rots, wilts and severe necrosis depending on individual host plants and some instances cultivars. For example, *B. cepacia* causes sour skin and bulb rot in onion, *B. gladioli* causes corm and pseudobulb rot in gladiolus, *B. gladioli* pv. *agaricicola* causes soft rot in mushroom, *B. glumae* causes sheath and grain rot in rice and *B. caryophylli* wilt and stem rot in carnation.

Candidatus Liberibacter

Nonculturable, phloem limited bacteria, based on 16S rRNA sequencing belong to the Gram negative alpha group of Proteobacteria, Gram negative and no flagella. Thin-section EM examination reveals elongated sinuous rods shaped with an uneven diameter of 0.15-0.25 μm. Round forms which are of larger diameter can also be observed in degenerating cells. Similar particles are observed in the haemolymph and salivary glands of both insect vectors. The pathogen is transmitted by insect vector *Diaphorina citri* Kuwayama, the Asian citrus psyllid. The bacterium is vectored by the psyllid from host plant to host plant, where it resides exclusively in the sieve tubes of the phloem vessels. Electron microscopy studies reveal that they also possess the characteristic double membrane cell envelope of the liberibacters. As the bacterium multiplies, it chokes off the supply of nutrients moving throughout the plant which weakens the plant and eventually killing it. The bacterium causes citrus greening (*Ca.* L. asiaticum. A heat tolerant from Asia and *Ca.* L. africanum: a heat sensitive from Africa) and papya bunchy top.

Candidatus Phytoplasma

Phytoplasmas were discovered in 1967 by Japanese scientists Doi *et al.* who termed them mycoplasma-like organisms. The organisms were renamed as phytoplasmas in 1994 at the 10th Congress of the International Organization for Mycoplasmology. It is a wall- less, phloem limited, single cell membrane, non culturable, no flagella, Gram positive Dienes' staining, rod shaped or pleiomorphic or filamentous shape, a diameter normally less than 1 μm and contains very small genome (~750 kb). Ca. Members of the genus '*Ca.* Phytoplasma' are sensitive to antibiotics of the tetracycline group, but not to penicillin types. According to results from buoyant density centrifugation, the G+C content in DNA of '*Ca.* Phytoplasma' is 23–29 mol per cent. Phytoplasm is transmitted by Leaf hopper. The Aster yellow pathogen alone infects over 300 hosts, with plant species occurring in 50 families. Symptoms include virescence loss of flower colour, phyllody, flower sterility, witches broom- like shoot proliferation, stunting and yellowing. Diseases like Little leaf of brinjal, Sesamum phyllody.

Clavibacter

The bacterium can be differentiated from other corynefrorm group of bacteria by the cell wall as peptidoglycan of this genus including plant pathogens is based on diamino butyric acid (DAB) and absence of mycolic acid. Predominated menaquinones contain 9 and 10 carbon chains (MK-9 and MK-10). The b+c ratio are 67-68 mol per cent. The bacterium is gram positive coryneform rods. Colonies appear on media (nutrient dextrose agar) for sub culture only, yeast peptone glucose agar, nutrient yeast dextrose agar after 20 days at 21°C incubation, pin point, cream and domed within 10 days due to slow growing. Growth rate are improved with sub culture. Typical colonies are-creamy white or ivory, round, smooth, raised convex (domed), mucoid - fluidal with entire edges and usually 1 - 3 mm in diameter. *Cliavibacter michiganensis* subsp. *michiganensis* grow on YGMA medium and colonies are convex, mucoid, pale yellow to yellow after four days of incubation. They are slow growing. The bacterium is catalase and esculin hydrolysis positive and oxidase, nitrate reduction, urease activity, H_2S production, indole production and nitrate utilization negative. Starch hydrolysis is negative or very weak. In case of lactose, it produces acid very weak or negative. They cause vascular and stunting symptoms *e.g. C. michiganensis* subsp. *michiganensis* causes bacterial canker of tomato; *C. michiganensis* subsp. *sepedonicus* causes ring rot of potato.

Curtobacterium

The bacterium is a Gram positive, irregular short rod. Lateral or sub polar flagella are found aerobic and catalase positive. The optimum temperatures for the bacteria to grow is 24 - 27°C. The colonies of sugar beet strains produces a pale yellow pigment. The bacteria utilize a wide range of organic compounds like hydrolyzed casein, starch, esculin and Tween 80 and releases H_2S from cystine and $Na_2S_2O_3$. $5H_2O$ but could not produce urease oxidase or indole. The DNA base compose is 68 - 75 mol per cent, G+C. The distinct cell wall peptidoglycan of this genus including plant pathogens is based on ornithine and no mycolic acids are present. MK-9 is the predominant respiratory menaquinane. The plant pathogens transferred to the single plant pathogenic species *viz., C. flacumfaciens* comprising the pathogens (*C. f. flacumfaciens,* C. *f. betae, C. f. orrtii and C. f. poinsettiae,* which cause vascular wilt and leaf spot in red beet, beans (*Phaseolus* spp. and *Glycine max*), tulip and *poinsetlia* respectively.

Enterobactor

The genus *Enterobacter* is a member of Enterbactericae. The bacterium is gram positive, rod shaped, measuring 0.3 –1.0 μm wide and 1.0–3.0 μm long, motile with peritrichous flagella. On TTC medium, after 24 h incubation at 30°C, dark pink, non-mucoid colonies with regular edges and translucent margins will be formed. These are facultatively anaerobic, catalase positive, oxidase negative. The bacterium causes canker and die back disease to the plants. For example species of *enterobacter* cause diseases as *E. cancerogenus* on *Populus canadensis, E. dissolvens* on *Zea mays, E. nimipressuralis* on *Ulmus* spp. and *E. pyrinus* on *Pyrus pyrifolia.*

Erwinia

The genus *Erwinia* is Gram negative, facultative anaerobes, peritrichous flagella rod shaped and acid produced from fructose, glucose, galactose and sucrose. All species exhibit motility. On CCT medium, *E .amylovora* colonies are translucent, smooth, large (4 - 7 mm), light blue with entire margins. On MS medium, member of Enterobacteriaceae produce red orange-colonies while P*seudomonas* and other bacteria are distinguished by the bluish colour of their colonies. Godfrey *et al.* (1998) suggest that pectolytic bacteria can be confirmed as *Erwinia* on the basis of being facultative anaerobic, catalase positive and oxidase negative. The Erwinia are broadly divided into two groups *i.e.* amylovora group (non soft rot group) and carotovora group (soft rot group). The most species of soft rot cause various blights, cankers, leaf spots, wilts and rots. For example fire blight of pear and apple caused by *E. amylovora* and bacterial wilt of cucurbits caused by *E. tracheiphila.* In carotovora group, *E. carotovora* subsp. *carotovora* and *E. c.* subsp. *atroseptica* cause soft rot of vegetables.

Ewingella

Ewingella americana is a Gram-negative rod and it is the only species in the genus *Ewingella.* It was first identified and characterized in 1983 under family Enterobacteriaceae. The bacterium is named in honor of an American biologist William H. Ewing, who contributed to modern taxonomy. *E. americana* is an organism with simple nutritional needs that can survive in water and citrate solution and preferably grows at 4°C. Domestic sources of water, including air conditioning units, ice baths and wound irrigation systems. It causes browning disorder called 'internal stipe necrosis on cultivated mushrooms (*Agaricus bisporus*). Phenotypically it is easy to distinguish from other enterobacteria, since it is negative for lipase and DNAase, arginine dihydrolase and lysine and ornithine decarboxylase tests, Voges-Proskauer positive. It produces acid from D-glucose but fails to produce acid from L-arabinose, ramnose, melibiose, D-sorbitol or sucrose. *E. americana* causes internal stipe necrosis disease in button mushroom.

Gibbsiella

The bacteriaum is Gram-negative, short rods (0.9 × 1–1.5 μm), facultative anaerobe, shows oxidase negative and catalase positive. Cells occur singly, in pairs or in groups of four and possess very fine fimbriae but flagella are absent. Colonies are white to cream on nutrient agar, convex, round and smooth with entire margins. Strains can grow at temperatures of 10-40 °C. Bacterium is positive for β-galactosidase activity but negative for H_2S, indole, urease, acetoin and gelatinase production. Citrate is utilized and nitrate is reduced to nitrite. Negative for arginine dihydrolase, lysine decarboxylase, ornithine decarboxylase and tryptophan deaminase activity. *Gibbsiella quercinecans* causes necrosis disease in oak.

Gluconobacter

Gluconobacter oxydans, previously known as *Acetobacter suboxydans* are Gram-negative rod or oval shaped bacteria of about 0.5 to 0.8μm x to 4.2μm. The name *oxy* from *Gluconobacter oxydans* in Latin for 'sharp' and 'acidic' and *dans* is 'giving'. Its abilities include partially oxidizing carbohydrates and alcohols through the process of oxidative fermentation and they can be used for synthesis of Vitamin C, D-gluconic acid as well asketogluconic acids. *G. oxydans* generally found in flowers, fruits, garden soil, cider, alcoholic beverages and soft drinks because they contain strains capable of growing in high concentrations of sugars and low pH values (optimal pH for growth is 5.5-6.0). They are able to grow in extreme conditions but their growth rate is slow and the concentration of mature cells is very low. *Gluconobacter oxydans* strains cause bacterial rot of apples and pears turning those into shades of brown. *G. oxydans* causes pink disease of pineapple fruit. Biosensors using *G. oxydans* can be used to measure substrate concentration as a biosensor.

Herbaspirillum

The genus *Herbaspirillum* of the *Betaproteobacteria* mainly comprises of diazotrophes with a potential for endophytic and systemic colonization of a variety of plants. After phylogenetic analysis of 16S rRNA gene sequence data the isolates could be allocated to the genus *Herbaspirillum* and 99.9 per cent similarity to the sequence of *Herbaspirillum lusitanum* P6-12^T was found. These places the three isolates as representatives of a novel species within the genus *Herbaspirillum*. A Biolog GN2 assay supported this conclusion. The major fatty acids were $C_{16:1}w7c$, $C_{16:0}$ and $C_{18:1}w7c$ and the DNA G+C content ranged from 60.9 to 61.5 mol per cent. *Herbaspirillum rubrisubalbicans* causes red stripe disease of sugarcane and mottle stripe disease of sorghum.

Janthinobacterium

Bacterium is Gram negative, polar flagella, rod shaped, size 1-0-1-5 x 1-8-2-5 µm and aerobic. It produces low convex, round, beige grey – white colony colonies on solid medium. There is no growth below pH 5 or in media containing 2-9 per cent NaCl. Bacterium is oxidase and catalase-positive and negative for indole and Proskauer - Voges. Nitrate and nitrite are not reduced. It utilizes citrate and ammonium ions as sole carbon and nitrogen sources and grows. The G+C content is 64.2 mol. *J. agaricidamnosum* causes soft rot disease in *Agaricus bisporus.*

Leifsonia

Davis *et al. (*1984) proposed new genus for 2,4-diaminobutyric acid-containing actinomycetes to accommodate *Corynebacterium aquaticum* Leifson 1962 and *Clavibacter xyli* subsp. *cynodontis* Davis *et al.* (1984). *Clavibacter xyli* Davis *et al.* (1984) with two subspecies in the new genus as *Leifsonia xyli* (Davis *et al.,* 1984) comb. nov., *Leifsonia xyli* subsp. *xyli* (Davis *et al.,* 1984) comb. nov. causes sugar cane ratoon stunting disease and *Leifsonia xyli* subsp. *cynodontis* (Davis *et al.,* 1984) comb. nov. The members of the proposed genus are characterized by coryneform shape, peptidoglycans based upon 2,4-diaminobutyric acid, the major menaquinone MK-11, phosphatidylglycerol and diphosphatidylglycerol as principal phospholipids, the high content of anteiso- and iso-branched saturated fatty acids and a DNA G+C base composition of 66-73 mol per cent. The bacterium is Gram +ve, cell size 5.0 x 02 µm. of bacteria appear on the medium after 7 -10 days as white in colour. No growth of bacteria on CB Colonies agar medium. They are negative in oxidase test, production of H_2S, Voges-proskauer test and methyl red test. Bacteria are not able to hydrolyzed starch, aesculin and gelatin and not able to produce acid from D- arabinose, D- galactose, salicin and D-sucrose. *Leifsonia xyli* subsp. *xyli* causes ratoon stunt diseases in sugarcane.

Lonsdalea

Lonsdalea, named for David Lonsdale in honour of his contributions to British forest pathology. The isolates tentatively identified as *Brenseria quercina* from oak trees in Britain later given name as *Lonsdalea quercina.* Cells are Gram-negative-staining short rods (0.5–1.06 x1.0– 2.0 µm) occurring singly or in pairs or in groups, motile by peritrichous flagella, facultative anaerobic, oxidase negative but catalase positive. Optimum temperature for growth is 28–30 °C. Colonies are white to cream in colour on tryptone soya agar, round, convex and smooth with entire margins. Negative in tests for b-galactosidase, arginine dihydrolase, lysine decarboxylase, ornithine decarboxylase and tryptophan deaminase activities. Citrate is utilized but no production of H_2S, urease, indole or gelatinase. Nitrates is not reduced to nitrites. The DNA G+C content ranges from 54.5 to 54.9 mol per cent. Members of this genus may cause diseases on oak trees including bark canker and drippy nut disease and may be associated with Acute Oak Decline in Britain.

Norcardia

Genus *Norcardia* is a genus of weekly staining from positive, catalase positive, rod shaped short filaments bacterium, no flagella. It forms partially acid test bladed branching filaments. They are strict aerobes with the ability to grow in a wide temperature range and produce reddish with some white aerial hyphae. *Nocardia vaccinii* causes gall on *Vaccinium* spp.

Pantoea

They are Gram negative, rod shaped, facultative anaerobic bacteria that are oxidase negative catalase positive. Most species of pantoea are motile by Peritrichous flagella. Acid is produced from arabenose, tructose, galatose, glucose, trehalose and N-acetyl glucusamine gas is not produced from glucose. The group is commonly differentiated from other *Erwinia* by the production of a yellow pigment. However *P. citrea* and a few strains of *P. agglomerans* do not produce yellow pigment. They do not degrade pectate, do not require growth factors, do not produce urease, lysine decarboxylase or ormithine decarboxylase and do not produce H_2S on Triple sugar iron agar. *Pantoea* spp. causes several diseases to the plant such as Stewart's wilt disease of maize (*Pantoea stewartii* subsp. *stewartii*), brown spot of Graminae (*P. stewartii* subsp. *indologenes)*, soft rot of pineapple (*P. ananatis*), pink disease of pineapple and orange (*P. citrea*).

Pseudomonas

They are gram negative, straight or curved rods, 0.5-1.0 x 1.5-4.0 mm in size. Cells are motile by are one to several flagella and to 6 + C content of the DNA catalase positive and strict aerobes except for a few which denitrify. They produce fluorescent pigments which becomes visible on iron-difficult media such as King's B medium. The genus includes many non-pigmented species but some produce fluorescein (yellow-green) and pyocyanin (blue-green) pigments, soluble in water, that with ageing may fade to brownish color diffusing into the medium. Except for *Pseudomonas corrugata,* no pseudomonad grows at 41°C, utilizes D-arabinose as sole carbon source and accumulates poly-B-hydroxy butyrate (PHB) as carbon reserve material. New molecular analyses and taxonomic rearrangement identified the rRNA group I species, including the type species *Pseudomonas aeroginosa* and other fluorescent species such as *P. fluorescens, P. putida, P. angulata, P. cickorii, P. marginalis* and *P. syringae* as a member of a phylogenetically homogenous group referred to as *Pseudomonas* based on 16S rRNA: DNA homology. 14 plant pathogenic pseudomonas species, of which *P. syringae* is the most economically important with more than 50 pathovars. Other includes *P. viridiflava*, a soft –rotting species and *P. marginalis*, a brown rotting species of lettuce, alfalfa and parsnip. *P. savastanoi* causes gall or tumour producing species. Additional phytopathogenic species of include *P. tolaasii* and *P. agarici* infect to the mushroom. Some other species of *Pseudomonas* have minor importance to cause disease in plants. *Pseudomonas* cause many diseases in plants such as *Pseudomonas syringae* pv. *syringae* (bacterial canker or blast (stone and pome fruits), *Pseudomonas syringae* pv. *garcae* (Bacterial blight of coffee), *Pseudomonas* syringae pv. *tomato (*speck disease of tomato*), Pseudomonas viridiflava* and *Pseudomonas cichorii* (Bacterial blight of tomato).

Ralstonia

Ralstonia solanacearum was categorized as non fluorescent *Pseudomonas* despite evidence that it was completely unrelated to the fluorescent pseudomonads. *R. solanacearum* is belong to rRNA homology group II, where as the fluorescent *Pseudomonas* in rRNA homology group I absorbed in 1992, Yabuuchi *et al. (*1992) transferred even species in rRNA homology group II including *P. solanacearum* into the new genus *Burkholideria.* Later Yabuuchi *et al. (*1995) established the genus *Ralstonia* to accommodate *R. solanacearum, R. eutropha* and *R. pickettii. Ralstonia solanacearum* is a gram negative, rod shaped, sized 0.5 - 0.7 x 1.5 -2.5 μm, oxidase and catalase positive, accumulate poly-B-Hydroxy butyrate (PHB) and reduce nitrate. There will be no growth at 40°C little or no growth in broth with 2.0 per cent NaCl and cultures shows negative for arginine dihydrolase, gelatine liquefaction and starch or esculine hydrolysis. Colonies are non fluorescent, but a defensible brown pigment is often produced on a complex media. Two type of colonies are observed on complex medium containing 0.5 per cent glucose; one is fluidal (Mucoid) due to copies production of extra cellular polysaccharides (EPS), where as the other is dry (butyrous) type. Flagella are polar when present. The disease caused by *Ralstonia solanacearum* is bacterial wilt of solanaceous members and other crops.

Rathayibacter

Rathayibacter (Ra. thay'. i. bac. ter. M.L. gen. n. *rathayi,* of Rathay, referring to E. Rathay, the Australian plant pathologist who first isolated members of the genus).A new genus, *Rathayibacter,* was been proposed to accommodate three species of gram-positive, aerobic, coryneform bacteria which were earlier placed in the genus *Clavibacter (Rathayibacter rathayi* comb. nov., *Rathayibacter tritici* comb. nov. and *Rathayibacter iranicus* comb. nov.), as well as six strains that were isolated from annual cereal grasses, probably responsible for ryegrass toxicity. The properties of genus *Rathayibacter* include coryneform shape, peptidoglycan based on 2, 4-diaminobutyric acid (type B2γ), predominant MK-10 type menaquinones and phosphatidylglycerol and diphosphatidylglycerol as basic polar lipids. The DNA base compositions range from 63-72 mol per cent G+C. Most of the *Rathayibacter* strains are associated with nematodes belonging to the genus *Anguina*. The *Rathayibacter* species differ from species belonging to related genera like *Clavibacter* and *Agromyces* species in the following characteristics: Irregular, pleomorphic, short rods (length, 0.5 to 1.8 μm and width, 0.4 to 0.8 μm;). Bacteria are V-forms and sometimes short cell chains are formed, Gram positive and not acid fast, non-motile, non-sporulating, obligately aerobic, chemoorganotrophic, possessing an oxidative type of catabolism. Reacts positive for catalase production. Bacteria are mesophilic in nature with optimal growth at temperatures between 24 and 28°C.Tween 60, Tween 65, Tween 80, starch, urea and xanthine are not hydrolyzed or decomposed.

Nitrate is not reduced to nitrite. Hydrolysis of hippurate shows negative. Methyl red reduction and Voges-Proskauer tests are variable in nature. Growth occurs at temperatures between 16 and 35°C, but not at or below 10 or at or above 37°C. Susceptible to 7 per cent NaCl and 0.02 per cent potassium tellurite but tolerates 0.01 per cent potassium tellurite. Some strains produce bacteriocins. The type species of the genus *Rathayibacter* is *R. rathayi*. *Rathayibacter tritici* causes yellow slime disease of wheat and in association with *Anguina tritici* it causes tundu disease of wheat.

Rhizomonas van

It is a Gram negative, rod shaped and belongs to the rRNA super family IV. It is naotile with one lateral sub polar or polar flagellum, most wrinkled. It grows at 20-33°C temperature with optimum temperature for growth is at 29-33°C. It is not grow at a pH of 5.1 or below. Bacterium is aerobic, metabolizes glucose oxidatively, oxidase positive, weakly catalase positive, reduces nitrate to nitrite and ammonia but not to nitrogen gas. The bacterium is mitrogenase and orginnedihydrolase negative. Ethanol is not converted to acetic acid by the bacterium and does not fluoresce on KB and utilizes very few carbon sources. The whole cell faulty acid composition is unique, escipcially the presence of relatively large amount of 2 tetradecanol (14:0.2 OH). B + C content of this bacterium are 59 per cent mol per cent. Bacterium grows on S-medium for three weeks at 29°C. Colonies are creamy in appearance and are wrinkled with undulated margins. *Rhizomonas suberifaciens* causes corky root disease of lettuce.

Rhodococus

It is Gram positive, aerobic, pleiomorphic, non motile. It does not form spores. When grown on the surface of agar plate, colonies are orange in colour and appear broth smooth or rough. The bacteria produce non defensible orange pigmentation on NBY agar medium. The bacterium grows on TTC medium and produced and for and sarbitol and negative in inseculin. Bacterium utilizes acetate and formate and does not hydrolyze casein and Insulin. The bacterium *Rhodococus fascians* causes leafy gall and fasciation of numerous plant species including tobacco, small fruits (caneberries, strawberries) and ornamental plants (butterfly flowers, *Primula*, kalanchoes, *Impatiens*, geraniums, carnations).

Spiroplasma

The bacterium is cultural, no cell wall, Gram positive, facultative anaerobic. They are Spiral pleiomorphic rod, minimum viable length of a helix is 2.0 x 0.1-0.2 µm, intracellular fibrils, flexing or rotation. Some strains are non-motile in nature and non-helical. *S. citri* is one of the very few plant pathogenic mollicutes that have been cultured so far, with transparent fried- egg colonies with granular margins. The pathogen *Spiroplasma citri* causes stubborn disease of citrus and *S. kunkelii* Whitcomb *et al.* causes corn stunt disease.

Streptomyces

The genus *Streptomyces* are filamentous and about 1mm in diameter and Gram positive bacteria. Initially colonies consist of substrate mycelium and are smooth (not shiny) and firm to rubbery in texture. The colonies produce fuzzy, white aerial mycelium several days after planting. Aerial mycelium is fragmentinti pigmented spore chains with a characteristic morphology is spiral, rectifications, straight *etc.* producing a colony with powdery appearances and a characteristic colour *i.e.* gray, white, pink, blue or green. They produce brown, red/yellow pigments on specific media. On the basis of morphological and physiological characters, species of *Streptomyces* can be differentiating. The plant pathogenic *Streptomyces* species cause diseases of underground structure of plants. *S. scabiei* causes common scab of potato and similar diseases of taproot crops (radish, turnip, carrot, rutabaga and parship). It the best studied of the pathogenic is characterized by necrosis of fibrous roots and cankers on storage roots at any stage of their development. Thaxtomin, appear to play a major role in most diseases caused by *Streptomyces* species. *Streptomyces scabies* or *Streptomyces scabiei* causes corky lesions on tuber and root crops as well as decreasing the growth of seedlings was also observed. The most common disease of potato is common scab of potato caused by *S. scabiei.*

Xanthomonas

Genus *Xanthomonas* is gram negative, rod shaped, measuring 0.4-0.7x0.7-1.8μm with a single polar flagellum (monotrichous). They are aerobic, catalase positive and weak producer of acids from carbohydrates. Colonies of most species and pathovars are mycoid, convex and yellow on YDC agar. They produce extra cellular polysaccharides, which is the source of xanthan gum' on media containing glucose. The unique yellow membrane –bound non-water soluble pigments produced by all but three species/pathovars, are brominated arylpolulene esters (Zanthomonadins) soluble in petroleum either, methanol and benzene. *Xanthomonas* species can be identified by the yellow xanthomonodin. They are oxidase negative, non-fermentative and no growth on 1.0 per cent TTC medium. They grow at 40°C, H_2S from cysteine positive, levan from sucrose positive. Xanthomonas are easily differentiated from other genera of aerobic, Gram negative rods such as *Pseudomonas, Acodovorax, Burkholderia* and *Ralstonia* and other yellow pigmented bacteria such as *Flavobacterium* and *Pantoea* spp. The following species of *Xanthomonas* such as *albilineans, axonopodis, campestris, citri, fragariae, maltophila, oryzae, phaseoli* and *populi* have been published in International Journal of Systemic Bacteriology and causes various diseases in the plants. For example, bacterial leaf blight of rice (*X. oryzae* pv. *oryzae*), citrus canker (*X. citri* pv. *citri*), bacterial blight of pomegranate (*X. citri* pv. *punicae*), black rot of crucifer (*X. campestris* pv. *campestris*).

Table 1. Phynotypic characters useful for differentiating *Xanthomonas* from *Pseudomonas* and other yellow pigmented bacteria like *Flavobacterium* and *Pantoea*

Character	*Xanthomonas*	*Pseudomonas*	*Flavobacterium*	*Pantoea*
Flagellation	1, polar	>1, polar	None	Peritrichous
Xanthmonadin	Yes	No	No	No
Fluorescence	No	Yes	No	No
Litmus milk	Alkaline[a]	Alkaline	Unchanged	Not determined
Growth at 40°C	Yes[b]	No	Yes	Yes
Levan from sucrose	Yes	Yes	No	No
H_2S from cysteine	Yes	No	No	No
Oxidase	Negative	Negative[c]	Positive	Negative
Fermentative	No	No	No	Yes
Growth on 0.1 per cent TTC	No	Yes	Yes	Yes

[a] *X. fragariae* is unchanged, [b] *X. fragariae* will not grow at 40°C, [c] *P cichorii* is positive.

Xylella

Bacterium is Gram negative, xylem limited bacteria. The bacterium was first discovered in association with pierces disease of grapevine in 1973 and was called ricktsia like bacteria. The first xylem limited bacteria was grown in axemic culture in 1978 (Davis *et al.,* 1978). The name *Xylella fastidisa* was proposed to establish a new genus with one species to include the all of the known strains of fastidious, gram negative, xylem limited bacteria. Cells of *X. fastidiosa* are very narrow and cannot be seen by bright field microscopy. They can only be observed by phase contrast or dark field microscopy. Since the cells are small 0.2-0.4x1.0-4.0 mm. their size becomes very useful in their identification. The bacterium is transmitted by sucking insects that feed on xylem sap, such as the harp-shooter leafhoppers but is not transmitted mechanically from plant to plant. Although *X. fastidiosa* strains are normally not host specific, they are cell or tissue specific. They are restricted to xylem vessels or inter cellular spaces of the xylem. Plant symptoms induced by *X. fastidiosa* include leaf scorch, leaf chlorosis, stunning dieback and general decline. Major plant diseases caused due to infections of *X. fastidiosa*, includes bacterial leaf scorch, oleander leaf scorch, coffee leaf scorch, alfalfa dwarf, phony peach disease and among them the economically important Pierce's disease of grapes and citrus variegated chlorosis.

Xylophilus

This genus contains a single plant pathogenic species *X. ampelinus* causes bacterial blight of grapes. The grape is the only known host. The organisms grow very slowly in general. *Xanthomonas a*nd Panagopoulos in 1969 identified as

Xanthomonas ampelinus sp. nov. and it was transferred by Willem *et al.,* in 1987 to a new genus *Xylophillus* gen. nov. as *Xylophillus ampelinus* (Panagopoulos, 1969). The bacterium is aerobic, Gram negative, rod shaped and 0.4 - 0.8 x 1.0 - 2.0 µm. Although *X. ampelinus* is yellow on NA, YDC and GYGP, the organism is quite easy to differentiate from *Xanthomonas* spp. because 10-12 days are required for colonies to reach 1mm diameter. The colonies are round and only slightly mucoid on YDC. The organism is urease positive whereas all *Xanthomonas* are urease negative. It has optimum and maximum temperature for growth of 24°C and 28°C respectively. It produces acid from arabinose and galactose but not from fructose and maltose. It does not hydrolyzed esculin or liquefy the gelatin. All strains are oxidase negative and produce H_2S from cysteine.

Fastidious Vascular Bacteria (FVB)

The fastidious vascular bacteria (earlier known as RLO) stimulates plant diseases which cannot be grown on simple culture medium in the absence of host cells. Phloem-limited fastidious bacteria were first observed during 1972 in the phloem of clover and later in citrus plants affected with the greening disease. In 1973, xylem-limited fastidious bacteria were noticed in xylem vessels of the grape plants affected with pierce's disease and alfalfa affected with alfalfa dwarf. Subsequently, similar organisms were identified in the xylem of plants affected over and above 20 other diseases. For example, phony peach disease of peach, ratoon stunting of sugarcane, plum leaf scald, almond leaf scorch and elm leaf scorch.

They are generally rod shaped cells having 0.2 to 0.5 µm diameter and 1 to 4 µm length. They are gram negative bacteria with no flagella, bounded by a cell membrane and a cell wall. Several such xylem limited bacteria have been placed in genus *Xyella,* while Bermuda grass stunting and sugarcane ratoon stunting are Gram positive and are classified as members of *Clavibacter*. So far, none of the phloem-inhabiting fastidious bacteria have been grown in culture but all the xylem-inhabiting fastidious bacteria can be cultured and all xylem inhabiting gram negative fastidious bacteria are transmitted by xylem feeding insects such as leaf hoppers, sharpshooter and spittlebugs. So far, no insect vector is known for the Gram positive xylem inhabiting fastidious bacteria.

The symptoms of diseases caused by xylem inhabiting fastidious bacteria often consists marginal necrosis of leaves, stunting, general decline and reduced yield. The most important diseases caused by forms of the *Xyella fastidiosa*, xylem-limited fastidious gram negative bacteria are pierce's disease of grape, citrus variegation chlorosis, phony peach disease, palm leaf scald and almond leaf scorch. Ratoon stunting disease of sugarcane is caused by *Clavibacter xyli* subsp. *xyli*, xylem-limited gram positive bacterium. Phloem limited bacteria are known to cause citrus greening disease and some minor diseases of periwinkle and clover.

Forms of Bacteria

Bacteria often produce variants that fails to produce cell wall. The progeny of such variants comprise populations of wall less bacteria called as L phase bacteria or L-form that are morphologically indistinguishable from the phytoplasmas observed in plants. Usually they are produced when penicillin or other substances that inhibit cell wall formation are added to the culture medium under laboratory conditions. They can also be developed apparently with antibiotics treatment in certain living organisms. L-form bacteria are unstable and revert back to the original form when the substances inhibiting cell wall synthesis is removed from the medium or they are stable, *i.e.*, they are unable to revert to the original bacterium. The plant pathogenic bacteria reported to produce L forms are *A. tumifaciens* (crown gall disease) and *E. carotovora* subsp. *atroseptica* (black leg of potato). The L-forms of *A. tumifaciens* retained the pathogenicity and produce tumours identical to those produced by the parent bacterium which can be reisolated and cultured from such tumors.

Chapter 5

Nutrition and Growth

Bacterial Nutrition

Bacteria have been divided into two groups on the basis of their existence in the presence or absence of the oxygen. Those which can live only in the presence of oxygen are known as AEROBIC and those which grow in the absence of oxygen as ANAEROBIC. On the basis of nutritional requirements, the bacteria can be grouped as AUTOTROPHIC and HETEROTROPHIC. Both are further divided into two categories. Thus, there are four major groups:

- ☆ **Photoautotrophs** – Those using light as the energy source and CO_2 as the principal carbon source. They include photosynthetic bacteria.
- ☆ **Photoheterotrophs** – Those using light as the energy source and an inorganic compound as the principal carbon source. This group includes some of the purple and green bacteria.
- ☆ **Chemoautotrophs** – Those using a chemical energy source and CO_2as the principal carbon source. Energy is derived by oxidation of reduced inorganic compounds. Only certain bacterial types belong to this group. Because of their ability to grow in strictly mineral media in the absence of light, these are also called *chemolithotrophs* (*lithos* = rock).
- ☆ **Chemoheterotrophs** – Those utilizing a chemical energy source and an inorganic compound as the principal carbon source. Most bacteria including phytopathogens belong to this group.

However, often the mode of nutrition within and between the groups overlap. The terms OBLIGATE and FACULTATIVE are often used meaningfully to know, identify and designate principally based on presence or absence of nutritional versatility. Plant pathogenic bacteria are basically facultative saprophytes because of the following reasons:

1. They survive as saprophytes on dead organic matter during unavailability of living host plants.
2. They can be grown on synthetic media,
3. The degree of saprophytic survival ability/competitive survival ability.

Culture Media for the Growth of Bacteria

Appropriate biochemical and biophysical environment is necessary for culturing of any bacteria. The biochemical environment is made available as a **culture medium a**nd depending upon the special needs of particular bacteria. A large variety and types of culture media have been developed with different purposes and needs. Culture media are employed in the isolation and maintenance of pure cultures of bacteria according to their biochemical and physiological properties and needs.

Culture media varies widely and it depends on manner in which bacteria are cultivated and the purpose. **Liquid media** are generally used for growth of pure batch cultures, while solidified media are used widely for the isolation of pure cultures, for estimating viable bacterial populations and a variety of other purposes. The usual gelling agent for solid or **semisolid medium** is **agar,** a hydrocolloid derived from red algae (ex. *Gellidium*). Agar is used because of its unique physical properties *i.e.* it melts at 100 degrees and remains liquid until cooled to 40 degrees, the temperature at which it gels) and it cannot be metabolized by most of the bacteria. Hence, as a medium component it is relatively inert in nature, it simply holds (gels) nutrients that are in aqueous solution.

Types of Culture Media

Culture media may be classified into several categories depending on their composition or need. A **chemically defined (synthetic) medium** is one in which the exact chemical composition of medium is known. A **complex (Undefined) medium** is one in which the exact chemical constitution of the medium is not known. Defined media are usually composed of pure bio-chemicals off the shelf, complex media usually contain complex materials of biological origin such as potato or blood or milk or yeast extract or beef extract, the exact chemical composition of which is obviously undetermined. It is said or defined to be **minimal medium** if it provides only the exact nutrients (including any growth factors) required by the organism for growth. The use of defined minimal media helps the investigator to know the exact nutritional requirement of the organisms in question. Chemically defined media has value in studying the minimal nutritional requirements of microorganisms, for enrichment cultures and for a wide variety of biochemical and physiological studies. Complex media usually provide the full range of growth factors that may be required by an organism so that may be more handily used to cultivate unknown bacteria or bacteria whose nutritional requirement are complex in nature (*i.e.* organisms that require a lot of growth factors, known or unknown).

Other concepts employed in the construction of culture media are the principles of selection and enrichment. A **selective medium** is one which possess a component (s) added to it which will inhibit or prevent the growth of certain types or species of bacteria and/or promote the growth of desired species. One can also adjust the physical conditions of a culture medium such as pH and temperature to render it selective for organisms that are able to grow under these certain growth conditions.

A culture medium may also be a **differential medium** if it helps the investigator to distinguish between different types of bacteria based on some observable trait(s) in their pattern of growth on the medium.

Bacterial Growth Curve

In the laboratory, under favorable conditions, active bacterial population doubles at regular intervals. Growth is by geometric progression manner: 1, 2, 4, 8, *etc.* or 2^0, 2^1, 2^2, $2^3.2^n$ (where n = the number of generations). This is called as **exponential growth.** In reality, exponential growth is only part of the bacterial life cycle and not representative of the normal pattern of growth of bacteria in Nature.

When a fresh favorable medium is inoculated with a specified number of bacterial cells and the population growth is monitored over a period of time and plotting the data will yield a **typical bacterial growth curve** (Figure 1).

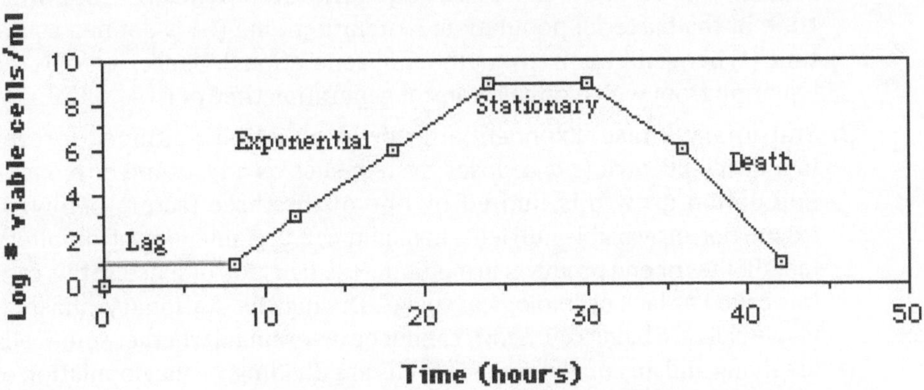

Figure 1. The Typical Bacterial Growth Curve.

When bacteria are grown in a closed system (also called as a batch culture), like a test tube, the population of cells almost always exhibits these growth dynamic patterns: cells initially adjust to the new culture medium (lag phase) until they can start dividing regularly by the process of binary fission (exponential phase or log phase). When their growth becomes limited, the cells stop dividing (stationary phase), until eventually they show loss of viability (death phase).

Note: The parameters on x and y axes. Growth is expressed as change in the number viable cells vs time. Generation times are calculated during the exponential phase of growth cycle. Time measurements will be in hours for bacteria with short generation times.

Four characteristic phases of the growth cycle are recognized.

1. **Lag Phase**. Immediately after inoculation of the bacterial cells into fresh medium, the population remains temporarily unchanged. Although there is no apparent cell division occurring, the cells may be growing in volume or mass or size, synthesizing enzymes, proteins, RNA, *etc.* and increasing in metabolic activity of bacterial cell. The length of the lag phase is apparently dependent on a wide variety of factors including the size/volume of the inoculum; time necessary to recover from shock or physical damage in the transfer; time required for synthesis of essential division factors; and time required for synthesis of new (inducible) enzymes that are necessary to metabolize the substrates present in the culture medium.
2. **Exponential (log) Phase**. The exponential phase of growth is a pattern of balanced growth where all the cells are dividing regularly by binary fission and are growing by geometric progression fashion. The cells divide at a constant rate depending upon the composition of the culture medium and the conditions of incubation. The rate of exponential growth of a bacterial culture is expressed as **generation time**, also the **doubling time** of the bacterial population. Generation time (G) is defined as the time (t) per generation (n = number of generations). Hence, G=t/n is the equation from which calculations of generation time derive.
3. **Stationary Phase**. Exponential growth cannot be continued forever in a **batch culture** (*e.g.* a closed system such as a test tube or a flask). Population growth is limited by one of the three factors below: 1. exhaustion of available nutrients in medium; 2. accumulation of inhibitory metabolites or end products in medium; 3. exhaustion of space, in this case it is called as lack of "biological space". During the stationary phase, if viable cells are being counted, it cannot be determined whether some cells are dying and an equal number of cells are dividing, or the population of bacterial cells has simply stopped growing and dividing. The stationary phase, like the lag phase, is not necessarily to be a period of quiescence. Bacteria that produce secondary metabolites, such as antibiotics, do so during this phase of the growth cycle (Secondary metabolites are defined as metabolites produced after the active stage of growth). It is during the stationary phase that spore-forming bacteria have to induce or unmask the activity of many genes that are involved in sporulation process.

4. **Death Phase**. If incubation continues after the population reaches stationary phase, a death phase follows, in which the viable cell population declines. During the death phase, the count of viable cells decreases geometrically (exponentially), notably the reverse of growth during the log phase. (Note, if counting done by turbidimetric measurements or microscopic counts, the death phase cannot be observed.).

Growth Rate and Generation Time

As cited above, bacterial growth rates during the phase of exponential growth, under standard nutritional and growth conditions (culture medium, temperature, pH, *etc.*), define the bacterium's generation time. Generation times for bacteria vary from about 12 minutes to 24 hours or more. The generation time for *Xanthomonas* in the laboratory is 12hours, but in the intestinal tract, the coliform's generation time is estimated to be 12 to 24 hours. For most of the known bacteria that can be cultured, generation times range about 15 minutes to 1 hour.

External Environment – Cell Responses

Natural Forces

Temperature, moisture, pressure, light *etc.* H-ion concentration do affect the "state of balance in cell's protoplasmic material". This leads to disturbed metabolic processes which results in poor growth, reproduction and reduced population.

Temperature

Bacteria can survive between 0 to 85°C or may be even more. Bacteria can be broadly grouped into three categories on the basis of growth temperature range:

1. Psychrophilic– Min: 0°C; Max: 30°C;Opt: 15°C
2. Mesophilic – Min: 2-25°C; Max: 30-50°C; Opt: 20-45°C
3. Thermophilic – Min: 20-45°C; Max: 60-93°C; Opt: 55°C.

Moisture

Bacteria are more aquatic than terrestrial and thrive under moist conditions. Water helps in absorption of nutrients and motility.

Other factors such as light intensity, pressure, chemicals, solutions, metabolites *etc.* affect biological activities both positively and negatively.

Reproduction and Genetics

Binary Fission

Under favourable conditions bacteria multiply rapidly by binary fission which is the most common method of reproduction found in them. The dividing cell elongate considerably and the genome divides into two (Figure 2). In the middle of the cell, two ingrowths appear followed by the cell wall formation. Finally two daughter cells are formed and these cells will further divide to form more daughter cells. Sometimes, cells may remain attached together and division continues forming a chain of cells. It has been estimated that within 6 hours,more than 700,000 new cells are formed.

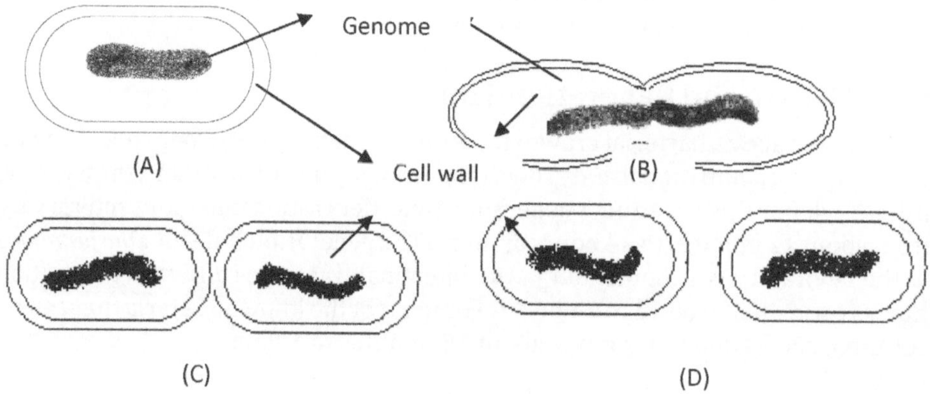

Figure 2. Binary Fission in Bacteria.

DNA Replication

The bacterial "chromosome" (circular DNA molecule) divides (replicates) resulting in the production of two circular chromosome. The molecule is duplicated at a site that transverses the whole molecule. This site is called as REPLICATING FORK. It is that point where the two strand become four in each of the two daughter circular "chromosomes", one strand is derived from the parent and the other one is new. This type of replication in which one strand is old and the other one is newly synthesized is called the SEMICONSERVATIVE REPLICATION. This mechanism of replication was proposed by Cairns in 1963 and is known as Cairns model.

Cell Division

A peripheral ring of plasma membrane invaginates and grows centripetally to form a double membrane septum (Figure 3). Wall material is deposited between the two membranes of the septum. Subsequently, septum formation and the cross wall synthesis occur simultaneously. The cell division is followed by separation of the daughter cells by dissolution of the matrix between the newly synthesized walls. Higgins and Shockman (1971) proved that mesosomes are involved in DNA replication and formation of septum and the wall.

Genetics

Bacteria have a great capacity to adapt to diverse environments through gene transfer, mutation or metabolic regulations. In this respect plasmids are particularly important as they transfer themselves or mediate transfer of chromosomal DNA to other bacteria and thus altering their genetic properties.

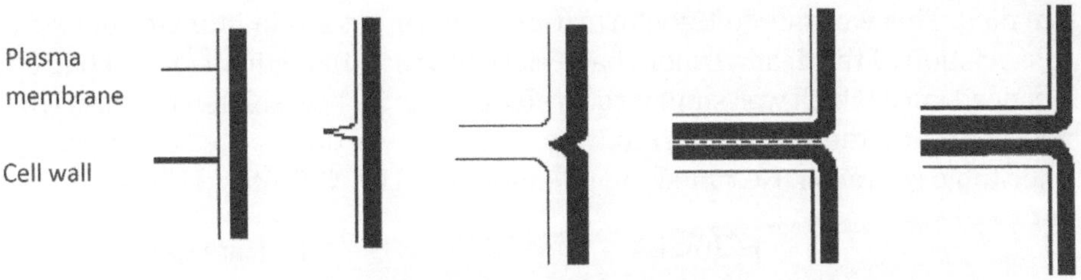

Figure 3. Septum and Cell Wall Formation during Cell Division.

Gene Transfer

Genetic materials are transferred between bacteria by CONJUGATION, TRANSFORMATION and TRANSDUCTION resulting in genetic re-combination in the recipient cell which acquires new characteristics.

Conjugation

Conjugation is a sexual process involving the exchange of nuclear material through a temporary area of fusion. The donor cell attaches to the recipient cell, usually with specific appendages called SEX PILI and transfers its gene by direct cell to cell contact. In *E. coli* strain K-12, the ability of the donor strain to produce conjugative pili is governed by the F (fertility) plasmid. The plasmid can be transferred from F+ to F- cells through DAN replication and converts the latter to F+ by its own F+ determinant. In Gram -ve bacteria, a pilus of the F+ cell attaches to the outer membrane protein of F– cell and retracts, resulting in wall to wall contact of both cells. The DNA is transferred through a transmembrane pore that is subsequently formed between the donor and recipient cell envelopes. In Gram + ve bacteria, conjugation may occur in the absence of pili.

The phenomenon was worked out in *E. coli* K 12 strain. Through "interrupted mating experiment" it was known that physical contact between cells involves, DNA passing from one cell to another through a conjugation tube, strains show sexual differences, one acting as male and other as female *i.e.* the donor and recipient. Conjugation between F+ (male) and F– (female) resulting in the conversion of female into male due to transferred fertility factor F is shown in Figure 4.

Transformation

Griffith observed transformation during 1928 in *Pneumococcus pneumoniae* for first time while working on the disease Pnemonnia and this initiated a biological revolution. He had both virulent and avirulent strains which differed in cultural characters. The virulent one was capsulated and formed a smooth colony on medium where in the avirulent was non-copsulated and formed rough colony. The virulent strain caused death of mice on injection while the avirulent was harmless. The heat killed virulent strain failed to kill the mice. However, when the living avirulent strain and the heat killed virulent strain were mixed and injected, the mice died. This was due to transformation of avirulent strain into virulent strain in association of the dead virulent bacterial cells. Griffith further isolated bacteria from dead mice which was similar to virulent bacteria. The isolated bacteria when injected to mice, the death occurred. This character was transmitted to the progeny as heritable character. He called the phenomenon TRANSFORMATION.

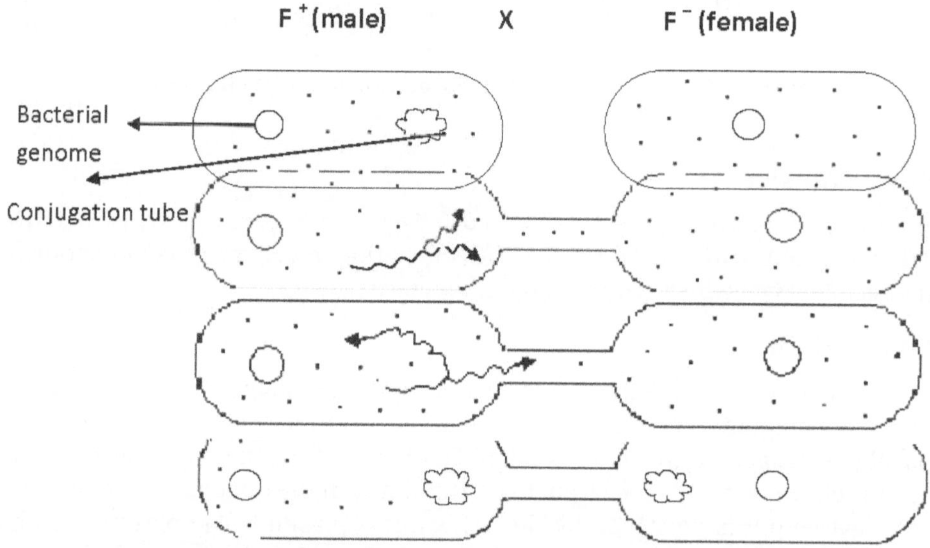

Figure 4. Migration of One F Factor into Female Cell.

The transformation process consists of three steps, (1) external binding of the DNA fragment onto the outer membrane (2) penetration of the DNA fragment through the cell envelop and (3) gene expression in the state of independent replicon or integrated into the chromosome.

Transduction

During prophage release from host chromosomes and initiation of independent replication, they may carry a very small portion of the host chromosome resulting in the assembly of temperate phase particles with the DNA consisting of phage and host cell DNA. They are defective phage as they do contain a full complement of the DNA's part that is left on the host chromosome when this phage infects other cells. Crossing over takes place between a fragment of chromosome of the donor cell and the homologous chromosome of the recepeient, resulting in alteration of the corresponding phenotypic properties. Such phage mediated recombination of chromosomes is called TRANSDUCTION.

Lysis of Bacteria

Plant pathogenic bacteria in nature remains in pure form only in plant tissues at initial stages of disease development. The lesions are soon invaded by various macroorganisms. The ability of the pathogens to survive in such a mixed population may be affected by synergism or antagonism with other organisms.

Bacteriophage

These are one of the groups of viruses that infect specific bacteria, usually causing their disintegration or disruption or dissolution. Phages are morphologically classified into three categories: tadpole, spherical and filamentous. The head of tadpole-shaped phages are actually icosahedral and measure 50-100 nm in diameter and the tails 15-170 x 10-20 nm. The size of spherical phages is generally 20-26 mm in diameter and that of filamentous phages is about 800-1400 nm in length and about 4-7 nm in width. Most phages of plant pathogenic bacteria have DNA but RNA phages are also known. The base composition of DNA varies with the phages although most phage NA consists of the usual bases of thymine, adenine, cytosine and guanine.

Lytic Cycles

Infection

The first step of phage infection is the adsorption of phage particles to receptor sites on the bacterial surface. The contact of phage particles and bacterial cells is purely physical and is determined by the density of both the phage and bacterial cells. Phage particles adsorb the receptors on the bacterial cell envelope with their tail. The tail fibers probe the adsorption sites and the spike fix the phage particles to the receptors. As isozyme located in the tail lyses the bacterial cell wall, the tail sheath contracts so that the core with an internal hole is penetrated, injecting phage DNA into the bacterial cell. The phage coat protein called "Ghost" remain outside the bacterial cell.

Reproduction

Phage progeny in the reproduction process do not acquire infectivity until mature phage particles are assembled in the infected host cell. The non infective form is called as vegetative phage. In the early stages of phage reproduction, infective phage particles cannot be recovered from bacterial cell lysates. This stage is called as eclipse phage. Replication of phage, NA formation of structural proteins and biosynthesis of enzymes required for synthesis of these viral components as well as the assembly of phage particles are perfomed on a regular time schedule according to phage genes. Some genes start working immediately after infection while others are transcripted late in the viral growth cycle. The phage head is likely assembled through the process in which viral NA is first pushed into prohead protein coat and then is pulled in as consequence of ATP hydrolysis.

Release

Bacterial cells containing mature particle usually burst to release phage progeny and called as LYSIS. Along with mass lysis of bacterial cells, turbidity sharply drops in liquid cultures and PLAQUES are formed on agar plates. The lytic cycle of virulent phages normally completed within an hour after infection.

Temperate Phage and Lysogeny

Temperate phages differ from virulent phages in that the phage DNA injected into the cells is integrated into the host chromosome and there after it replicates synchronously with the host genome. The phage genome in this state is called PROPHAGE and the phages that have the capacity to undertake a prophage state arec alled TEMPERATE PHAGE. Such relationships occurring between temperate phages and the bacteria are called LYSOGENIC BACTERIA.

Lysogenic bacteria successively transfer their prophage DNA to the daughter cells during cell divisions without releasing any free phages. The prophage is however, occasionally released from the host chromosome in some host cells and initiates the lytic cycle as in virulent phages. This is called as SPONTANEOUS INDUCTION. Thus the cultures of lysogenic bacteria usually contain a small number of phages.

Chapter 6

Survival and Dispersal of Bacteria

Ecology

It is a science where interactions within in, between and/amongst the populations and communities *i.e.* all biotic and a biotic factors, is studied. During evolution process, each species, genus, populations, communities, flora, fauna *etc.* adapt to certain given conditions. There are several systems in which certain groups interact and play a role of energy conversions and its transfer at different trophic levels of ECOSYSTEM.

In relation to soil where majority reside we have broadly two groups: Soil inhabitants and soil invaders. In the latter category all those bacteria that are active, those that attack plants and live as parasite, then on refuge mixed in soil are included. This cycle is repeated almost seasonally. SOIL INHABITANTS are those which entirely and exclusively carry their activities within the soil system either actively or passively. Some have made plant's root zone as their home and are called RHIZOSPHERE BACTERIA. Rhizosphereis the zone where roots release/exude/secrete energy rich organic compounds and this attracts microorganisms. These bacteria grow fast, breed fast, produce varieties of enzymes and toxins, produce antibiotics, tolerate metabolites and naturally dominate the root zone to grab the nutrients. Bacteria have these properties and therefore have approximately over 40 per cent of the total rhizosphere populations. *Pseudomonas, Xanthomonas, Erwinia, Agrobacterium, Bacilli* are dominant ones. Some groups are found invariably on plants foliage particularly on leaves. They are PHYLLOSPHERE BACTERIA. The leaf exudates promote certain group to grow as epiphytes.

Life Cycles and Dispersal

Like fungi, bacteria over season survive and disperse through different means and infect when host and environment are favourable. Thus, primary inoculum (primary infection) and secondary inoculum (secondary infection) are produced (Table 3 & 4). The chain of events (pathogenesis) forming disease cycle includes survival, dissemination and infection.

Table 3. Economically Important Seed-borne Phytopathogenic Bacteria

Sl.No.	*Bacterium*	*Disease*
1.	*Clavibacter michiganensis* sub sp. *michiganensis*	Canker of tomato
2.	*Curtobacterium flaccumfaciens* pv. *flaccumfaciens*	Silvering of red beet
3.	*Pantoea stewartii* sub sp. *stewartii*	Bacterial leaf blight and bacterial wilt of corn
4.	*Pseudomonas savastanoi* pv. *glycinea*	Bacterial blight of soybean
5.	*Pseudomonas savastanoi* pv. *phaseolicola*	Halo blight of bean
6.	*Pseudomonas syringae* pv. *aptata*	Bacterial blight of sugar beet
7.	*Pseudomonas syringae* pv. *atropurpurea*	Halo blight of bromegrass
8.	*Pseudomonas syringae* pv. *lachrymans*	Angular leaf spot of cucumber
9.	*Pseudomonas syringae* pv. *pisi*	Bacterial blight of pea
10.	*Pseudomonas syringae* pv. *tabaci*	Angular leaf spot of tobacco
11.	*Rathayibacter tritici*	Tundu, yellow ear rot, yellow slime of wheat
12.	*Xanthomonas citri* pv. *malvacearum*	Angular leaf spot, Black arm, Bacterial blight of cotton
13.	*Xanthomonas phaseoli* pv. *phaseoli*	Common bacterial blight of bean
14.	*Xanthomonas citri* pv. *vignicola*	Bacterial blight, Bacterial Canker, Bacterial leaf spot of cowpea
15.	*Xanthomonas campestris* pv. *campestris*	Black rot of crucifers
16.	*Xanthomonas campestris* pv. *sesami*	Bacterial spot and blight of sesamum
17.	*Xanthomonas hortorum* pv. *carotae*	Bacterial blight of carrot
18.	*Xanthomonas oryzae* pv. *oryzae*	Bacterial leaf blight of rice
19.	*Xanthomonas oryzae* pv. *oryzicola*	Bacterial leaf streak of rice
20.	*Xanthomonas translucens* pv. *translucens*	Back chaff, leaf blight of cereals and grasses
21.	*Xanthomonas euvesicatoria*	Bacterial spot of tomato

Table 4. Survival and Dispersal of Bacteria through different Modes {Source}

1. Survival in planting materials
Clavibacter michiganesis subsp. *sepedonicus* – Ring rot of potato
Burkholderia gladioli pv. *gladioli* – Scab of gladiolus
2. Survival as epiphytes
Erwinia amylovora – Fire blight of pear and apple
E. carotovora subsp. *carotovora* – Soft rot of cabbage
Pseudomonas syringae pv. *glycinea*– Soybean blight
Ps. syringae pv. *syringae*– Brown spot of bean
Xanthomonas citri pv. *malvacearum* – Cotton blight

Infection Process

Bacteria establish physiological synchrony as well as genetical synchrony to derive energy and parasitizes. Each bacterium has a foci of infection and place to over a season. Thus, both primary inoculum as well as secondary are existing in the ecosystem. The first step in infection chain is the contact. Several passive agencies (wind, water, soil movement *etc.*) and also through the active movement *i.e.* flagellar movement, they reach the site of infection. A mere contact does not guarantee infection. Both must RECOGNIZE each other. Elicitors facilitate recognition. Elicitors are signals of biological origin, located both on host cell wall and parasite cell wall. There could be compatible (onset of parasitism) or incompatible reaction. Thereafter, the bacterial cells find entry inside the cells of their hosts to derive nutrients. Ingress inside host is called as ENTRY OR PENETRATION. Bacteria cannot penetrate as the required force, energy, structures *etc.* are absent. Therefore there is no direct entry.

Bacteria enter the hosts either through wounds and/or natural openings such as stomata, hydothodes, nectaries, *etc.* Wounds are invariably caused/created by several physical, chemical and biological factors operating continuously.

Different chemical weapons are produced by invading bacteria like other organisms particularly fungi. They include enzymes, toxins, polysaccharides, growth factors *etc.* Among the weapons the enzymes play major role. They produce pectolytic enzymes for dissolution of pectin, pectic acid and release the polymers. These hydrolytic enzymes include methyl esterases, polygalacturonase transeliminases. Proteinases that digests the protein, cellulase/hemicellulase that digests the cellulose/hemicellulose *etc.* are produced. Among the toxins, both host specific and host non-specific are produced. *e.g.,* TABTOXIN or WIDLFIRE toxin produced by *P. syringae* pv. *tabaci.* It is a potent toxin. Besides enzymes and toxins, growth regulators, polysaccharides, ethylene *etc.* are also produced and play important role in pathogensis.

After the host tissue colonization, signs and symptoms appear. Bacteria, in general produce following specific symptoms –

1. Crown gall bacterium – *Agrobacterium* spp. crown gall, twig gall, cane gall, hairy roots, *etc.*
2. *Clavibacter* spp. – Ring rot, canker, wilt, spots, fasciation *etc.*
3. *Erwinia* spp. – Blight, wilts, soft rots *etc.*
4. *Pseudomonas* spp. – Leaf spots, galls, wilt, blight, cankers *etc.*
5. *Xanthomonas* spp. – Leaf spots, rot, cankers, blights *etc.*
6. *Streptomyces*– Scab, soft rots *etc.*

Chapter 7

Diagnosis of Bacterial Diseases

Diagnosis

The process of identification of a symptom is called diagnosis and the science (theoretical and practical) is called diagnostics. Quick and correct identification of the disease(s) is the first step in devising and applying effective control measures.

I. Field Diagnosis

Distribution of disease in the field, record of the problems appeared in the past, cultural practices adapted, application of agro-chemicals, varieties/genotypes, rotations are the necessary inputs for reaching some conclusion to the identity of the disease.

II. Plant Diagnosis

1. Symptoms and Signs

- ☆ Classify the symptoms into wilting, necrotic spot, blight, soft rot, hypertrophy, malformation *etc.*
- ☆ Examine for water soaked affected tissues.
- ☆ Examine for exudation and ooze.

Symptoms of Plant Diseases Caused by Bacteria

Plant pathogenic bacteria cause symptoms on plants can be divided in to seven groups such as leaf spots, excrescences and galls, tumours, wilting (vascular diseases, necrosis and cankers rotting and bacteria embedded in slime. These types of symptoms, however can also be caused in a number of cases by fungi, mycoplasmas, viruses or nematodes, due to the fact that plants have a limited number of possible reactions to infection. Visual judgment in most cases is in sufficient to make reliable diagnosis. We can only make a preliminary or presumptive diagnosis. The symptoms and related causal organism is given below-

1. Bacterial Blight Symptoms on Leaves

Leaf Spot

Bacterial leaf spots can often be distinguished from those caused by other organisms by a chlorotic halo which is formed under the influence of toxins followed by a water soaked zone formed by EPS, a brownish necrotic part and a brown papery dry centre. The veins of infected leaves and stems and also roots turn black as the multiplying bacteria plug the flow of water and nutrients. When leaf spots are bordered by larger lignified veins they are angular in case of the dicots. Longitudinal in the case of monocot plants. For the rest, they are circular or irregular.

Bacterial Blight of Rice Caused by *Xanthomons oryzae* pv. *oryzae*

On the leaf blade pale green to yellowish water soaked stripes develops, often near to the tip of the leaf. The stripes later increase in length and width and often show wavy necrotic margin (Figure 1).

Figure 1. Symptoms of Bacterial Blight of Rice

Halo Blight of Beans (*Pseudomonas syringae* pv. *phaseolicola*)

Hydrolytic spots that become necrotic. Angular necrotic lesions which remain small and conspicuous yellow halos often one and a half inches across develop early around the brown necroses (Figure 2).

Figure 2. Halo Blight of Beans

Common or Fuscous Blight of Beans

The initial symptoms usually occur on the leaves. Infected leaves develops water-soaked or pale green spots that later turn brown and dry. These spots may coalesce to form irregular blotches of varying sizes. However, each spot or blotch invariably has a narrow chlorotic margin surrounding i (Fig. 3a).

(a) (b)

Figure 3. Symptoms of Common Blight of Beans Caused by *X. phaseoli* pv. *phaseoli* (a) and Fuscous Blight of Beans caused by *X. phaseoli var. fuscans* (b)

In fuscous blight, Infected leaves develop water-soaked, pale green spots that later turn brown and dry and each spot or blotch invariably has a narrow chlorotic margin around it (Figure 3b).

Bacterial Blight of Pomegranate

On leaves, bacterial blight symptom appear as small (2 to 5 mm), irregular, water soaked spots with necrotic centre, which later turns light to dark brown with prominent water soaked margins. The individual spots coalesce giving a blighted appearance. On fruits, irregular necrotic, brown to black L and Y shaped cracks appear on pericarp of the fruits (Figures 4a and b).

(a) (b)

Figure 4. Bacterial Blight Symptoms on Leaves (a) and Fruit (b).

2. Excrescences and Galls

The filamentous bacteria cause excrescences called scab on potato, sugar beet, radish and carrot. The *Streptomyces scabies* doesn't produce toxin but produces pectinases which kill cells and enable the hyphae to grow between cells. The cork layer meristem tries to localize the infection and forms cork cells and parenchymatous cells which form a raised scab. Scab symptoms may superficial, deep – pitted or netted russet depends on species of *Streptomyces* cause the disease (Figures 5a, b and c).

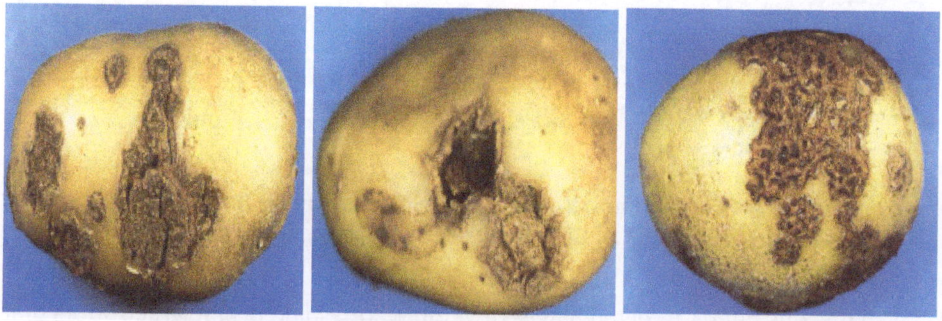

(a) Superficial Scab **(b) Deep – Pitted Scab** **(c) Netted/Russet Scab**

Figure 5. Symptoms of Common Scab of Potato Caused by *Streptomyces scabies*.

3. Galls

Bacterial galls are tissue swellings generated by hypertrophy and hyperplasia, which is caused by the hormone balance disturbing influence of bacteria. The growth of the galls only takes place when the pathogen is present. Bacteria cause two types of galls-

a) **Histoid galls:** The galls consist of many newly formed parenchymal cells with spontaneously formed island of xylem vessel tissue and bacterial cavities surrounded by cork layers. The galls develop under the influence of hormones indole acetic acid and cytokinins, which is secreted by bacteria *e.g.* galls of olive tree caused by *Pseudomonas savastanoi* pv. *savastanoi.*

b) **Organoid galls:** The bacterium stimulates resting meristematic tissue in buds and bulbs form abnormal growth and sprout formation due to cytokinines and indole acetic acid, which is produced by bacteria *e.g.* organoid galls on *Lilium* sp. caused by *Rhodococcus fascians.*

4. Tumours

Small, white and soft malignant swelling at the base of the plant stem caused by mostly *Rhizomium radiobacter* (*Agrobacterium tumefaciens*). The bacterium is a wound parasite which only can change plant cells to cancer cells when there are living, dividing cells at the base of the margin of the wound (Figure 6). Virulent strains of *A. tumefaciens* carry large Ti (tumour inducing) plasmid with a size of 150 -250 kb. Another species *Agrobacterium rhizogenes* causes syndrome of abnormal root proliferation, which is called hairy root. It contains Ri (root- inducing plasmid) plasmid and hair like adventitious roots are developed near the infection site.

(a) Crown Gall Symptoms **b) Galls Appear on Vine of Grape**

Figure 6. Symptoms of Tumour Caused by *Rhizobium radiobacter* (*Agrobacterium tumefaciens*) (a) and *R. vitis* (b)

5. Vascular Disease and Wilting

Bacterial wilt of solanaceous crops is caused by *Ralstonia solanacearum.* First visible symptoms of wilt is initial drooping of tips, twigs and upper leaves and later whole upper ground portions droops, yellowing and drying of the plant (Figure 7).

Ring rot of potato is caused by *Clavibacter michiganensis* subsp. *sepedonicus.* First symptoms appear on the plants are slow wilting, initially limited to the margins of leaves low on the stem. Later leaves show chlorotic, yellow to orange,

Figure 7. Symptoms of Bacterial Wilt of Tomato under Field Condition.

inter coastal areas. Infected leaflets, leaves and even stems may become necrotic and eventually die (Figures 8a and b).

(a) (b)

Figure 8. Symptoms of Bacterial Ring Rot on Potato Plant, (a) advance stages of infection, rolling of leaf margins, mottling and necrotic tissue, (b) Vascular ring appear only as a broken, sporadically, appearing dark line or a continuous yellowing discolouration.

Bacterial canker of tomato is caused by *Clavibacter michiganensis* subsp. *michiganensis.* In vascular infection, unilateral wilting of leaflets and leaves can be seen and at the end severe wilting occurs. Vascular tissue shows a yellow brown discouloration and tissue at nodes are soft (Figures 9a and b).

Black rot of crucifers is caused by *Xanthomonas campestris* pv. *campestris.* Symptoms appears on leaves, V- shaped yellow sectors with blackened veins develop along the margin. Later it become brown and necrotic, papery and may

(a) Marginal Browning of Leaves (b) Fruit Spots Caused by Bacterial Canker

Figure 9. Bacterial Canker Symptoms Appear on Leaves (a) and Fruit Spot (b) on Tomato.

cover the whole leaf. Yellow spots and distortions may be observed on cotyledons. When stems are cut, black vascular bundles are present (Fig. 10).

Figure 10. Symptoms of Black Rot showing V-shaped and Blackening of Veins Symptoms on Cabbage Leaf.

6. Necrosis and Cankers

When bacteria spread rapidly through the tissues and kill them, large areas of sunken, dead, brown to black necrotic tissue are formed. Bark, leaf and internal necrosis can be distinguished. For example, bark and leaf necroses are caused by *Erwinia amylovora* in apple, pith necrosis of tomato caused by *Pseudomonas corrugata* and phloem necrosis of walnut is caused by *E. rubrifaciens.*

Citrus Canker

First symptoms appears as small spots on leaves, shoots, twigs, as well as fruits and these spots becomes raised pustules or blister like eruptions later. On leaves, lesions develops up to 10 mm in size and appears brown and necrotic, with a depressed center and surrounded by a yellow halo. On fruits, necrotic lesions develop without yellow halo (Figure 11) and some time can be mistaken for scale insects.

Figure 11. Canker Symptom on Citrus.

7. Rotting

The typical symptom of rotting is necrosis in combination with exudation of fluid and bad smell. Bacteria can cause rotting by rapid degradation of the middle lamella *i.e.* maceration and cell content by pectolytic, cellulolytic and proteolytic enzymes. Example, soft rot of carrot is caused by *E. carotovora* subsp. *carotovora*.

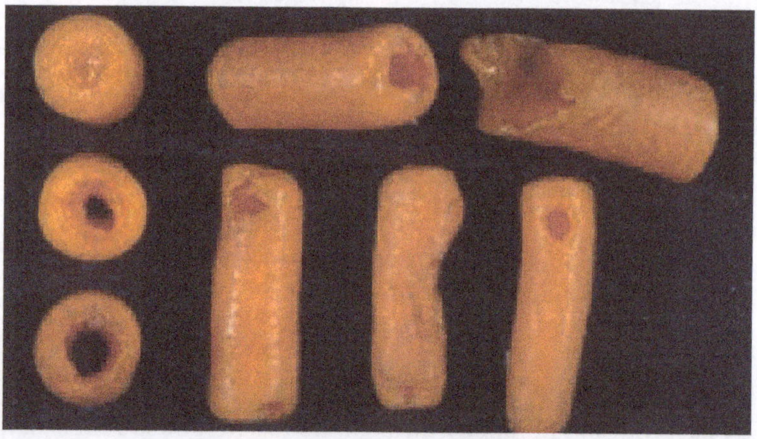

Figure 12. Soft Rot of Carrot.

8. Bacteria Embedded in Slime

Bacteria are visible in masses in slime, which protrude from the tissue, either spontaneously or after cutting or wounding of the tissue. In brown rot of potato, this slime is one of the diagnostic values. Bacterial slime may be present as very thin threads, so called strands. The colour of the slime is often grey, but also yellowish or orange. For example, Bacterial slime, which contasin bacteria and can be dispersed by wind on water soaked lesions caused by *Pseudomonas syringae* pv. *porri* on leak (*Allium porrum*).

9. Symptoms of Fastidious Bacteria

i. *Xylella fastidiosa* (Xylem Limited Bacteria)

The bacterium *Xylella fastidiosa*, a Gram-negative xylem-inhabiting, vector-transmitted, very slow growing bacterium, was cultured and properly described for the first time in 1987 in the USA as the cause of Pierce's disease (PD) of grapevine, phony peach disease in peach and citrus variegated chlorosis or citrus X disease. Symptoms appear on grapevine caused by *X. fastidiosa* are sudden drying of large parts of a green leaf. These parts become brown necrotic and the surrounding tissues become yellow to red. The necrosis is often present at the leaf margins. Scorched (burnt-like) leaves usually drop from the distal and not from the usual basal end of the petiole, leaving bare petioles attached to canes, often well after normal leaf fall.

ii. *Candidatus* liberobacter (Phloem Limited Bacteria)

Candidatus liberobacter asiaticus*:* The bacteria Symptom ca n occur throughout the tree, especially if the infection occurs at or soon after propagation. If infection occurs later the symptom and the causal organism are often partially confined. Earlier reports suggesting that the organism's movement it's restricted to new growth with little downward movement are challenged by recent findings that it can move 30-50 cm downwards in 12 months. Infected trees or branches suffer heavy leaf drop followed by out-of-season flushing and blossoming, with dieback occurring in severe cases. In general leaf symptoms are of two types: Primary symptoms are characterized by yellowing of normal-sized leaves along the veins and sometimes by the development of a blotchy-mottle. With secondary symptoms the leaves are small, upright and show a variety of chlorotic patterns resembling those induced by zinc and iron deficiencies. Analysis of symptomatic leaves shows a higher potassium content and lower calcium, magnesium and zinc concentrations. Infected fruit are small, lopsided and have a bitter taste, probably because of higher acidity and lower sugars. Many fall prematurely, while those that remain on the tree do not color properly, remaining green on the shaded side, hence the name of the disease. Any seeds in severely affected fruit are often abortive.

iii. *Candidatus* Phytoplasma

A common symptom caused by phytoplasma infection is phyllody, the production of leaf-like structures in place of flowers (Sesamum phyllody). Other symptoms, such as the yellowing of leaves, are thought to be caused by the phytoplasma's presence in the phloem, affecting its function and changing the transport of carbohydrates. Phytoplasma-infected plants may also suffer from virescence, the development of green flowers due to the loss of pigment in the petal cells. Photoplasma-harboring plants which are able to flower may nevertheless be sterile. Many plants infected by phytoplasma gain a bushy or "witches' broom" appearance due to changes in their normal growth patterns. Most plants show apical dominance, but phytoplasma infection can cause the proliferaztion of auxiliary (side) shoots and an increase in size of the internodes is observed. Such symptoms are valuable in the commercial production of poinsettia. The infection produces more axillary shoots, which enables the production of poinsettia plants with more than one flower. Phytoplasmas may also cause many other symptoms that are induced because of the stress placed by infection on the plant rather than specific pathogenicity of the phytoplasmas. Photosynthesis, especially photosystem II, is inhibited in many of the phytoplasma infected plants. Phytoplasma infected plants often show yellowing which is caused by the breakdown of chlorophyll, whose biosynthesis inhibition also observed. For examples, Aster yellows (*Ca.* Phytoplasma asteris, *Ca.* Phytoplasma japonicum), Coconut lethal yellowing, (*Ca.* Phytoplasma palmae, *Ca.* Phytoplasma castaneae, *Ca.* Phytoplasma cocosnigeriae), Rice Yellow Dwarf (*Ca.* Phytoplasma oryzae), Sugarcane Grassy Shoot Disease (*Ca.* Phytoplasma oryzae), Clover proliferation (*Ca.* Phytoplasma trifolii) diseases are caused by different group of phytoplasmas.

iv. *Spiroplasma*

Citrus Stubborn disease is caused by *Spiroplasma citri* arises from the persistence of the original characters of a tree when it is 'top-worked' with healthy bud wood. Affected trees are more or less stunted in appearance. Leaves become shorter and broader ('little leaf'), abnormally upright, cupped, sometimes mottled or chlorotic. Shoots will be abnormally bunched and development of multiple axillary buds may give rise to witches' brooms appearance. Fruiting will be suppressed in infected plants. Fruits may be stunted, lopsided or acorn-shaped (*i.e.* with thick rind at the base and thin rind at the tipportion) and may show colour inversion (peduncular end discolours while stylar end remains green). Another example corn stunt disease is caused by *Spiroplasma kunkelii.* The initial symptoms of corn stunt show characteristic small chlorotic stripes at the leaf bases of young plants after about 25-30 days. The chlorotic stripes become fused and extend further towards the tip of the leaves in the older leaves with green spots and stripes on a chlorotic back-ground. The infected plants have much shorter internodes and a proliferation of secondary shoots in leaf axils is observed. Reddening on leaves varies depending on the corn genotype and environmental conditions prevailing.

2. Microscopic

By using scraping, teasing or sections, alongwith proper staining, the presence of bacterial cells will help in identification of the cause and disease.

3. Isolation

Isolate the bacterium following standard technique, purify the isolate, multiply, prove pathogenicity and simultaneously study cellular characters and Gram staining.

III. Diagnosis by Bacteriophage

This technique is useful for examination of seed materials. Macerate seeds in a liquid medium to release bacterial cells. Add known numbers of phage particles and incubate overnight. A significant increase in the numbers of phage particles indicates the presence of host bacterium.

IV. Serological Methods

Among various procedures immuno fluorescence and ELISA have been most widely used for the diagnosis of phytopathogenic bacteria.

V. Gene Diagnosis

Rapid and accurate identification is possible by NA hybridization using dot-blot or colony hybridization techniques.

Detection of Bacterial Pathogens

Detection deals with establishing the presence of a particular target organism within a sample, with special emphasis on symptomless individuals. Diagnosis relates to the identification of the nature and cause of the disease problem, thus concerning plants exhibiting symptoms. Accurate routine disease detection requires high levels of specificity, sensitivity accuracy and speed. In this context, specificity is defined as the capability of a technique/practice or equipment to detect the organism of interest in the absence of false positives and negatives. Sensitivity relates to the lowest number of pathogens reliably detected per assay. For bacteria, the sensitivity levels of different techniques for detection in plant material are as follows (in colony forming units (cfu)/ml)]: isolation, about $10–10^2$; immunofluorescence (IF), 10^3; conventional PCR, $10^3–10^4$ and ELISA, $10^5–10^6$.

1. Bioassay

The wilting symptom of tomato crops can be induced by many factors like bacterial and fungal pathogen, root-knot nematode and deficit or excess of soil moisture. The most characteristics symptoms of bacterial wilt of tomato, chilli and brinjal are very rapid wilting, especially when the plants are young and succulent. The flowering stage is the most critical stage where a plant shows sudden wilting. The first visible symptom is the flaccid appearance of the youngest leaves. Other primary indications of infection are stunting, downward curling of leaves and petioles. Generally infected plants collapse quickly, but where this fails to happen there will be development of blackening of vascular system at the junction between stem and leaf. Further down the stem, the whole vascular system may be completely blackened and creamy bacterial slime oozes cut end.

2. Bacterial Ooze Test

It is common test to diagnose the disease caused by bacteria. It is the important sign of a bacterial diseases. Oozing is most prominently observed in vascular diseases caused by bacteria. A common sign of bacterial wilt of tomato caused by *R. solanacearum* can be observed at the surface of freshly-cut sections from severely infected stems is sticky ends with milky-white exudates, which indicates the presence of dense masses of bacterial cells in infected vascular bundles and particularly in the xylem tissues, which is responsible for transportation of water and nutrients from roots to aerial parts of the plant. Ooze also may accumulate on the cut surfaces of the infected surface. The ooze is usually an almost pure culture of *R. solanacearum,* which can be cultured on standard, low ionic strength specified bacteriological media. This water streaming test (Ooze test) is of presumptive diagnostic value in the field because this only occurs with bacterial wilt and not with any other type of pathogen or abiotic cause.

3. Bacteriophages

A specific recent research thrust is investigating the use of bacteriophages (phages) as recognition elements. These phages identify their host by specific receptor molecules present on the outside of the bacterial cell, infect the cell and use it as a factory for their own replication. The advantage of phages is their high specificity and sensitivity to bacteria. Moreover, these small groups of viruses are harmless to humans and animals and much less expensive and faster to produce than antibodies and also will have a far longer shelf life. Phage-based recognition receptors also show very high thermal stability. Furthermore, they can be immobilized onto the sensing device in the same way as antibodies or nucleic acid does. In contrast to other novel recognition elements, phages can also be immobilized by simple dip coating, where a physical adsorption mechanism is employed. For example, *Pseudomonas cannabina* pv. *alisalensis* PBSPCA1 bacteriophage was isolated from a commercial broccoli raab field that was typical symptomatic for bacterial blight. Previous studies showed that PBSPCA1 infects all *P. cannabina* pv. *alisalensis* isolates in various geographical locations and from various symptomatic crucifers but PBSPCA1 does not infect *P. syringae* pv. *maculicola,* which is important because this pathovar is difficult to distinguish from *P. cannabina* pv. *alisalensis,* which is a causative of pepper spot on crucifers that is often confused with earlier. Consequently, phage lysis assays employing PBSPCA1 have been utilized to distinguish *P. cannabina* pv. *alisalensis* from other pathovars.

3. Morphological, Biochemical and Physiological Characters

Plant pathogenic bacteria are characterized by using classical methods like Gram staining, shape and size, KOH test and microscopically study. Determination can be growth at different temperatures (4-37°C), thermal death point (50-55°C for 10 min in liquid medium), growth at different levels of NaCl (2, 5 and 7 per cent *etc.*). The patter of antibiotic resistance may also be determined. Screening for toxin production and ice nucleation activity will yield additional characteristics useful in the identification of plant pathogenic bacteria specially those belonging to the *Pseudomonas syringae* group and *Clavibacter* species. Detection of these substances nowadays often takes place via molecular detection of the responsible genes. In biochemical tests, the expression of the genetic material by the diverse enzymes systems of the bacterium is determined the tests have been developed to determine if bacteria are able to decompose certain C sources (Sugars, alcohals, organic acids, glycosides) and or N sources (*e. g.* amino acids). This may be done by inoculation of tubes containing a sole C or N source in minimal medium, including a pH indicator. If the bacterium is able to decompose the C or N a source to acid or alkali products, the pH indicator will change (change the colour of the medium)when inoculating two tubes with glucose as the carbon sources and covering the medium in one of them with sterile paraffin the oxidative (O) and/or fermentative (F) metabolism of glucose can be determined. This is important test for differentiation between genera (*Pseudomonas, Xanthomonas* O^+ F^-, *Erwinia* O^+ F^+. the action of these enzyme systems can be checked. Other biochemical tests determine the formation of certain end products by the bacteria *e.g.* Formation of H_2S for cysteine, indole from tryptophane, NO_2 or NH_4 from nitrate (NO_3) are used. Effect of enzymes can be also be visualized directly or after staining on certain agar medium containing the substrates for the enzymes like hydrolysis of pectin, gelatin, starch and casein by cellulose or fats/fatty acids by lipases (esterase) respectively. Besides these some tests oxidase and catalase, accumulate poly-B-Hydroxy butyrate (PHB) are also used for identification of bacteria.

'The Biolog technique is one of the methods which rely on measurements of utilizing different carbon sources by microorganisms. Measurements of carbon source utilization enable qualifying microbial metabolic capabilities and hence functional diversity of a microbial community. This method is inexpensive, reliable, fast, easy to use and can provide insight into the physiological ecology of the soil microbial community. It can be and has been, used in a wide variety of environments such as soil, sludge reactors, subsurface aquifers and mine tailings. The isolated strains will be characterized by determining their substrate utilization profiles with BIOLOG GN2 or BP2 on a micro titer plate, which is designed to test the ability of an organism to oxidize 95 different types of carbon sources. The results will be interpreted according to the instructions of the manufacturer (Biolog Inc., California, USA). For this determination, the isolates will be grown for 24 hr at 28 C on medium BUG (Biolog Inc.) supplemented with 5 per cent sheep blood.

Criticized for several reasons:

1) It is useful only for the bacterial portion of the community (fungi cannot reduce the tetrazolium dye).
2) It is biased toward fast-growing, aerobic, heterotrophic organisms that are culturable under laboratory conditions.
3) The results are not easily statistically analyzed, nor interpreted with respect to their field-relevance.

Phenotype MicroArrays' (PMs)

Growth phenotypes allow the microbiologists to describe and thereby differentiate cells. The power of phenotypic description of bacteria was demonstrated in a systematic manner by the doctoral dissertation of L.E. den Dooren de Jong, worked first under M.W. Beijerinck and completing the project under A.J. Kluyver at the Technological University at Delft, the Netherlands. A method for global analysis of cell phenotypes includes a standard set of nearly 2000 assays that could be used productively to differentiate a very wide range of bacterial species. The assays can be performed in 100-μL cultures in 96-well culture plates. The set consists of about 200 assays of C-source metabolism, 100 assays of P-source and S-source metabolism, 400 assays of N-source metabolism, 100 assays of biosynthetic pathways, 100 assays of pH effects and pH control with deaminases and decarboxylases, 100 assays of ion effects and osmolality and 1000 assays of chemical sensitivity. In the chemical sensitivity assays, there will be 240 diverse chemicals, each at four concentrations.

In this system selected chemicals which are toxic to most of the microorganisms and that are toxic by interfering with diverse cellular pathways, like DNA replication, RNA transcription, translation, cell wall synthesis, cell membrane synthesis, nutrient transport, *etc.* are incorporated into this set, tests to measure the sensitivity of bacteria to numerous inorganic chemicals, such as cations (Na+, K+, Fe3+, Cu2+, Co2+, Mn2+, *etc.*) and anions (chloride, sulfate, chromate, phosphate, vanadate, nitrate, nitrite, tellurite, *etc.*).

4. Serological Methods

In isolating required bacterial colonies can be confused with saprophytic colonies. Since it is very easy to examine cells from isolated colonies on the medium, agglutination tests can be made and allow a more efficacious screening of the bacterial colonies. Otherwise, agglutination tests can be used directly with exudates from several infected plant tissues. In this way, the bacterium can be directly detected in the host. The most commonly used assays for bacteria detection and identification are agglutination, enzyme-linked immunosorbent assay (ELISA), immunofluorescence, lateral flow strip tests or flow-through assays immunodiagnostic assays using *R. solanacerum* specific. Antibodies (also known as immunoglobulins) are proteins which are used by the immune system to identify and neutralize foreign objects like bacteria. A more subtle detection technique is immunofluorescence, which provides a means for detecting the bacterium in the plant, water as well as soil. The IF technique can be also used in detecting presence of the bacterium in the host-plant and in studying the rate of spread of the pathogen.

A simple method is required for rapid detection of bacterial pathogens in plant for testing huge number of samples by non- experienced personnel. For this purpose, tissue print ELISA and lateral flow devices have been designed for detecting bacteria. Although detection specificity is very high by using the appropriate monoclonal antibodies. The sensitivity is low for detecting bacteria and they are more appropriate and reliable for analyzing plants with symptoms. The lateral flow devices kits are based on existing technology similar to a pregnancy test kit and such kits are available only for few bacteria. A lateral flow device kit developed by Central Science Laboratory, U. K., permits detection of *R. solanacearum* in 3 min, in a single step. Rapid ImmunoStrip® is available from Agdia Inc. Immunofluorescence.

Cellular Analysis and Notification of Antigen Risks and Yields (CANARY)

B lymphocytes which are previously engineered to produce a cytolsolic bioluminescent protein are transfected with a plasmid containing a gene coding for a membrane-bound antibody specific to a pathogen. When these cells are exposed to the target pathogen, they emit the light within seconds. CANARY offers a combination of speed and sensitivity neither available with current immunoassays nor with PCR-based protocols, detecting as little as 50 CFU of some bacteria within few seconds. An even more sensitive technique has emerged that is based on immunoassay. Silica- based nano particles are filled with a fluorescent dye and conjugated to an antibody specific to a surface antigen of the microbe of interest. The fluorescence of a single bacterial cell can be sufficient for detection using this technique. CANARY has been applied to plant pathogens, but as per current knowledge, bio conjugated nano particles have not yet been used to detect plant pathogens.

6. Phospholipid Fatty Acid Analysis

Phospholipid fatty acid analysis (PLFA). It is a common bacterial community assay. This analysis quantifies fatty-acids present in the membrane structures belonging to different taxonomic groups in soil (Archaea- and Eubacteria, fungi and other eukaryotic organisms). The presence and relative abundance of fatty-acids unique to these particular groups allows the generation of whole-community profiles based on a taxonomic parameter, rather than functional. PLFA generates a large multivariate data set. The advantage in using PLFA is that it yields a broader range of data, including all microorganisms in the sample and allows comparisons of the relative abundance of different groups of microorganisms present in the sample. However, PLFA data sets are often difficult to interpret and do not yield specified information on organism activity or function. In addition, the assay is sensitive to temperature variation.'Thermoadaption' of membrane lipids by microorganisms in an existing community can falsely appear to be a change in community composition. Individual fatty acids are found in the membranes of nearly all microorganisms in nature, but because the relative abundance of each fatty acid varies among organism groups, lipid profiles are useful in looking at the presence and relative abundance of Gm-, Gm+, actinobacteria and fungi in soil communities. Thus, while all bacteria have the capability of synthesizing branched, saturated, or unsaturated fatty acids, Gm+ membranes tend to be predominantly composed of branched and/or saturated fatty acids, with less than 20 per cent of their lipids being monoenoic in nature. In contrast, Gm- bacteria dominantly use monoenoic, cyclopropane and hydroxy-fatty acids in their membranes. One can use these as general biomarkers for Gm+ and Gm- bacterial abundance. Eukaryotes, including fungi, have a high proportion of polyenoic, unsaturated fatty acids. The 10Me18:0 fatty acid is commonly used as an actinomycete indicator (Table 1).

Table 1. Table of Fatty Acid Indicators

i,a 15:0, I, a 17:0	Gm+ Bacteria
16:1w7c, 18:1w7c, cy17:0, cy19:0	Gm- Bacteria
10Me18:0	Actinomycetes
10Me16:0, i17:1w7c	Sulfate reducingbacteria
16:1w8c, 18:1w8c	Methane oxidizers
18:2w6, 18:3w6, 18:3w3	Fungi
20:3w6, 20:4w6	Protozoa
18:1w9c, 18:1w11c, 18:3w3, 20:5w3	Plants

Polymerase Chain Reaction (PCR)

Polymerasechain reaction is developed by Kary Mullis in the year 1983. PCR is based on using the ability of DNA polymerase to synthesize new strand of DNA complementary to the available template strand. Because DNA polymerase can add a nucleotide only onto a preexisting 3'-OH group, it needs a primer to which it can add the first nucleotide. This requirement makes it possible to delineate a specific region of template sequence that the researcher wants to amplify. At the end of the PCR reaction, the specific sequence will be accumulated in billions of copies (amplicons).

The PCR reaction requires the following components:

- ✰ **DNA template** - the sample DNA that contains the target sequence. At the beginning of the reaction, high temperature is applied to the double-stranded DNA molecule to separate/denature the strands from each other.
- ✰ **DNA polymerase** - a type of enzyme that can synthesize new strands of DNA complementary to the target sequence. The first and most commonly used polymerase is *Taq* DNA polymerase (from *Thermis aquaticus*). *Pfu* DNA polymerase (from *Pyrococcus furiosus*) is also used widely because of its higher fidelity while copying DNA. Although these enzymes are subtly different, they both have two capabilities that make them suitable for PCR: 1) they can generate new strands of DNA using a DNA template and primers and 2) they are resistant to heat.
- ✰ **Primers** - short pieces of single-stranded DNA that are complementary to the target sequence. The polymerase begins synthesizing new strand of DNA from the end of the primer.
- ✰ **Nucleotides (dNTPs or deoxynucleotide triphosphates)**-called as "building blocks" these are single units of the bases A, T, G and C, which are essentially for synthesis of new DNA strands.

- ☆ **Buffer solution** - It provides a suitable chemical environment for optimum activity and stability of the DNA polymerase and reaction.
- ☆ **Divalent cations, magnesium or manganese ions** - Generally Mg^{2+} is used, but Mn^{2+} can be used for PCR-mediated DNA mutagenesis, as higher Mn^{2+} concentration increases the error rate during DNA synthesis.

Procedure of PCR

Typically, PCR consists of a series of 20-40 repeated temperature changes for a specified time, called cycles, with each cycle commonly consisting of 2-3 discrete temperature steps, usually three. The cycling is often preceded by a single temperature step (called *hold*) at a high temperature (>90°C) and followed by one more hold at the end for final product extension or brief storage. The temperatures used and the length of time they are applied in each cycle depend on a variety of parameters and conditions. These include the enzyme used for DNA synthesis, the concentration of divalent ions and dNTPs present in the reaction and the melting temperature of the primers. The steps involve in PCR reaction is as follows-

- ☆ **Initialization step:** It is only required for DNA polymerases only that require heat activation by hot-start PCR. This step consists of heating the reaction to a temperature of 94–96 °C (or 98 °C if extremely thermostable DNA polymerases are used), which is held for 1–9 min. This step is not a compulsory in all in PCR.
- ☆ **Denaturation step:** This step is the first and regular cycling event and consists of heating the reaction to 94–98 °C for 20–30 seconds. It causes DNA melting of the DNA template by disrupting the hydrogen bonds between complementary bases, yielding two single-stranded DNA molecules.
- ☆ **Annealing step:** The reaction temperature is lowered to 50–65 °C for 20–40 seconds allowing annealing of the primers to complementary bases on the separated single-stranded DNA template. Typically the annealing temperature is about 3-5° C below the melting temperatures of the primers used. Stable DNA-DNA hydrogen bonds are formed only when the primer sequence very closely matches the template sequence. The polymerase binds to the primer-template hybrid and begins DNA synthesis.
- ☆ **Extension/elongation step:** The temperature at this step depends on the DNA polymerase used in the reaction. Taq polymerase has its optimum activity temperature at 75–80°C and preferably at 72 °C. At this step the DNA polymerase synthesizes a new DNA strand complementary to that of template strand by adding dNTPs that are complementary to the template in 5' to 3' direction, condensing the 5'-phosphate group of the dNTPs with the 3'-hydroxyl group at the end of the nascent (extending) DNA strand. The time required for extension step depends both on the DNA

polymerase used and on the length of the DNA fragment to be amplified. As a rule-of-thumb, at its optimum temperature, the DNA polymerase can polymerize a thousand bases per minute. Under optimum conditions, *i.e.*, if there are no limitations due to limiting substrates or reagents, at each extension step, the amount of DNA target is doubled, leading to exponential (geometric) amplification of the specific DNA fragment.

- ☆ **Final elongation:** This single step is occasionally performed at a temperature of 70–74 °C for 5–15 minutes after the last PCR cycle to ensure that any remaining single-stranded DNA is fully extended.
- ☆ **Final hold:** This step at 4–15 °C for an indefinite time may be employed for short-term storage of the reaction product.

Detection of bacteria in a given sample by PCR is not solely dependent on the performance on the PCR assay itself, but also on the efficiency of the procedure followed to extract the nucleic acids from the sample. The sensitivity of detection can be reduced by inhibitors that may be present in the extract of nucleic acids. To check for substances that may interfere with the amplification process, internal controls can be designed for each pair of primers, or real-time PCR can be employed instead. PCR efficiency is controlled by many parameters, such as type of polymerase, buffer composition, stability, purity and concentration of dNTPs, cycling parameters as well as the quality abd characteristics of the starting template. Several expensive commercial integrated systems allow for the automated extraction and analysis of nucleic acids from microorganisms, but they are not efficient with all types of plant material and sample used and they need to be evaluated before they can be adopted for routine detection practice. The specific primers can be designed based on either the amplification of specific genes from the chromosome or plasmids, or on different approaches such as sequences selected from RAPD differential bands obtained by subtractive hybridization. Some specific primers have been developed for detection of phytopathogenic bacteria from soil, seeds and planting material as well as for characterization are given in Table 2.

Table 2: List of Primer Sets for Detection and Identification of Phytopathogenic Bacteria

Phytopathogenic Bacteria	*Primer Sequences*	*Product of Primer*
Acidovorax avenae	RST49: 5'- GATGGCCGT GCC CTT CTT CAT CCT CG-3' RST51: 5'- CAT GGC CAC GAT GAG GAT CG-3'	390bp
Acidivorax avenae sub sp. *citrulli*	WFB 1: 5'- GAC CAG CCA CAC TGG GAC-3' WFB2: 5'- CTG CCG TAC TCC AGC GAT-3'	390bp
Agrobacterim strains	Primer VCF: 5'- ATC ATT TGT AGC GACT-3' Primer VCR:5'- AGC TCA AAC CTG CTT C-3'	730bp
Burkholderia gladioli	CMG-23-1: 5'- ATA GCT GGT TCT CTC TCC GAA-3' CMG-23-2: 5'- CCT ACC ATG CAY ATA AAT-3'	388bp

Phytopathogenic Bacteria	*Primer Sequences*	*Product of Primer*
Clavibacter xyli subsp. *xyli*	RSD33: 5'-CTG GCA CCC TGT GTT GTT TTC-3' RSD292: 5'-TTC GGT TCT CAT CTC AGC GT-3'	265bp
Candidatus liberobacter asiaticus	O11: 5'- GCG CGT ATG CAA GAG CGG CA-3' O12C: 5'- GCC TCG CGA CTT CGC AAC CCA T-3'	1160 bp
Clavibacter michiganensis sub sp. *michiganensis*	CMM5: 5'- GCG AAT AAG CCC ATA TCA A-3' CMM6:CGT CAG GAG GTC GCT AAT A-3'	614bp
Erwinia amylovora	EaF:5'- GCGCAGTAAAGGGTGACAG CCCCGTACACAAAAAGG CAT-3' EaR:5'- CCCTAGCCGA AACAGTGCTCTACCCCCGG-3'	565bp
E. carotovora sub sp. *atroseptica*	Y45: 5'- TCA CCG GAC GCC GAA CTG TGG CGT-3' Y46: 5'- TCG CCA ACG TTC AGC AGA ACA AGT-3'	438bp
Leifsonia xyli sub sp. *xyli*	Lxx82F: 5'- TTC AAC GCA GAA ATT GTC CAG G -3' Lxx22R: 5'- CAA GCA GGC GTA CTA AGT TCG A-3'	139 bp
Pantoea stewartii	Es1: 5'- GGC AGC GAA CTT GGC AGA GAT GCC-3' Es2 : 5'- TTG GTG CCT TCG GGA ACC G TG A-3'	46 bp
Pseudomonas	Ps-for: 5'- GGT CTG AGA GGA TGA TCA GT-3' Ps-rev: TTA GCT CCA CCT CGC GGC-3'	1018bp
P. fluorescens	Ph12a: 5'-GAG GAC GTC GAA GAC CAC CA-3' Ph12b: 5'- ACC GCA GCA TCG TGT ATG AG-3'	745bp
P. syringae	D21: 5'- AGCCGTAGGGGAACCTGCGG-3' D22: 5'-TGACTGCCAAGGCATCCACC-3'	558bp
P. syringae pv. *phaseolicola*	HM6: 5'- CGTGTCCGTGGATAAAAGC-3' HM13: GTTGAATTTCACTACCCG-3'	1900bp
Ralstonia solanacearum	Y2: 5'- CCCACTGCTGCCTCCCGTAGGAGT-3' OLII: 5'- GGGGGTAGCTTGCTACCTGCC-3'	288bp
R. solanacearum	759: 5'- GTCGCCGTCAACTCACTTTCC-3' 760: 5'- GTCGCCGTCAGCAATGCGGAATCG-3'	281bp
Spiroplasma citri	Spiralin F: 5'- GTCGGAACAACATCAGTGGT-3' Spiralin R: 5'- TGCTTTTGGTGGTGCTAATG-3'	675 bp
Xanthomonads	RST2: 5'-AGG CCC TGG AAG GTG CCC TGG A-3' RST3: 5'- ATC GCA CTG CGT ACC GCG CGC G-3'	840bp
X. citri pv. citri	4: 5'- TGT CGT CGC TTG TAT-3' 7: 5'-GGG TGC GAC CGT CAG GA-3'	467bp
X. fragariae	XF9: 5'- TGG GCC ATG CCG GTG GAA CTG TGT GG-3' XF11: 5'- TAC CCA GCC GTC GCA GAC GAC CGG-3'	537bp

Real-Time (Quantitative) - PCR

Quantitative real time PCR has become possible by the development of detectors that can detect and measure fluorescence that is emitted during the PCR cycle. This method is based on the 5' —> 3' exonuclease activity of the Taq DNA polymerase, which results in the cleavage of fluorescent dye- labeled with various probes (TaqMan®)during PCR. The exponential nature of PCR in theory allows the amount of starting material to be calculated from the amount of product at any point in the course of reaction. In practice, however, reaction conditions can interfere with exponential amplification and affect product concentration. Early attempts at quantization involved stopping the PCR reaction at various points to generate standard curves, which resulted in a laborious, low-throughput process. Several plant pathogenic bacteria have been detected in plants including *Clavibacter michiganennsis* subsp. *spedonicus*, *Agrobacterium* strains from plants and *Xyllella fastidiosa* in asymptomatic grape vines by using several real –time PCR, *R. solanacearum* in potato tubers, (Figures 13 and 14). It was also found severe reaction inhibitors in potato tubers in a real time TaqMan® assay for *C. michiganensis* subsp. *sepedonicus*. For detection of plant pathogenic bacteria, various techniques used in real time PCR as described below:

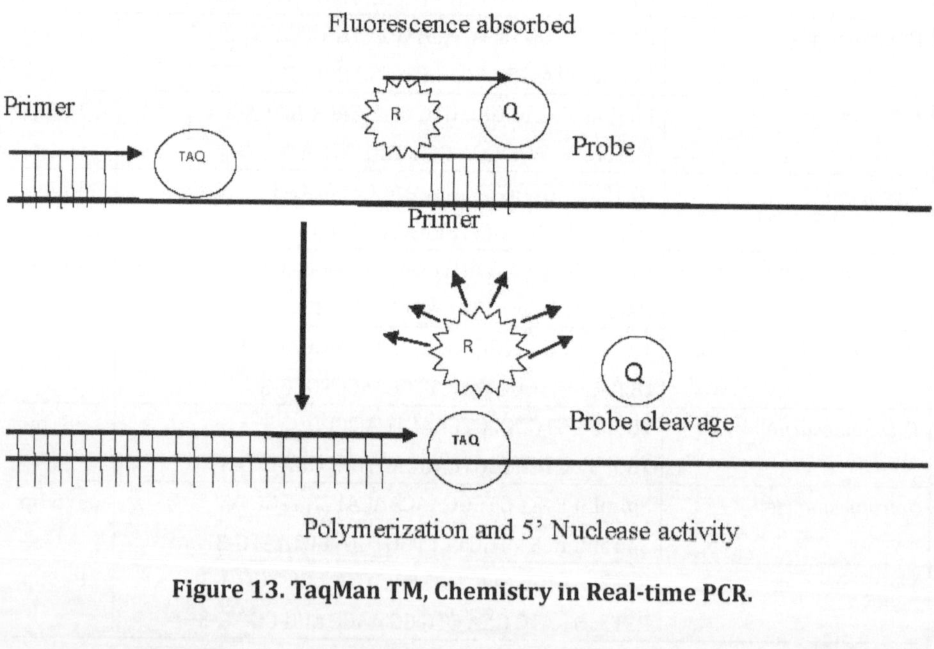

Figure 13. TaqMan TM, Chemistry in Real-time PCR.

i. TaqMan Probe

In this systems, an oligonucleotide probe sequence of approximately 25 -30 nucleotides in length is labeled at the 5' end with a fluorochrome, usually 6- carboxyfluorescein (6- FAM) and a quencher fluorochrome, usually carboxytetramethyl- rhodamine (TAMRA), at the 3' end. The probe is degraded by the 5'-3' exonuclease activity of the Taq polymerase as it extends the primer during each PCR amplification cycle and the fluorecent chromophore is being released. The amount of fluorescence is monitored during the cycle of amplification and it proportional to the amount of PCR product generated in the reaction. Real-time reaction monitoring with specific instruments and fluorescent probes combines amplification, detection and quantification in a single step. TaqMan probes are single-stranded oligonucleotides that are complementary to one of the target strands. A fluorescent dye adorns the 5' end and a quencher is bound to the 3' end of the probe, when the polymerase replicates a template on which a TaqMan probe is bound and the 5' exonuclease activity cleaves the probe and fluorescence occurs. They have been proposed for detection of *R. solanacearum*. Other probes are based on stem-loop hairpin structures, molecular beacons being the simplest hairpin probes generally used. Scorpion probes covalently couple the stem-loop structure to a PCR primer and, because hybridization of probe sequence to amplicons is intra molecular.

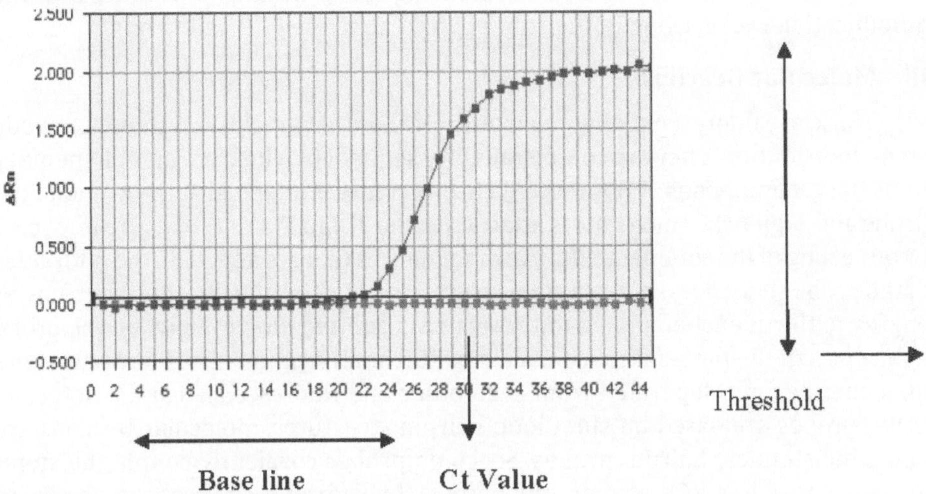

Figure 14. Amplification Plot from Positive and Negative Real Time PCR Reactions.

SYBR Green I became the most widely used DNA dye for real-time PCR applications because of cost efficiency, generic detection of amplified DNA and its ability to differentiate PCR products by melting curve analysis. Several protocols in the present compilation utilized SYBR. It is easy to adapt existing conventional PCR protocols to a real-time PCR assays by using SYBR® Green Master Mix (Qiagen, USA) and utilizing them for identification of bacterial cultures. However, there are disadvantages with the use of SYBR Green I, such as inhibition of PCR amplification in a concentration- dependent manner, effects on DNA melting temperature and preferential binding to certain DNA sequences. The drawback of using SYBR Green

I for melting curve analysis is that the melting temperature is highly dependent on the concentration of the dye and the DNA.

ii. Fluorescent Resonance Energy Transfer (FRET)

The probes require labeling of two oligonucleotide probe sequences that is adjacent to each other within the PCR target region. Probe 1 contains fluorescein label at its 3' end, whereas the probe 2 contain label at its 5' end with another label (Light Cycler red 640). It is imperative that the two probes are designed in such a way that when they hybridize to the amplified PCR product, they are aligned head to tail to bring the two fluorescent dyes in close proximity to each other. The fluorescein dye attached to the Cycler ™ or RAPID (ruggedized advanced pathogen identification device) unit emits a green fluorescent light at a slightly longer wavelength. When the second probe is in close proximity, the energy emitted by the first probe excites the light Cycler Red 640 dye attached to the second probe and red fluorescent light at longer wavelength is now emitted that can be detected at 640 nm. Fluorescence is measured during the annealing step of each of the amplification cycles when both probes hybridize to the PCR amplicons.

iii. Molecular Beacons

These are fluorescent oligonucleotide probes that are designed and it include stem- loop folding. They are complementary sequences which are complementary to the target amplicons. A fluorescent chromophore is attached at the 5' end of the probe and a quencher molecule is attached at the 3' end. A stem structure is formed by annealing of the complementary arm sequences that are added on the both sides of the probe sequences. When a stem structure is formed, the fluorophore transfer energy to the quencher and no fluorescence is emitted. However, when the probe hybridizes to the target amplicons during PCR amplification, the fluorophore and quencher become separated from each other and fluorescence can be detected. Other probes are based on stem-loop hairpin structures, molecular beacons are being the simplest hairpin probes. Scorpion probes covalently couple the stem-loop structure to a PCR primer and, because hybridization of probe sequence to amplicon is intramolecular, Scorpion probes are more efficient when compared to binary systems such as molecular beacons.

Nested-PCR

Sensitivity and specificity problems associated with conventional PCR and RT-PCR can be lowered by using nested PCR-based methods, which is based on two consecutive rounds of amplification. Usually, the products of the first amplification are transferred to another tube before the nested PCR is being carried out using one/two internal primers. The potentiality of nested-PCR in plant pathology has been already reported. Sensitivity is increased by two orders of magnitude reaching about 10^2 bacterial cells/ml of extract and reduces the negative effect of inhibitors. However, the two rounds of amplification in different tubes also increase the risk of contamination, especially when the method is used on regular and in a large scale. One limitation of the nested PCR approach concerns the need to accurately establish the ratio between external and internal primers and the use of limiting amounts of external primers to avoid interference during the process of second amplification. Several interesting alternatives with single closed tubes have been developed recently in order to avoid risk of contamination. The development of nested-PCR in a single closed tube using a single compartmentalized Eppendorf tube with a pipette tip is another interesting and viable alternative. This method, however, requires the use of 0.5 ml Eppendorf tubes and cannot be employed in rapid reaction capillary tubes.

Co-operational-PCR

It is a new PCR, which has high sensitivity for the amplification bacterial targets from plant material/samples. The Co-PCR (co-operational amplification) technique can be performed easily in a simple reaction based on the simultaneous action of three or four primers. The reaction process consists of the simultaneous reverse transcription of two different fragments from the same target, one internal to the other, the production of four amplicons by the combination of the two pairs of primers, one pair external to the other and the co-operational action of amplicons for the production of the largest fragment. Metal block and capillary air thermal cyclers have been employed for the detection of a bacterium, but by using three primers only, which shows the possibilities of this new approach. The low amount of reagents (ten times less than in conventional PCR) probably increases susceptibility to inhibitors. However, this step was not necessary when analyzing the presence of *R. solanacearum* in water samples. Co-PCR requires only one reaction, minimizing manipulation and reducing risk of contamination.

Bio-PCR

In this PCR, it combine the viable enrichment of mostly growth media with an enzymatic amplification, which is known as BIO-PCR. The target bacterium is enriched in liquid or solid media and detected at extremely low levels in seeds and other propagative materials. The BIO-PCR assay includes the following very simple steps as- i) extracting a sample ii) plating a sample onto agar/liquid media (half are retained for visual recovery or adding sample to liquid medium iii) incubating for 15 -72h, depending on the growth of target bacterium; iv) washing plates to remove bacteria from agar media or centrifuge liquid medium and using only 1 or 10 µl for direct PCR. The samples can be stored at – 20°C. This techniques has several advantages over tradition PCR including increased sensitivity of the primer, elimination of PCR inhibition and detection of only viable cells. It can be used in traditional PCR or real time PCR. The key to developing a successful real –time BIO-PCR protocol are choose suitable semi-selective or selective liquid or solid medium for bacterium and determine precisely the time for enrichment. In solid medium only pin point size of colony of bacterium should be selected for PCR because larger size (1 mm) of colony may allow to grow other saprophytic bacteria. The most plant pathogenic bacteria are take 24 -48 h to grow pin point colonies.

However, *R. solanacearum* growing fast and require only 10 -15h. A single pinpoint size colony normally contains over 1000 cells/ml, which is the threshold for the PCR. Mucoid region of the colony of bacterium should be avoided for PCR. In Bio –PCR, semi selective medium as SM-1 was developed by and further modifying the medium as MAS-1 used for *R. solanacearum*. The detection limit of this medium for through this technique is 100 cfu/ml from soil based on tested artificially or naturally infested samples. The advantages of bio- PCR include the elimination of false negatives due to potential PCR inhibitors in sample extracts, elimination of false positives resulting from dead cells or free DNA and increase of detection sensitivity due to an increase in the number of the target cells. In general, bacterial assays can be very fast since DNA extraction step is not needed; a simple 2 to 3 min hot start is adequate to break the cells. Also, direct PCR reduces the chances of cross contamination. However, even when used as a direct assay, real-time PCR has several disadvantages compared with isolation techniques when used to detect very low levels of bacteria as often occurs in seed samples. False negatives can result from using small volumes (1–10 μl) and presence of PCR inhibitors is also a problem. False positives can result from presence of dead cells or free DNA. To avoid these inherent disadvantages and to utilize the best of both techniques, PCR can be used in combination with pathogen isolation on agar media.

BIO-PCR has several advantages over traditional PCR, including increased sensitivity, elimination of PCR inhibition and detection of only viable cells. The keys to developing a successful real-time BIO-PCR protocol are (i) choose an appropriate semi- selective medium and (ii) determine precisely the time required for enrichment. For solid media, the time for pinpoint-size colonies to appear is the critical step for the success of BIO-PCR. Too long an incubation time permits saprophytes to grow and interferes with PCR. Ideally, pinpoint-size colonies of the target organism should be visible before the saprophyte colonies reach 1-mm diameter or greater. If such a medium is not available for the target organism, a new one should be designed as per requirements. For most plant pathogens, 24 to 48 h is sufficient time to form pinpoint colonies.

Multiplex-PCR

Multiplex-PCR allows the simultaneous detection of different DNA targets in a single reaction and also it is very sensitive in detecting. On the other hand, PCR detection protocols can be designed to verify the presence of more than one pathogen in plant material by looking for common specific sequences in two or more of them, or to detect related bacteria on multiple hosts. Multiplex PCR is useful in plant pathology because different bacteria and/or RNA viruses frequently infect a single host and consequently sensitive and immediate detection is needed for the propagation of pathogen-free plant material. There are several examples in plant pathology for simultaneous detection of several targets and the amplification by multiplex PCR. Multiplex-PCR protocols using specific primers have also been set up for simultaneous detection of two/more genes of the same bacterial pathogen, thus limiting false positives, or allowing amplification of several pathogenic bacteria in seed or plant material in a single run.Ex. one bacterium and four viruses were detected in olive plants.

A multiplex nested RT-PCR in a single closed tube has been developed for simultaneous and sensitive detection of the viruses CMV, CLRV, SLRSV and ArMV and the bacterium *Pseudomonas savastanoi* pv. *savastanoi* from olive plants in a single run using20 compatible primers in a compartmentalized tube. This newly developed method combines the advantages of various techniques of PCR like multiplex RT-PCR with the sensitivity and reliability of nested RT-PCR carried out in a single closed tube. It enables the simultaneous detection bacterial DNA targets in a single analysis, performed with woody plants. It also saves time and reagent costs because it can be performed in a single reaction, although accurate design of compatible primers is the critical step. The compartmentalization of a single Eppendorf tube with a pipette tip allowed multiplex-PCR and nested PCR to be combined effectively. During the first amplification reaction there is no interference of the external with internal primers because they are physically separated from the initial reaction mixture. Once the multiplex RT-PCR ends, the internal primers are mixed with the products of the first reaction before proceeding to the nested multiplex PCR. The sensitivity achieved by multiplex nested RT-PCR for the bacterium *P. savastanoi* pv. *savastanoi* is similar to that reached by applying monospecific nested PCR after an enrichment step, which was demonstrated to be 100-foldmore sensitive than conventional PCR.

Various PCR – based techniques for rapid, specific and sensitive detection of plant pathogenic bacteria have been improved and increased in the last few years. To reach higher sensitivity in serological or molecular detection, two complementary strategies can be applied together: amplification of the target and/or amplification of the signal. The pathogen or target amplification or enrichment can be carried out in vitro, preferentially with selective media and at the optimum growing conditions for each bacterial pathogen, or by molecular techniques including PCR. Among molecular techniques, multiplex PCR is increasingly used because it improves the efficiency and time saving in diagnosis. In the near future multiplex PCR will probably be adapted for the simultaneous detection of viruses and bacteria of one particular crop and for the simultaneous detection of other major plant pathogens like viruses, viroids, bacteria and fungi in the same reaction. The objective is real-time analysis or immediate detection of microorganisms in their natural environment without culturing or amplification. Qualitative and quantitative detection data generated with new technologies should provide a more complete picture of the life cycle of plant pathogens.

Restriction Fragment Length Polymorphisms (RFLP)

Restriction fragment length polymorphisms (RFLP) of DNA representing selected genes/DNA sequence can be used to identify pathogen species. This approach to species identification depends on having an excellent database on existence of variability in fragment length polymorphisms that may be found among isolates of individual species, since nonspecific isolates may differ in the presence or absence of particular restriction sites, consequently changing the RFLP banding profile/pattern. Total DNA can be analyzed and fragments visualized in an agarose gel after staining with ethidium bromide or any DNA staining agent. RFLP analysis is discriminative at a low taxonomic level, often strains level and is now often combined with PCR, *e.g.* Digestion after PCR of the 16S rRNA gene.

Amplified Fragment Length Polymorphism (Aflp)

It is a modification of the RFLP technique, which is used for species identification and more commonly to examine genotypic diversity within a population. Basically genomic DNA from a sample is digested typically with two different restriction enzymes in order to produce a large number of fragments. Specific oligonucleotide adapters of 25 – 30 bp are ligated to the restricted DNA fragments. It is useful in discerning the geographic origin of pathogens associated with new disease outbreaks. Although AFLP analysis is a powerful diagnostic tool, it is laborious, technically demanding and not suitable for routine use in diagnosis.

Repetitive Extragenic Palindromic (Rep) - PCR

It is also a fingerprinting technique that uses repetitive sequences which are of mostly unknown function that are interspersed throughout the DNA present in the genomic DNA of bacteria. Primers used are which specifically anneal to these sequences and multiply the DNA between two repetitive sequences. The primers are 18 – 22 mers. About 10 – 40 fragments are generated ranging in size from 200 bp to 6kbp. These fragments are separated by size in electrophoresis unit on a gel stained with ethidium bromide and visualized. REP- PCR is discriminative at very low taxonomic level, usually strain level.

Fluorescent *In situ* Hybridization (FISH)

In this method 20 – 30 mers oligonucleotide probes against 16S or 23S rRNA/DNA being used. These nucleotide probes can be used for in situ hybridization, because they are able to diffuse through the cell wall of microorganisms that are present in thin sections of tissue or in plant or soil extracts fixed on a microscopic slide. Fluorescent in situ hybridization (FISH) with probes targeted to 23S rRNA have been used to detect *R. solanacearum* race 3 biovar 2 from potato (peel). A probe specific for *R. solanacearum* was used to confirm the identity of the pathogen in potato samples that were simultaneously and independently tested by a pathogen-specific immunofluorescece assay. Use of two independent assays in a rapid (1-day) test immediately increases the confidence limits for a positive result, which is especially important for large-scale screening of propagative materials for major crops.

Micro/Macroarrays

Oligonucleotide arrays offer greatly expanded multiplexing capability compared to conventional PCR, real-time PCR and other more commonly used techniques. Arrays consist of discrete spots of pathogen-specific oligonucleotide sequences immobilized onto a solid surface such as a glass slide or nylon membrane. The sample DNA is amplified by PCR, labeled and then hybridized to the array. Because the sample DNA is amplified by PCR, sensitivity of arrays is typically high and the number of pathogen-specific tests that can be run on the array is essentially unlimited. Having the capability to use arrays to simultaneously test diseased plant tissue for dozens or hundreds of pathogens would be most useful in cases when the diagnostician may not know which species-specific detection assay to apply. Arrays will also be valuable because plants are often infected with several pathogens, some of which may act in concert to cause a disease complex in the host. Macro arrays may be more useful in plant disease diagnosis than micro arrays because of their greater sensitivity and freedom from the need for highly specialized equipment and personnel. While arrays are superior for multiplexing, they are not well-suited for quantification. They may be limited when high-throughput assays are required.

PCR/Electrospray Ionization-Mass Spectrometry (PCR/ESI-MS)

Earlier, it is known as "TIGER" (Triangulation Identification for the Genetic Evaluation of Risks). It is a new technique for rapid identification of specific bacteria. It is based on the use of mass spectrometry to characterize the base compositions of PCR products generated using a suite of conserved primers. For species determinations, the resulting base compositions are compared to a database of microbial base composition signatures. In contrast to probe-based detection of microorganisms, mass spectrometry does not require one to anticipate and test for particular microorganism.

Loop-Mediated Isothermal Amplification (LAMP)

The LAMP method proved to be the best approach for amplifying nucleic acid with high specificity, efficiency and rapidity without the need for thermocycling. The principle is based on strand displacement activity followed by amplification using a unique DNA polymerase and a set of four specially designed primers that recognize a total of six distinct sequences on the target DNA. Species-specific primers were designed by targeting the *R. solanacearum* fliC gene coding for flagellar proteins. Amplification performed for 60 min at 65 C resulted in production of magnesium pyrophosphate, which increased the turbidity of the solution, permitting visual assessment. The LAMP product was detected only in samples containing R. solanacearum and not from other species. It also successfully amplified target DNA in mixed cultures containing other bacterial species. *Rhodococcus fascians*is a phytopathogenic actinobacterium which causes leafy galls and other plant distortions that result in economically significant losses to nurseries producing ornamental plants. The LAMP assay accurately detected pathogenic *R.fascians* in 26 of 28 (93 per cent) naturally infected plants and did not react with 23 asymptomatic plants. The LAMP primers also amplified productfor DNA extracts of 40 of 41 bacterial strains isolated from plants with leafy galls. The detection limit of LAMP assays was approximately 103 CFU/30-µl reaction. The LAMP assay in particular is a significant advancement in rapid *R. fascians* diagnostics and enables those with limited laboratory facilities to confirm the presence of this pathogen in infected plants.

Recombinase Polymerase Amplification

In 2006 a new isothermal amplification method for the detection of specific DNA sequences was reported (Piepenburg *et al.*, 2006). Like LAMP technology, recombinase polymerase amplification (RPA) works at a constant temperature (usually 37–42 °C), making it suitable for in-field applications, as it does not require the global melting of the template and further steps usually done by a thermocycler. The isothermal amplification is achieved using oligonucleotide primers and three core enzymes: (i) recombinase enzyme, (ii) single-stranded DNA-binding proteins and (iii) a strand displacement polymerase. The recombinase enzyme forms complexes with the primers that 'scan' the target sequence, displacing the template strand which is subsequently stabilized by the single-stranded DNA-binding proteins, avoiding the displacement of the primer. Finally, the strand displacement polymerase starts the amplification of the target sequence. Since its first report, RPA has been widely used for diagnosis of human and animal pathogens. However, there are only a few reports of its application for plant pathogen detection, including fungi, bacteria and viruses. These studies have reported better results than common PCR-based diagnosis in terms of sensitivity and time (10–15 min of amplification are enough for detection). Simple and rapid diagnosis has been achieved for plum pox virus and little cherry virus 2 using high affinity lateral flow strips to detect terminally labelled RPA amplicons. Simple DNA extraction methods have been used in these studies, making the technology accessible to anyone (Zhang *et al.*, 2018). Recombinase polymerase amplification has also been used for multiplex detection of plant pathogens when combining the technology with surface-enhanced Raman scattering (SERS) labelled nanotags. SERS is a physical phenomenon that occurs on metal nanoparticle surfaces when they are excited by a laser. This results in enhanced Raman scattering patterns that are characteristic for each adsorbed molecule. In-field detection with the technique was demonstrated for *Botrytis cinerea, Pseudomonas syringae* and *Fusarium oxysporum* in tomato samples using a portable Raman after following a simple sampling protocol.

Microsphere Immunoassays (MIA)

This technique which is based on microspheres coded with different fluorophores, emerged as a promising alternative for microbial detection. Microspheres are conjugated with specific antibodies to capture the pathogen antigen and a magnet is used for separating bound and nonbound microspheres followed by the addition of specific antibodies linked with a reporter fluorophore to the microsphere-sample mix. A Luminexx MAP analyser excites the internal dyes of the microspheres with a red laser and the reporter fluorochrome, captured by the antigen, with a green laser, allowing the specific detection of multiple analytes at the same time (Charlermroj *et al.*, 2013). There were no reports of detecting plant pathogen using multiplex MIAs till 2008. The development of a multiplex MIA was first reported for potato virus X, potato virus Y, potato leaf roll virus and for the detection of *Acidovorax avenae* subsp. *citrulli*, chilli vein-banding mottle virus (poty virus), melon yellow spot virus (tospo virus) and watermelon silver mottle virus (tospo virus serogroup IV) (Charlermroj *et al.*, 2013). Multiplex MIA allowed the detection of the pathogens with a higher sensitivity and take approximately 1 h when compared with ELISA (4h) (Charlermroj *et al.*, 2013). MIAs are described as grower-friendly approach of diagnosing pathogen, *i.e.* 100 parameters can be tested in a single run. Agricultural companies can have their own diagnosis station provided with appropriate staff training and well calibrated equipment for fluorescence reading without the need of sending the samples to specialized laboratories. However, the technology costs are still high for small-scale farmers. Rettcher *et al.* (2015) suggested simple portable magnetic immunoassay for rapid detection and sensitive quantification as an alternative to overcome the high cost and complexity of fluorescence. In this study, magnetic beads coated with specific antibodies for the grapevine fan leaf virus capsid protein were used to enrich an immune filtration column with the same antibodies immobilized in a polyethylene matrix. After the enrichment, the column is placed in a magnetic hand held reader device. During measurement, the coated particles undergo a magnetic saturation by applying a strong low frequency magnetic field. The frequency signal measured in a resonant detection coil of the device allows the detection and quantification of the pathogen in less than 30 min. Even though the technology developed by Rettcher *et al.* (2015) has still not been validated for the detection of multiple pathogens in the same run, the authors conclude that this goal can be achieved by arranging consecutive matrices coated with different antibodies and adding a readout that is able to discriminate between different signals. Even more simple assays have been developed for plant pathogen detection using multiplex immune detection techniques. Safenkova *et al.* (2016) developed a multiarray in a test strip (MATS), combining immune chromatography with array techniques to obtain a rapid and high-throughput test. The test allowed the visual and instrumented detection of eight important potato viruses by spotting gold nanoparticles conjugated with specific antibodies in a nitrocellulose strip. This test can be performed by untrained staff during on-site inspection (Safenkova *et al.*, 2016).

Aptamer-based Diagnosis

Aptamers are single-stranded (ss) DNA or RNA ligands of less than 100-mer, selected to bind particular target molecules. Usually these ligands are identified following a protocol known as SELEX (systematic evolution of ligands by exponential enrichment). For aptamer identification, the target molecule is screened against an initial random pool of oligonucleotides (usually around 1000 sequences). Nonbinding oligonucleotides are discarded by repeated washes, leaving only the desirable aptamers binding to the target molecule followed by sequencing of these desirable oligonucleotides and used in the next SELEX rounds (usually 12–16 rounds) until the final aptamers are selected (Sefah *et al.*, 2010). To date, specialized companies like Admera Health, AptaGen, AM Biotechnologies, Creative Bio-gene.

Aptamer Sciences offer services of aptamer discovery and development, thus saving considerable time regarding the SELEX aptamer discovery cycles. Discovery of an aptamer service may costs around $5000. Once discovered, the production cost is lowered making it a cheap technology, in comparision with immunodetection-based ones as simple oligonucleotides are required. Most diagnostic approaches of aptamers are based on ligand-induced conformational changes. These changes can be detected by dye binding or fluorescence quenching methods. As Aptamer are small in size their affinity to their targets is comparable, or even higher than commonly used monoclonal antibodies. Aptamers are chemically stable, cost effective, viable alternatives to antibodies in many applications and used mainly in human disease diagnosis or food security. In the plant pathology field, only few studies have focused on the detection of plant pathogens. In the case of fungi, fumonisin (toxin produced by Fusarium species) detection (McKeague *et al.*, 2010) aptamers have been reported with its application in food-safety approach. Today, portable SPR imaging devices are available. In future these instruments, together with technologies such as the reports of detection of ASPV, could have a promising potential in-field plant pathology using aptamers.

Alternating Current Dielectrophoresis Technology

Dielectrophoresis (DEP) explains the movement of a polarizable particle induced by non-uniform alternating current (AC) electric field. The force exerted on the particle allows it to move towards high or low electric field density regions (Adekanmbi and Srivastava, 2016). Both the target particle and medium in which the particle is suspended undergo changes in their conductivity and permittivity under different AC frequencies, which can be determined by the Clausius– Mossotti (CM) factor. The moment of the particle either towards the maximum or to a weak electric field was represented by different CM factors. AC frequency at which the CM factor equals zero is known as the crossover frequency (cof) whose values differentiate one type of particle from another. Since then, it has been used widely for manipulation and separation of biological or artificial particles, including bacteria, yeasts, stem cells, viruses, mitochondria and DNA, while it is mainly used for separation and concentration of microorganisms in diagnosis field prior to their detection by traditional or molecular techniques. For example, DEP has been used to selectively isolate three different bacteria by using a microchannel; cells (including live/dead cells) with different electrical properties become trapped at different positions of the microchannel (Pysher and Hayes, 2007). AC-DEP has been used to develop simpler and portable methods for microorganism detection using DNA samples. For example, Gagnon *et al.* (2008) reported a novel pathogen detection method using AC-DEP to detect DNA–DNA hybridizations by imaging the study, silica nanoparticles were oligonucleotide-functionalized and the assays were performed using a quadrupole electrode array attached to a function generator via coppe rtape and wire leads. The nanoparticles were found to sensitively affect the cof at nanomolar target ssDNA (obtained by asymmetric PCR) concentrations as ssDNA hybridization modifies the particle surface conductivity. In recent years, the advances in micro-/nanoscale microelectrodes have allowed them to be easily integrated into small chips, making it possible to combine DEP-based diagnosis with other functions of portable or in-field diagnosis chips (Gagnon *et al.*, 2008). For example, devices like the NESDEP IU from F-cubed LLC promise to produce accurate test results in any location, without a full laboratory or special training in only 2 h, with a cost of $25–50 per analysis depending on the sample volume (www.foodqualitynews.com).manipulation and separation of artificial or biological particles, including bacteria, yeasts, stem cells, viruses, mitochondria and DNA (Gagnon, 2011). In the diagnosis field it has been used mainly for separation and concentration of microorganisms prior to detection by traditional or molecular techniques. For example, DEP has been used to selectively isolate three different bacteria by using a microchannel; cells with different electrical properties (including live/dead cells) become trapped at different positions of the microchannel. In an effort to develop simpler and portable methods, AC-DEP has been used for microorganism detection using DNA samples.

Portable Nanopore Sequencing

Rapid enlightment of next-generation sequencing (NGS) technologies in modern era has dramatically reduced the time and cost of identifying pathogen by metagenomic approaches. NGS has been widely used in the field of plant pathology, especially for virus detection, without culturing and using environmental samples directly. Majority of NGS assays are performed in specialized laboratories, meeting the requirements of expensive equipment or service provider, a great deal of sample preparation and complicated algorithms for processing the data. Nanopore sequencing, the fourth-generation DNA sequencing technology, is based on the philosophy of applying an external voltage that make the electrolyte ions in the solution to move electrophoretically from one side of a nanopore to another thus generating a quantifiable ionic current. When DNA is added into the solution under the applied external voltage, negative charged DNA will block the nanopore, thus interfere the ion current. The characterization of physical and chemical properties of the analyte is based on amplitude and duration of the temporary blocking (Feng *et al.*, 2015). Beta-testing of the MinION, the first portable nanopore sequencing device that works with consumable flow cells where the nanopores are immobilized can be operated simply by connecting it to a standard USB port was started by Oxford Nanopore Technologies during 2014. Each flow cell of the device can generate approximately 5–10 Gb of sequencing information with additional benefit of working in remote locations because of its small size and easy connection. The device has been tested for applications in the International Space Center. The MinION technology measures the rate at which a DNA strand moves through biological nanopores when applying an electric field, detecting the translocation of the DNA molecule with a single-nucleotide precision. For base-calling, Metrichor an online cloud-based analysis platform integrated with remote nanopore sensing devices which also allowed the real-time analysis of sequences read by the MinION. Workflows like 'What's in my pot' (WIMP) that combined both MinION technology with a realtime species identification bioinformatics application have been developed to achieve faster identification of species alon with classification of bacteria, fungi and viruses present in a sample in approximately 3.5 h. Using Metrichor for base-calling, the analysis requires an internet connection. 'Nanocall', an open source base-caller for Oxford Nanopore sequencing data was developed in 2016, thus providing users with an offline and private analysis advantage for MinION generated data, even in remote locations. MinION device also allows accurate identification and differentiation of microorganisms in biological samples. Most diagnosis applications of the device have been reported for medical, food safety and even animal viral disease diagnostics, with promising results. For example, the MinION was demonstrated to comprehensively identify pathogens directly from urine samples, without culturing in a timeframe of 4 h from sampling to results. Multiplex detections can also be done, making it possible to detect 96 different barcodes in a single sample using the barcoding kits offered by the Oxford Nanopore technologies. For simplifying the detection method in resource-limited situations the device can also be used in different combination with some of the technologies mentioned above. To date there are no reports of its application for the detection of plant pathogens. High cost of the technology, is the greatest barriers for implementing NGS in diagnostic settings, with the exception of high-impact pathogens which may justify the use of the fast and reliable system at higher costs. Currently, a starter pack of Nanopore Technologies including the MinION device can be acquired for $1000. However, supplies like the consumable flow cells with the nanopores and all necessary reagents are still expensive when considering routine diagnosis analysis. However the current technology still needs improvement to reduce error rates, the portability of the device, together with new simplified workflows like Nanocall or WIMP and the fact that NGS technologies and costs of reagent are decreasing rapidly, thus making portable nanopore sequencing devices a promising alternative for the future of point-of-care diagnosis, including plant pathogen detection.

Chapter 8

Management of Bacterial Diseases

Diseases caused by bacterial plant pathogens are major constraints in crop production and cause significant annual losses on a large scale. The attainment of consistent effective management of these can be extremely difficult and it often affected by grower dependence on highly disease-susceptible cultivars because of high yielding nature and consumer preferences. New and emerging bacterial disease problems (glume blotch of wheat) and established problems in new geographical regions (*e.g.* bacterial canker of kiwifruit in New Zealand) grabbing the concern, but the list of bacterial disease problems with few effective management options is long. The ever-increasing global human population requires the continued stable production of a safe food supply with greater yields because of the shrinking areas of cultivable land. One major facet in the maintenance of the sustainability of crop production systems with predictable yields involves the identification and deployment of sustainable disease management solutions for bacterial diseases.

In world scenario many countries banned the usage of antibiotics of animal and human usage in plant disease management, recently India had also taking steps to phase out utilisation of Streptomycin sulphate and oxytetracycline in plant disease management in all forms by 01.01.2024. In addition, the identification of novel management tactics has also come to the fore because of the increasing evolution of resistance to existing bactericides. A number of central research foci, involving basic research to identify critical pathogen targets for control, novel methodologies and methods of delivery, are emerging that will provide a strong basis for bacterial disease management into the future.

In this scenario factors to be considered are:

- ☆ Near-term solutions are desperately needed. Are there replacement materials for existing bactericides that can provide effective disease management under field conditions?
- ☆ Experience should inform the future. With prior knowledge of bactericide resistance issues evolving in pathogens, how will this affect the deployment of newer compounds and biological controls?
- ☆ Knowledge is critical. A comprehensive understanding of bacterial pathosystems is required to not only identify optimal targets in the pathogens, but also optimal seasonal timings for deployment.
- ☆ Host resistance to effectors must be exploited, carefully and correctly. Are there other candidate genes that could be targeted in transgenic approaches? How can new technologies (CRISPR, TALEN, *etc.*) be most effectively used to add sustainable disease resistance to existing commercially desirable plant cultivars?
- ☆ We need an insider's perspective on the management of systemic pathogens. In addition to host resistance or reduced sensitivity, are there other methods that can be used to target these pathogen groups?
- ☆ Biological systems are variable. Can biological control strategies be improved for bacterial disease management and be made more predictable in function?

By considering above queries various researchers tried to answer them by their experimental studies quoted under various sections in this chapter.

Many general characteristics of plant bacterial diseases which favour disease incidence, pathogen dissemination and pathogen survival and confound management strategies in agricultural ecosystems had to be considered for devising any management strategy an. Examples of these include: (i) the growth rate of bacterial pathogens under optimal environmental conditions can quickly result in population sizes that favour both increased levels of infection and the potential for rapid spread to new infection sites, driving disease epidemics (ex. fire blight of apple by *Erwinia amylovora*); (ii) epiphytic growth on symptomless leaf surfaces can result in the attainment of sufficient population size for disease occurrence (ex. bacterial brown spot, *Pseudomonas syringae* pv. *syringae*); (iii) some bacterial pathogens are soil-borne and their association with other soil organisms like nematodes in soil are extremely difficult to manage(bacterial wilt, *Ralstonia solanacearum*); (iv) many bacterial pathogens predominantly colonize internal locations within plants that are inaccessible to most spray-applied chemical and biological controls that target plant surfaces (ex. Huanglongbing (HLB), 'Candidatus Liberibacter spp.'); (v) populations of insect vectors, especially of destructive xylem- and phloem-dwelling pathogens, are as difficult to manage as the diseases (vi) pathogen survival and growth on crop debris and on alternative hosts defeat strategies of inoculum reduction (ex. Goss's wilt, *Clavibacter michiganensis* subsp. *nebraskensis*); and (vii) asymptomatic survival of pathogen cells on planting material for extended periods of time can result in disease dissemination to new geographical locations (ex. bacterial blight of cassava, *Xanthomonas axonopodis* pv. *manihotis*).

Management strategies for plant bacterial diseases require a thorough knowledge onhost- pathogen interaction in order to identify the most appropriate timing for the targeting of pathogen populations and to determine when the critical host tissues are vulnerable to infection. An integrated management approach, including the use of host plant resistance/growth of less susceptible cultivars, intervention with chemical and/or biological controls and cultural practices aimed at reduction of inoculum, typically represents the best overall strategy for effective and sustainable disease management. However, integrated approaches depend on the availability of suitable host plant cultivars, efficacious chemical and biological controls and cultural practices that are physically achievable and economically suitable, such that growers will deploy them. In recent years, genomic advances have facilitated a rapid increase in our understanding of bacterial pathogen–host interactions and have led to the invention of novel techniques to incorporate resistance into high yielding desirable plant cultivars. Thus, the management of bacterial plant diseases remains a formidable 'grand challenge' for plant pathologists. Various methods are used to manage bacterial diseases of crops as described below:

1. Cultural Practices

a) **Eradication** - Eradication methods are employed to the host plants or alternate or collateral or alternative hosts against the pathogen. Practical procedures include fumigation, heat treatment, flooding ,solarization, burning or removal of residues.

b) **Field hygiene** - Removal of diseased plants or plant parts from field is important to reduce inoculum density. Defoliated leaves or pruned twigs of fruit trees and straw of cereals are preferably burnt. Infected woody stems such as tomato, brinjal and tobacco should not be ploughed into the soil because they may remain partly undecomposed and thus harbour the inoculum.

c) **Disinfection of seeds and planting materials** - Several bacterial pathogens survive in or on propagative materials including seeds. Necessary physical (temperature, heat, radiation), chemical or biological methods should be used.

d) **Disease Escape** - By proper selection of cultivars and some modifications in cultivation period particularly date of sowing/planting, many bacterial pathogens can be managed. It is now proved that cultivation of early maturing varieties and early enough torrential rain will increase the susceptibility of plants to bacterial leaf blight.

e) **Rotation** - This is an unique method to reduce the soil sickness particularly nutritional and microbiological. Rotation is ineffective against bacteria having wide host range or when the organism is soil inhabitants (*e. g. R. solanacearum, Agrobacterium, Erwinia*) or when the bacterium is both seed and soil borne.

f) **Irrigation** - Controlled irrigation both with respect of amount and timings provide good management. In common scab of potato (*Streptomyces scabies*), irrigation for several weeks after tuber initiation is effective in reducing the disease. It is experimentally proved that maintaining the soil moisture deficit of 0.6 for 3-4 weeks in the tuber initiation period controls the disease.

g) **Plant density and nutrition** - Adequate plant density and balanced fertilization are obviously important for growing healthy plants. High N levels and plant density are congenial conditions for infection as pathogen gets predisposed and proliferates. Dense plant canopy reduces air movement and drying of foliage.

2. Chemical Control

Several chemical compounds such as streptomycin, kasugamycin, oxytetracycline, novobiocin, copper compounds *etc.* are available as commercial spray materials to control bacteria infecting above ground plant parts, sodium hypochlorite and oxolinic acid as seed disinfectant and, chloropicrin and thiram as soil disinfectant.

Copper and Antibiotics

The potential for the chemical management of any disease can be largely driven by many factors like availability of effective modes of action, convenience to access the pathogen on plant surfaces, the susceptibility of the pathogen to that chemical, the economic importance of the crop and ease of production from an industrial perspective. Compared with fungicides relatively a very few chemicals targeting plant bacterial diseases have been available I markets for these decades. Probably the most important reason for this is that the best antibacterial compounds are antibiotics and mot of the historically have been developed for use in clinical medicine and not for agriculture purpose. Initially lot of compounds were tested against the, among these copper compounds (introduced in the 1880s) and the antibiotic streptomycin (1950s) proved to be the most effective in managing plant bacterial diseases (McManus *et al.,* 2002). Although, in general, these bactericides have been relatively more reliable and successful disease management tools, but use of copper and streptomycin both has been impacted by the evolution of resistance in populations of plant pathogens (McManus *et al.,* 2002). Which might happened mainly due to the extensive use of copper and antibiotic sprays over multiple years high numbers of applications within individual seasons. We now know that horizontal gene transfer, a mechanism for the acquisition of resistance determinants by bacteria even from phylogenetically distinct species, has driven the antibiotic resistance crisis which currently affecting the human population also (Perry and Wright, 2013). The major gene responsible for copper resistance is encoded by a copper-inducible operon (copABCD) in *P. syringae* pv. *tomato* (Mellano and Cooksey, 1988) and by related variants in other *P. syringae* pathovars (Gutierrez Barranquero *et al.*, 2013), *X. campestris* pv. *juglandis* (Lee *et al.*, 1994) and in the citrus pathogens *Xanthomonas citri* pv. *citri* and *X. alfalfae* pv. *citrumelonis* (Behlau *et al.*, 2011). In fire blight pathogen *E. amylovora* for the first time, the first streptomycin resistance were conferred by a chromosomal mutation altering the ribosomal protein target of the antibiotic (Moller *et al.*, 1981). Recent analysis on streptomycin resistance in major pathogens like *E. amylovora, P. syringae* and *X. vesicatoria* have shown that it is conferred by the strAB genes, which encode aminoglycoside phosphotransferase enzyme that modify streptomycin to a non-toxic compound, this gene is known to be located on the transposon Tn5393 and related variants.

Along with streptomycin a few additional antibiotics have been used as in various pathosystems. Oxytetracycline has been used on pome fruit trees to control fire blight (*E. amylovora*) in the USA and Mexico, on peach and nectarine targeting bacterial spot (*X. arboricola* pv. *pruni)* in the USA (McManus *et al.*, 2002). Gentamicin has been used in Latin American countries to control fire blight and various vegetable diseases and oxilinic acid has been used in Israel to control fire blight (Shtienberg *et al.*, 2001). In these cases, gentamicin-resistant bacteria were recovered from lettuce in Costa Rica and oxolinic acid resistance in *E. amylovora* was detected in Israel 1–3 years after its introduction (Rodriguez *et al.*, 2006). Kasugamycin has been registered for use in the USA to target fire blight, especially in orchards containing streptomycin-resistant *E. amylovora* (McGhee and Sundin, 2011), was used in European Union countries for more than three decades until antibiotic use was banned and has been used in Japan as a seed treatment for bacterial diseases. Kasugamycin does not have any applications outside of these plant agricultural uses and does not appear to be as broad spectrum as streptomycin. However, the use of antibiotics has generally been discouraged in few regions of the world because it's potential impact on the transfer of antibiotic resistance into clinical pathogens. In addition, experience with horizontal gene transfer and the acquisition of streptomycin resistance genes by plant pathogens in many ecosystems suggests that antibiotic resistance evolution is an eventuality with any new antibiotic deployed. Thus, the main reality of using antibiotics for crop diseases is that any strategy that selects for resistance in the target pathogen is not sustainable.

Nanotechnology

Nanotechnology is the leading branch of science in recent years with its immense applications in various sectors. Nanotechnology has substantially advanced in field of medicine and pharmacology but hadn't received among agricultural applications. The application of nanotechnology in agriculture is currently being explored in plant hormone delivery, seed germination, water management, nanosensors, transfer of target genes, nanobarcoding and controlled release of chemicals. Nanoparticles (NPs) are particles with at least one dimension within the 1–100-nm range. Particles within this size range have unique physical and chemical properties, including large surface to mass ratio, high reactivity and unique interactions with biological systems. Some of the above properties make them excellent antimicrobials and some properties make them ideal carriers/ delivery systems for other antimicrobials. The evaluation of NPs for plant disease management to date has mainly focused on oomycete and fungal diseases but promising results have also been obtained for bacterial diseases (discussed below). The antimicrobial activities of most NPs result from three major aspects: photocatalysis, physical damage to the bacterial cell envelope and the release of toxic metal ions. NPs with photocatalysis activity are mostly metal oxides, such as CuO, TiO_2, ZnO and Fe_3O_4 and metals, such as Ag (Choi and Hu, 2008). During photocatalysis, reactive oxygen species (ROS) are generated in the form of hydrogen peroxide, hydroxyl radicals and peroxide. These ROS are toxic to bacteria, as they can damage cellular constituents, such as DNA, lipids and proteins, resulting in bactericidal and bacteriostatic effects.

In plant disease management, an NP formulation of titanium dioxide (TiO2) induced photocatalysis, resulting in antimicrobial effects against the bacterial spot pathogen *Xanthomonas perforans* (Paret *et al.*, 2013). Interestingly, doping the TiO2 NPs with Ag and Zn significantly increased the photocatalytic activity against *X. perforans*. Treatment of *X. perforans*-infected tomato plants with TiO2/ Zn NPs at approximately 500–800 ppm significantly reduced bacterial spot severity compared with untreated and copper controls (Paret *et al.*, 2013). ZnO and Ag NPs also exhibited promising antimicrobial activity against *E. amylovora* with minimum inhibitory concentrations of 83.3 and 108.3 ppm, respectively. Some NPs can exert antimicrobial activity through the release of ions, such as Ag1, Zn21 and Cu21, which are toxic to bacteria. For example, the release of Ag1 greatly contributed to the antimicrobial activity of Ag NPs (Lok *et al.*, 2007). Some modifications and improvements can also enhance the antimicrobial activities of NPs: for example, the association of NPs with a support material has recently shown promising antimicrobial effects. Bare NPs tend to form agglomerations that weaken the antimicrobial activity (Ocsoy *et al.*, 2013). Grapheme oxide (GO) has been used as support material to grow Ag NPs, significantly reducing agglomeration and enhancing the antimicrobial activity against the bacterial spot pathogen *X. perforans* (Ocsoy *et al.*, 2013). In a glasshouse assay, control of bacterial spot on infected tomato transplants treated with 100 ppm of the Ag@dsDNA@GO NPs occurred at a similar level as conventional copper treatment. Prabhu *et al.* (2015) have reported the antibacterial activity of iron oxide nanoparticles against *Xanthomonas* sp. and *Proteus vulgaris* at 50 mg/ml. Vinay *et al.* (2018) proved the antibacterial activity of green synthesized *P. fluorescens*-based ZnNPs against *X. oryzae* pv. *oryzae* and *X. axonopodis* pv. *punicae* at 500–1250 ppm and no phytotoxicity was observed. Silver nanoparticles synthesized from cotton stem showed antibacterial activity against *X. citri* pv. *malvacearum* and *X. campestris* pv. *campestris*. Furthermore,

AgNPs exhibited strong antioxidant activity and these findings suggest that cotton stem extract could be efficiently used in the synthesis of AgNPs. Vinay and Narguad (2019) reported an antibacterial activity of green synthesized water-soluble chitosan-based iron nanoparticles (Ch-FeNPs) against *X. oryzae* pv. *oryzae* and *X. axonopodis* pv. *punicae* at 500–1250 ppm. They noticed bacterial growth inhibition due to treatment of iron nanoparticles. Bryaskova *et al.* (2011) observed the significant antibacterial activity of ZnO NPs against *Staphylococcus aureus, Streptococcus pyogenes, Bacillus cereus, Pseudomonas aeruginosa, Proteus mirabilis* and *Escherichia coli.*

In addition to antimicrobial activities, NPs have also been suggested to be efficient delivery systems for many other antimicrobial compounds. Multiple NP delivery systems, such as hydrogel, dentrimers, liposomes, carbon nanotubes, micelles and microand nanoemulsions, have been studied for the delivery of various active ingredients. Among them, nanoemulsions have shown promising delivery of an agriculturally important herbicide and a nanoemulsion formulation was able to enhance the permeability of the antibiotic ampicillin through the citrus cuticle into the phloem via a foliar spray targeted against Huanglongbing disease (Yang *et al.,* 2015). As the plant cuticle (wax, cutin and pectin) acts as the major barrier preventing antimicrobial compounds from penetrating into plant tissues and as many plant-pathogenic bacteria infect the phloem and xylem tissues, the development of antimicrobial delivery technology which penetrates through the cuticle may have wide applications in the control of bacterial plant diseases.

In summary, NPs possess many desirable traits that may make them excellent antimicrobials for the management of bacterial plant diseases in the future. The application of antimicrobial NPs in agriculture could significantly enlarge our selection of antimicrobials. In addition, most NPs display antimicrobial activities through different mechanisms, which could possibly reduce the risk of resistance development. The application of NPs as delivery systems for pesticides also opens up new opportunities to enhance the efficacy of currently available antimicrobials targeting pathogens inside plants. However, limitations, such as cost, material preparation, particle agglomeration and efficient active ingredient delivery, still need to be resolved. Moreover, the environmental fate of NPs needs to be determined before large scale application and the potential for persistence of NPs in food products must be determined because of toxicity effects.

3. Biocontrol Control

In nature microorganisms coexist in soil in association with plant roots and also interfere with plant physiological functions. It is estimated that bacteria occupies 7 to 15 per cent of the total root surface area. Of these, some bacteria positively affect plants and have been designated as plant growth promoting rhizobacteria (PGPR) (Kloepper 1978). In present scenario biocontrol of phytopathogenic bacteria has been considered as one of the major and potential control strategies (Table 1). The use of biocontrol strategies offers several advantages over the chemical control, since it is environmentally safe, economical, self-perpetuating and usually free from residual side effects. However, in reality, it will not immediately nor totally replace chemicals, but the use of biocontrol agents can significantly enhance quality of life, the environment and agricultural productivity in long run. Biological agents reduces the disease severity by various means like production of lytic enzymes, chelating agents like siderophores for nutrients competition and various antibiotics with different modes of action *etc.* Bacterial leaf blight (BLB) caused by *Xanthomonas oryzae* pv. *oryzae* is one of the destructive diseases in rice. Phenazine antibiotic produced by some fluorescent pseudomonads has excellent antibacterial activity against *X. oryzae* pv. *oryzae*. Different *Bacillus* spp. applied to rice plants as seed treatment before sowing; a root dip prior to transplantation and two foliar sprays prior to inoculation could suppress 59 per cent of bacterial leaf blight (Vasudevan *et al.*, 2002). Solanaceous crops other than potato and other vegetatively propagated crops were protected biologically from bacterial wilt caused by *R. solanacearum* by dipping the root system of seedlings before transplanting; biocontrol of is mostly based on antibiosis and competition. Other method to deliver bacterial antagonist is by seed treatment, either by seed dressing, seed coating or seed pelleting. Treatment of tomato seed with water suspension of *P. putida*-Pf-20 suppressed bacterial wilt into some degree (Asrul *et al.*, 2004). Crown gall, caused by *Rhizobium radiobacter* (*Agrobacterium tumefaciens)*, is a very serious threat for many crops at nursery and field level, especially in stone fruit trees. The method of control is by inoculation of planting material with non-pathogenic *A. radiobacter* strain K84 immediately before sowing or planting. For over 15 years, crown gall on many different host plants has been successfully controlled by this non-pathogenic strain in many countries. Recently *A. radiobacter* strain K1026 has been popularised for particular areas. The inhibition of the pathogen by Agrocin 84 and 1026 (a bacteriocin produced by K84 and K1026 respectively), is mainly due to competition for site and certain nutrient. Biological site competition and competition of a certain nutrient that common these bacteria. Endophytes occur almost everywhere in various parts of the world with diversified climatic conditions and their association with all groups of plant kingdom like algae, bryophytes, pteridophytes, gymnosperms and angiosperms has been studied in detail. Thangavelu and Gopi (2015) evaluated 71 endophytic bacteria from root and corm tissues of banana and 37 bacteria from the rhizospheres of banana. Rajamanickam *et al. (*2018) tested the antibacterial activities of two endophytic bacterial strains by biohardening plantlets against rhizome rot disease with soil application of talc-based formulation of both, individual and mixture of endophytic *Bacillus* strains *viz.*, PP and CL3 under glasshouse conditions. The result revealed that the banana plantlets biohardened with *Bacillus subtilis* PP strain + soil application of a consortium of *B. subtilis* strains (PP and CL3) @ 10 g/plant at the time of planting recorded the minimum disease incidence of 19.56 per cent which accounted for about 75.55 per cent reduction over the control. They concluded that biohardeing could be one of the best way for vegitatively propagated crops like banana *etc.*

Table 1. Microbial Anatagonist Used for Control of Bacterial Diseases of Crops

Microbial Antagonists	*Disease and Host*	*Pathogen*
Bacterial antagonists		
Actinomycetes	,	*Ralstonia solanacearum*
Azotobactor chroococum	,	*R. solanacearum*
Pseudomonas putida W4P63	Soft rot of potato	*Erwinia carotovora*
P. fluorescens	Leaf light of rice	*Xanthomonas oryzae* pv. *oryzae*
Serratia J2, *Bacillus* BB11, *Pseudomonas*	Bacterial wilt of tomato	*R. solanacearum*
B. cereus, B. pumilus, B. lentimorbus	Black rot of cabbage	*X. compestris* pv. *compestris*
P. fluorescens	Wilt of brijnal	*R. solanacearum*
P. putida	Wilt of potato	*R. solanacearum*
Stenotrophomonas maltophilia	Wilt of potato	*R. solanacearum*
B. amyloliquefaciens	Wilt of tomato	*R. solanacearum*
B. subtilis	Wilt of tomato	*R. solanacearum*
Paenibacillus elgii JCK-5075	Wilt of tomato	*R. solanacearum*
Enterobacter aerogenes	Wilt of tomato	*R. solanacearum*
Streptomyces corchorusii and *S. mutabilic*	Wilt of tomato	*R. solanacearum*
P. glumae	Wilt of tomato	*R. solanacearum*
Bacillus subtilis var. *amyloliquefaciens* (FZB 24)	Bacterial leaf blight of rice	*X. oryzae* pv. *oryzae*
P. fluorescens RR8	Bacterial leaf blight of rice	*X. oryzae* pv. *oryzae*
Bacillus subtilis (FZB 24)	Bacterial leaf blight of rice	*X. oryzae* pv. *oryzae*
B. subtilis UiTMB1	Bacterial leaf blight of rice	*X. oryzae* pv. *oryzae*
Streptomyces toxytricini VN08-A-12	Bacterial leaf blight of rice	*X. oryzae* pv. *oryzae*
Lactobacillus plantarum CC100, PM411 and TC92 and *Leuconostoc mesenteroides* CM160 and CM209	Bacterial canker of kiwifruit Bacterial spot of stone fruits Angular leaf spot of strawberry.	*P. syringae* pv. *actinidiae* *X. arboricola* pv. *pruni* *X. fragariae*
Fungal antagonists		
Pichia guillermondii	Bacterial wilt of tomato	*R. solanacearum*
Candida ethanolica	"	*R. solanacearum*
Glomus mosseae	"	*R. solanacearum*
G. faxiculatum	"	*R. solanacearum*
T. asperellum	"	*R. solanacearum*
T. viride, T. harzianum	"	*R. solanacearum*
T. harzianum	Bacterial leaf blight of rice	*X. oryzae* pv. *oryzae*

Targeting Bacterial Biofilms

Biofilm formation provides bacteria with effective protection from environmental stresses, antimicrobial drugs and host defence mechanisms and it is also a is a critical virulence factor of many plant pathogenic bacteria. Thus, inhibiting the biofilm formation to control bacterial infection has become an attractive research area. Currently, most of the developed anti-biofilm inhibitors, such as D-amino acids and indole derivatives, have been applied to prevent early biofilm formation, resulting in reduced disease symptoms caused by animal and plant pathogens and increased bacterial vulnerability to antibacterial chemicals (Worthington *et al.,* 2012). Bis-(30,50)-cyclic dimericguanosine monophosphate (c-di-GMP) is a bacterial secondary messenger that participates in the regulation of multiple cellular behaviours and virulence in many bacterial species. Recent studies have demonstrated that a decrease in intracellular c-di-GMP levels can trigger dispersal of biofilm, which has been proposed as a promising strategy for suppression of biofilm. One compound showing excellent promise is 2-aminoimidazole (2AI), which inhibits biofilm formation by the bacterial spot pathogen *Xanthomonas euvesicatoria* under *in vitro* conditions. When applied in mixtures with copper hydroxide and alone 2AI suppressed bacterial spot infection in field experiments significantly (Worthington *et al.,* 2012). Two other compounds, 3-indoylacetonitrile and D-leucine, known to inhibit biofilm formation by the citrus canker pathogen *X. citri* pv. *citri* and reduced disease incidence on citrus plants.

Bacteriophages

Bacteriophages (phages) are obligate intracellular parasites that multiply inside bacteria by using few or all of the host biosynthetic machinery. Phages are very specific which making them an attracting alternative for managing phytobacterial diseases. The use of bacteriophage for bacterial disease management is another biological control method that has been the subject of research activity at varying times over the past four decades. Bacteriophages are abundant in most if not all ecosystems on Earth and phage capable of lysing plant pathogenic bacterial species can be readily isolated from host plant tissue or soil. Inundative application strategies are typically used for disease management. The two main limiting factors affecting bacteriophage efficacy are stability in the environment and spontaneous resistance mutations in target bacterial pathogens. Phages specific to *X. oryzae* pv. *oryzae* are found in the water of rice field, irrigation canal and rivers. The population density of bacteriophage is correlated with the number of its bacterial host. However, the problem in using of bacteriophage in the biocontrol of this pathogen was its inactivation by UV light, variable bacterial sensitivity and the rapid development of resistance to the phage. Application of bacteriophages as biologicals exploits the ability of lytic phages to kill the pathogen. The results of past field trials were variable from failed trials to 50–65 per cent efficacy. Tomato bacterial wilt caused by *R. solanacearum* is difficult to control. The phage that had widest spectra against several isolates of *R. solanacearum* was then chosen for the next experiment. A bacteriophage from a region in Central Java showed the wide host range compared with the others (Arwiyanto *et al.,* 2019). A commercially available formulation of bacteriophage in the name of AgriPhage™ is being released by company Omnilytics against *X. vesicatoria* and *Psuedomanas syringae* of tomato. These are also considered as environmentally safe from USDA.

Interest in bacteriophage for disease control is undergoing a resurgence, particularly in clinical medicine, as antibiotic resistance becomes more problematic. Likewise for plant bacterial diseases, there is interest in using phage, but significantly more field research and the optimization of formulations and spray timings are necessary to facilitate the development of successful applications. Ultimately, a thorough understanding of the ecology of potential antagonists, interactions with plants, positive and negative interactions with the plant microbiome and interactions with pathogens is needed for improved and sustainable deployment in commercial systems. Functional genomic analyses appear to represent a solid route to obtaining knowledge on current gold standard biocontrol agents, but this work needs to be coupled with significant, dedicated field efforts, such that the behaviour of these organisms can be assessed under a wide range of environmental and plant host combinations.

4. Resistant Variety

One effective tool in disease prevention is the use of resistant cultivars/varieties. At the same time, the development of resistant cultivars/varieties through plant breeding is an industrious and time-consuming effort to combine resistance and desired commercial traits. Also, there is not any one plant cultivar/variety that is completely resistant to all disease threats. The cultivars/varieties having the marking of resistance against a disease have higher level of resistance than those labeled tolerant.

i. Grafting Technique

Grafting onto resistant rootstocks provides an alternative and effective solution to manage soil-borne bacterial pathogens in Solanaceous and Cucurbitaceous crops. Tomato and eggplant rootstocks are used for grafting tomatoes worldwide. However, eggplant rootstocks are preferred over tomato rootstocks in many parts of the world due to their stronger resistance to bacterial wilt and tolerance to flooding. Presently many studies have been conducted to evaluate the bacterial wilt resistance, as well as the agronomical potential and the efficiency of the newly identified eggplant accessions in the field as rootstocks for tomato grafting. The graft of susceptible cultivars of tomato on resistant rootstocks belonging to *Solanum* spp. is found effective to manage the wilt disease. The bacterial disease was controlled by grafting susceptible tomato cultivars onto resistant eggplant cv. Pusa Purple Cluster rootstocks. In grafting, the wilt incidence in Ponderosa scions was less on the rootstock of Hawaii 7996 as compared to the most resistant rootstock (LS-89) and Hawaii 7996 may be used as an alternative genetic source for breeding for resistance against bacterial wilt diseas. Bacterial wilt resistant varieties of eggplant used as rootstocks: Eg 203, EG 195, S 69 and moderately resistant cultivars 89-002 and A 300. Moreover, Five bacterial wilt resistant eggplant accessions *i.e.* VI041809A, VI041943, VI041945, VI041979A and VI041984 from the World Vegetable Center were evaluated as rootstocks for grafting with two different fresh market tomato cultivars (Victoria and TStarE) as scion under open field conditions in Taiwan. Graft compatibility using the tube grafting method as well as bacterial wilting percentage, disease index, fruit yield and quality parameters were assessed. Yield for the eggplant rootstock grafted tomato plants was higher compared to the non-grafted tomatoes and self-grafted tomato. It has also been reported that the fruit quality was not affected by grafting, but the differences in antioxidant activities were observed. The new eggplant rootstocks can be considered as alternatives to the rootstocks currently used for commercial production of tomatoes during the hot-wet season (Manickam *et al.*, 2021). The pepper scion ('Nokkwang') grafting onto breeding lines ('PR 920' and 'PR 921' and 'PR 922') resistant to bacterial wilt showed greater rate of survival when they were inoculated with *R. solanacearum.* The mechanism involve in controlling bacterial disease is the avoidance when the rootstock is resistant to the pathogen. In the case of bacterial wilt resistance, it has been shown that resistant rootstocks physically limit the movement of bacteria from the soil to the scion it was later shown that when the resistant rootstock was used as the scion to graft onto susceptible rootstock, the resistant plant succumbed to the disease.

ii. CRISPR/Cas9-Novel Tool for Management of Bacterial Plant Diseases

CRISPR (Clustered Regular Interspaced Palindromic Repeats)/Cas9 (CRISPR associated protein) is a great breakthrough in gene editing technology. It can be utilized to exploit defensive mechanism in plants against plant pathogens attack with recognition and degradation of the invading pathogenic genes by bacterial immune system. CRISPR/Cas9-mediated genetically engineered resistance can be inherited to further generation of crop plants after segregation of Cas9/sgRNA transgene in F1 generation. CRISPR/Cas9 proves itself as a fascinating tool to revolutionize plant breeding for developing a disease-resistant cultivars (Ghimire, 2017). Phytopathogenic bacteria are difficult to control, mainly because of undetected latent infections and also the lack of suitable sufficient agrochemicals. CRISPR/Cas9 mutagenesis of OsSWEET13 has been performed in rice to achieve resistance to bacterial leaf blight (BLB) disease caused by γ-proteobacterium *Xanthomonas oryzae* pv. *oryzae* (Zhou *et al.*, 2015). OsSWEET13 is a susceptibility (S) gene encoding a sucrose transporter which is plays major role in plant-pathogen interaction. *X. oryzae* produces an effector protein, PthXo2, which induces OsSWEET13 expression in the host and the consequent condition of susceptibility. In a previous work concerning OsSWEET14 promoter mutagenesis adopting a TALEN approach, the disruption of this gene lead the *X. oryzae* effector unable to bind OsSWEET14 and ultimately resulted in resistance against BLB (Li *et al.*, 2012). Zhou *et al.* (2015) obtained a null mutation in OsSWEET13 in order to better explore PthXo2-dependent disease susceptibility and resultant mutants were resistant to bacterial blight. Jia *et al.* (2016) generated canker resistant mutants by editing the PthA4 effector binding elements in the promoter region of the Lateral Organ Boundaries 1 (CsLOB1) gene in Duncan grapefruit. Mutated lines showed a decrease in typical canker symptoms 4 days post inoculation with *X. citri* subsp. *citri* (Xcc) and no further phenotypic alterations were detectable. Peng *et al.* (2017) confirmed the link between CsLOB1 promoter activity and citrus canker disease susceptibility in Wanjincheng orange (*Citrus sinensis* Osbeck). The complete deletion of the EBEPthA4 sequence from both CsLOB1 alleles induced resistance enhancement to citrus canker which is achieved by employing CRISPR/Cas technology. This CRISPR/Cas9 genome editing technology has been successfully employed for the genetic editing of single or multiple gene targets in several plants it has been shown to be efficacious for the generation of mutations in the TaMLO-A1 allele of bread wheat, which encode a protein that represses plant defence against powdery mildew. This modification conferred resistance to the pathogen (Wang *et al.*, 2014), demonstrating that CRISPR/Cas editing can be successfully employed in the engineering of durable resistance, even at different levels of ploidy

Antimicrobial Peptides in Management of Phytobacterial Diseases

Antimicrobial peptides (AMPs) are the short polymers of amino acids with a peptide bond having 50 amino acids possessing broad-spectrum antimicrobial activity against various groups of plant pathogens. Antimicrobial peptides are also called as host defence peptides and cell-penetrating peptides. These are generally isolated from many organisms, like amphibians, insects, humans, microorganisms and plants. AMPs are part of the nonspecific host defence system and are active against different types of microorganisms including phytopathogenic bacteria. Microscopic analysis revealed wide-scale damage to the bacterial membrane, in addition to inhibition of pathogen growth. In planta, potent antibacterial activity is also demonstrated. Treatment with the lipopeptides of Arabidopsis leaves infected with *P. syringae* efficiently and rapidly reduced the number of bacteria with no toxicity on the plant tissues (Makovitzki *et al.*, 2007). The ultrashort lipopeptides could serve as native-like antimicrobial agents economically feasible for use in plant protection. The synthetic LF chimera showed potential antibacterial activities against wide range of phytopathogenic bacteria such as *X. campestris, P. carotovorum, E. amylovora, R. solanacearum* and *P. syringae* (Chahardoli *et al.*, 2017). The synthetic antimicrobial peptide, *i.e.* Shiva-1, was isolated from the silk moth and introduced in transgenic tobacco (Jaynes *et al.*, 1993). Lactoferrin is another iron-binding glycoprotein known to possess antibacterial properties. The expression of a human lactoferrin gene in tobacco delayed the onset of symptoms caused by *R. solanacearum* from 5 to 25 days. This resistance appears to be due to the truncation of lactoferrin, resulting in a smaller peptide with strong antibacterial activity (Mitra and Zhang, 1994).

Chitosan an Organic Elicitor in Management of Plant Bacterial Diseases

The polysaccharide chitosan represent a renewable source of natural biodegradable polymers and meet with the emergence of more and more food safe problems. Chitosan is a deacetylated derivative of chitin that is naturally present in the insect body wall, in crustacean shells and in fungal cell wall from which it can be extracted easily. Chitosan has been reported to have antimicrobial activity. Chitosan also behaves as a resistance elicitor inducing both local and systemic plant defence responses in plants (Orzali *et al.,* 2017). Chitosan has been emerging as one of the most promising polymers for the efficient delivery of various agrochemicals and micronutrients in nanoforms (Kashyap *et al.,* 2015). Chitosan and its derivatives inhibited the growth of a wide range of bacterial plant pathogens (Liu *et al.,* 2001). Gram-negative bacteria are more susceptible to chitosan (Du *et al.,* 2009). They suggested that hydrophilicity in Gram-negative bacteria is significantly higher than that in Gram-positive bacteria, which makes them more sensitive to chitosan (Chung *et al.,* 2004). The highly charged nature of lipopolysaccharides inouter wall will confer an overall charge of negative to the cell wall of the Gram-negative bacteria. Therefore, Gram-negative bacteria with high electronegative charge will interact more effectively with the polycationic chitosan compared to Gram-positive bacteria. Chitosan was evaluated against several bacterial pathogens and effective concentration for suppression of growth of bacteria is presented hereunder. The *in vitro* antibacterial effect of chitosan and its ability in protection of watermelon seedlings from *Acidovorax citrulli* were evaluated by Li *et al.* (2013). The disease index of watermelon seedlings planted in soil and the death rate of seedlings planted in perlite were significantly reduced upon treating with chitosan at 0.40 mg/mL. Chitosan solution at 0.10 mg/mL inhibited the growth of one of the most important pathogenic bacteria *Xanthomonas.* Pre-treatment of tomato seedlings with 10 μg/mL chitosan against *P. syringae* pv. *tomato* DC3000 (Pto DC3000) inoculation significantly decreased bacterial damages in cotyledons compared with control. It is also known to affect already established biofilms (Mansilla *et al.,* 2013). A wider comprehensive knowledge of the mechanism of action of chitosan in pathogens and plants will increase the chance of its successful application to control disease spread in plants.

Transgenic Approaches in Mitigating Phytopathogenic Bacteria

Many molecular approaches have been proposed to enhance plant resistance to bacterial pathogens like *P. syringae.* These strategies include the use of genes producing antibacterial proteins from different insect vectors and their transformation in plants for development of disease resistance (Huang *et al.,* 1997) and inactivation of various virulence factors resulted the immunity of plants against the bacterial species (Anzai *et al.,* 1989). The resistant non-bacterial genes can also be introduced by transgenic approaches for broad-spectrum resistance against the devastating plant pathogens. Recent years the utilization of RNAi has emerged as an important tool to counter the bacterial genome at transcriptional and post-transcriptional level. Small interference RNA has been proved effective against the crown gall disease caused by *Rhizobium radiobacter* in plants like *Arabidopsis, Nicotiana* and *Lycopersicum* species by transformation of inverted repeats of this pathogen genes ipt and iaaM to encode precursors of biosynthesis for two important phytochromes auxin and cytokinins.

Targeting Quorum Sensing

Quorum sensing (QS) is a cell-to-cell communication process involved in the regulation of various traits in response to cell density via extracellular signalling molecules, such as acyl homoserine lactones (AHLs) or diffusible signal factors (DSFs) *etc.* (Rutherford and Bassler, 2012). QS regulates critical traits involved in host–pathogen and pathogen–vector interactions in most of plant pathogenic bacteria (Helman and Chernin, 2015). Because QS relies on bacterial perception of an external signalling molecule, this process represents an excellent potential target for manipulation, either through signal degradation or overproduction of the signalling molecule this comes under a term known as Quorum quenching, For example, AHL degrading enzymes have been reported in many bacteria and this strategy has been used in disease management efforts, either as a biological control strategy via inundative application of AHL degrading organisms or through the expression of AHL-degrading enzymes in transgenic plants (Dong *et al.*, 2001; Molina *et al.*, 2003). In other situations, the synthesis of AHL molecules in transgenic plants has been used, as the stimulation of QS-controlled traits at inappropriate timings can cause malfunctions in the orchestration of pathogenesis through a process this is also termed as 'pathogen confusion' (Lindow *et al.*, 2014).

The first QQ enzyme was identified and purified from the Gram positive *Bacillus* sp. strain 240B1 (Dong *et al.*, 2000). The gene encoding the enzyme was named aiiA (for AI inactivation) and its protein product was shown to inactivate AHL signals by hydrolysis of the lactone ring. Compared with control plants, transgenic plants expressing the aiiA lactonase from *Bacillus* sp. exhibited a significant reduction in the maceration area of tobacco leaves or potato tubers when infected with *Pectobacterium carotovorum* (Dong *et al.*, 2000).

Molina *et al.* (2003) were the first to report the use of bacteria expressing native or ectopic AHL-degrading enzymes as biocontrol agents. The authors examined two biocontrol agents: (i) Bacillus sp. A24, natively expressing AHL-degrading enzyme; and (ii) genetically engineered P. fluorescens with the plasmid pME6863, carrying the aiiA gene from Bacillus sp. A24. Against phytopathogenic bacteria, *P. carotovorum*. Co-inoculation of Pe. carotovorum infected potato plants with wild-type Bacillus sp. A24, or the genetically engineered *P. fluorescens* P3/pME6863 strain, resulted in a significant decrease in tissue rot symptoms compared with the controls.

Targeting the Type Three Secretion System

The T3SS is a needle-like apparatus that enables plant pathogenic bacteria to translocate effector proteins directly from the bacterial cytoplasm into the host cell cytoplasm. These secreted effectors play critical roles during bacterial infection and inhibit the host immune response (Stavrinides *et al.*, 2008). As the T3SS is required for pathogenesis in most of the Gram-negative bacterial pathogens, this system represents an obvious target for the inhibition of pathogenesis in the designing of novel disease management strategies. Two phenolic compounds namely o-coumaric acid (OCA) and t-cinnamic acid (TCA), have been identified as inducers of the T3SS genes in plant pathogen *Dickeya dadantii* (Yang *et al.*, 2008). A collection of phenolic compounds was screened by Li *et al.* (2009) for the ability to inhibit the expression of hrpA, which encodes the structural component of the T3SS pilus in *D. dadantii*. This approach enabled the identification of a new phenolic compound, p-coumaric acid (PCA), as a T3SS inhibition. The identification of PCA as a potential inhibitor of T3SS was further employed to improve the inhibition efficiency through compound modification; subsequent screening of a library of PCA derivatives resulted in the identification of one compound (trans24-hydroxycinnamohydroxamic acid), exhibited an eightfold higher inhibitory potency towards T3SS structural and regulatory genes relative to PCA (Li *et al.*, 2015). Other studies have also indicated that several plant phenolic compounds are able to modulate T3SS gene expression in pathogens like *E. amylovora* and in the multi host pathogen *Pseudomonas aeruginosa* (Yamazaki *et al.*, 2012), similar to salicylidene acylhydrazide family compounds, which not only exhibit an inhibitory role over T3SS related genes, but also suppress amylovoran an exopolysaccharide production of *E. amylovora* (Yang *et al.*, 2014). In addition, many of the above-mentioned PCA derivatives showed little effect on bacterial growth *in vitro*, suggesting that these compounds do not have as high a selective pressure for resistance development as traditional bactericides.

5. Quarantine

Many bacterial pathogens introduced into abroad have caused serious epidemics in new areas where they were previously unknown. For example fire blight of apple and pear, which was prevalent in the US as early as 1780's was introduced in U.K. during 1950's and caused first outbreak in 1958. The disease was subsequently introduced into continental Europe in 1960's. Citrus canker was introduced in 1910's from Asia to Gulf states of North America and caused a severe epidemic in Florida. An eradication programme was launched from 1914 to 1934. It was one of the most intensive programmes in the history of plant pathology. Categories of seed borne bacteria on the basis of potential hazard in the international quarantine are given below:

Category 1

Definition: They are dangerous pathogen and were not known in regions of introduction before. It has considerable epidemic potential.

Precaution: Complete prohibition against introduction.

Examples: *Curtobacterium flaccumfaciens* pv. *flaccumfaciens* (French bean, soybean) *Clavibacte ryathayi* (cocksfoot), *X. campestris* pv. *papavericola* (poppy), *X. campestris* pv. *sesame* (sesame)

Category 2

Definition: Dangerous pathogen, not present, moderate epidemic potential.

Precaution : Seed testing, tested seeds must be free.

Examples: *P. syringae* pv. *glycinea* (soybean), *X. campestris* pv. *campestris* (crucifers), *X. campestris* pv. *vesicatoria* (tomato)

Category 3

Definition : Other plant pathogens of importance to the planting value of seed.

Precaution: Adequate testing.

Examples: *Clavibacter michiganensis* subsp. *michiganensis* (tomato), *P. Syringae* pv. *lachrymans* (cucumber), *P. syringae* pv. *phaseolicola* (French bean), *P. syringae* pv. *pisi* (pea), *X. campestris* pv. *phaseoli* (French bean)

Chapter 9

Bacterial Diseases of Fruit Crops

1. Grapes (*Vitis vinifera*)

Bacterial Spot

In India, the disease was noticedfor the first time on *Vitis vinifera* cv. Anab-e-Shahi at Tirupati (Andhara Pradesh) during 1960. The disease appeared in an epiphytotic form during 1984 in Sangli and Solapur districts of Maharashtra on Thompson Seedless grapes. Now the disease has been noticed in Karnataka, Maharashtra and Tamil Nadu. The disease appeared in severe form during 1999-2002 in north Karnataka.

Symptoms

Initially at lower surface of the leaf disease appear as small water-soaked spots surrounded by yellowish halo. Later these spots enlarge in size and become dark brown in color and angular in shape. Sometimes these spots coalesce to form larger patches. Leaves also show vein infection after drying infected leaves remain attached to the stem. Lesions are brown to black in color (necrotic), elongated and cankerous on petioles and canes (Figure 1). In advance stages of infection stunting, cracking and abnormal growth of canes are seen. Berries show brown to black cankerous lesions. Severely affected ones become small and shrivelled.

Causal Agent

The disease is caused by *Xanthomonas citri* pv. *viticola* (Gama *et al.,* 2018) (Syn. *Xanthomonas campestris* pv. *viticola* (Nayudu) Dye. The bacterial colonieson nutrient agar medium are white, round, smooth, glistening with entire margins.

The bacterial produce a clear starch digestion zone on SX medium. Bacterium is Gram negative, rod in shape with round ends, motility by single polar flagellum and measures 0.4 to 1.2 x 2 to 3 μm. The bacterium is negative in potato soft rot, arginine dihydrolase, tobacco hypersentitivity, nitrate reduction and production of fluorescent pigment on King's medium B. The bacterium is positive in starch hydrolysis, growth at 5°C and 35°C oxidative mode of nutrition.

Figure 1. Symptoms of Bacterial Spot on the Leaves and Canker on the crop vine.

Epidemiology

Temperature 25 to 30°C is favourable temperature for the disease development. Free water from dew, rain or irrigation on leaf surface is more important factors for pathogenesis. High relative humidity above 80 per cent indirectly helps to pathogen by retention of dew on the leaves for longer period. The free water helps in the localized infection and inoculum build up. The disease frequency is positively correlated with number of rainy days. The alternate hosts are mango, neem and *Phyllanthus madraspatensis*. The bacterial pathogen can survive in basal buds and it can also remain alive in the infected, dry leaves up to 65 days. However, under buried moist soil conditions it was reduced to 25 days. The pathogen may carry over from summer pruning to winter mainly with the previously infected canes and dormant buds. The disease is transmitted from one place to another place mostly by the infecting planting materials such as rooted cuttings or infected cuttings for nursery. Secondary spread of the disease is by wind-splashed rain. Disease spreads by diseased cuttings.

Management

- ✰ Regular inspection and destruction of infected plant materials.
- ✰ Use of cuttings which are disease free and late October pruning are recommended for its management.
- ✰ Pruning near bud break is strongly recommended due to more rapid wound healing later in the season.
- ✰ Use of resistant grape species such *Vitis cineria, V. longii, V. riperia, V. palmata* and *V. parviflora.*

- ✰ Spraying with streptocycline 300 ppm starting from two leaf stage up to 70 days at an interval of 15 days is also effective.
- ✰ Use of biocontrol such as *Erwinia herbocola* and *Bacillus* spp. for control of the diseases

2. Pomegranate (*Punica granatum*)

Bacterial Blight

Bacterial blight caused by *Xanthomonas citri* pv. *punicae* is a major disease of pomegranate. The bacterial blight disease in pomegranate has become a very serious threat to the growers in India and it leads to deceased yield (nearly by 90 per cent) and quality of fruits affecting the marketing value of fruits. It has become an increasingly serious threat for pomegranate growers of the states Andhra Pradesh, Maharastra and Karnataka State of the Indian Subcontinent. The disease was most severe in Karnataka and recorded in a range of 60 to 90 per cent incidence. The increase in day temperature (38.6°C) and afternoon relative humidity of 30.4 per cent along with cloudy weather and intermittent rainfall favoured the disease initiation and further spread of the disease.

Symptoms

Symptoms appear as small irregular, water-soaked spots on the leaves. Spots vary from 2 to 5 mm in diameter with necrotic center of pinhead size. Spots are translucent, later turn light brown to dark brown in color and are surrounded by water-soaked margins. Spots coalesce to form larger patches and severely infected leaves fall off. Bacterium attacks stem, branches and fruits. On the stem, the disease starts as brown to black spots around the nodes leading to girdling and cracking of nodes. Finally the branches break down. The spots on fruits are raised and oily in appearance. Brown to black spots appears on the pericarp with L or Y shaped cracks. In severe cases, there will be extensive breaking of fruits (Fig. 2a-e).

a. Foliar Infection

b. Disease Severity on the Branches

c. **Stages of Infection in the Fruits**

d. **The Cracking of the Fruits due to Bacterial Infection**

e. **Devastated orchard due to bacterial blight infection**

Figure 2. Bacterial Blight Disease Symptoms on Foliar, Branches, and Fruits.

Causal Agent

The disease is caused by *Xanthomonas citri* pv. *punicae* (Hingorani and Singh) Vauterin *et al.* [Syn. *Xanthomonas campestris* pv. *punicae* (Hingorani and Singh) Dye)]. Bacterium is a Gram negative, rod in shape, motility by single polar flagellum and 0.5 to 1.0 × 2.5 μm. It is non-acid, fast and aerobic.

Disease Cycle

Bacterial cells survives in soil for >120 days. Survive in fallen leaves during off season. The primary infection is through infected cuttings and secondary spread is through wind-splashed rains. The bacteria causing blight to pomegranate will enter the plant tissues either through natural openings or injuries. The disease incidence is more severe during the fruiting time.

Epidemiology

The disease is favoured by low humidity (> 50 per cent) and high temperature (25-35 ° C) favours. The bacterium survives on the tree. The pathogen survives for 120 days on the fallen leaves during off-season. Irregular rains with cloudy weather conditions are most favourable for the fast development of bacterial blight disease in pomegranate. The continuous growing of pomegranate over three seasons has lead to increased susceptibility of the crop in this part of the geographical area. Deficiency of nutrients makes the plants weak and more susceptible for disease development.

Management

- ☆ Clean cultivation and strict sanitation in the orchard help to reduce the disease incidence.
- ☆ Use of secature disinfected with sodium hypochlorite solution (1.0 per cent).
- ☆ Paste the infected branches with Bordeaux paste (2 per cent) and remove severely infected branches, fruits and leaves and burn them.
- ☆ During severe incidence of bacterial blight, take hasta bahar crop (by pruning during September-October) and give rest to the crop from December to May which reduces the percentage of disease.
- ☆ Before pruning spray with 1 per cent Bordeaux moisture to reduce the severity of disease and then spray with Ethrel to defoliate the plants and destroy foliage by burning.
- ☆ After pruning, smear the basal trunk of the infected plants with a mixture of streptocyclin or streptomycin sulphate (0.5gm) or bactinash @ 0.5 gm + copper oxychloride @ 3gm/L which reduces the incidence. Use red soil for better smearing/application.
- ☆ As soon as the disease appears, spraying with streptomycin sulphate @0.5g/L + copper oxychloride @2g/lt and repeating the spray combination at 10 days interval for 4-5 times will control the disease. Spraying with streptocycline (0.5 per cent) was found effective followed by K-cycline.
- ☆ Since, the disease is aggrevated in deficient stages of micronutrients, farmers are advisedto apply zinc sulphate (100g), magnesium sulphate (50g) and boron (20g) per plant as soil application. Additional spray should be taken up in the events of unforeseen rains.
- ☆ The plant extracts of Tulasi, Patchouli and Miswak were moderately effective under field conditions in the management of the disease (Manjula *et al.*, 2003).
- ☆ Prophylactic application of slicylic acid @ 3 gm/L will improve the systemic acquired resistance of plants against the disease.

3. Citrus (*Citrus* spp.)

Canker

The disease was first reported in 1933 on herbarium of *Citrus medica* in England and the United States, collected at Dehradun during 1827-1831 (Reddy, 1968). In India, citrus canker was first reported from Punjab (Luthra and Sattar, 1940). Disease is reported from Assam, Andhra Pradesh, Karnataka, Madhya Pradesh, Rajasthan, Tamil Nadu and Uttar Pradesh (Chand and Pal, 1982). Now, the disease is prevelant in almost all the citrus growing areas of India. Acid lime is highly susceptible to citrus canker. Heavy losses have been reported in epidemics due to leaf drop and fruit loss due to premature fruit drop and fruits with spots that can not be marketed or start to rot and have to be destroyed. Furthermore quarantine measures such as burning of trees and destruction of fruits may add to these losses.

Symptoms

Symptoms are observed on all the parts of grown up trees *viz.*, leaves, twigs, thorns, older branches and fruits.

On Leaves

Initially disease appears as small, watery, translucent spots of yellow colour where in surrounding tissues become raised convex surfaces. The spots may appear on both side of the leaf but are usually on the lower surface. As the spots matures, the surface becomes white or grayish in color and finally ruptures in the center giving a rough, hard, corky and crater like appearance (Figure 3). The lesions are circular at young stage and become irregular at old stage. The old lesions are light brown color in grape fruit, dark in sweet orange, mandarin orange and trifoliate orange and almost back on lime and lemon leaves. The spots increase in size from 1.0 to 10.0 mm in diameter and may coalesce to form elongated lesions. The size abundance of lesions vary with the kinds of citrus trees and conditions of growth. Lesions are largest on grape fruit leaves nearly 13 mm in diameter and on leaves of lime and lemons, they are much smaller, often not more than 3mm in diameter.

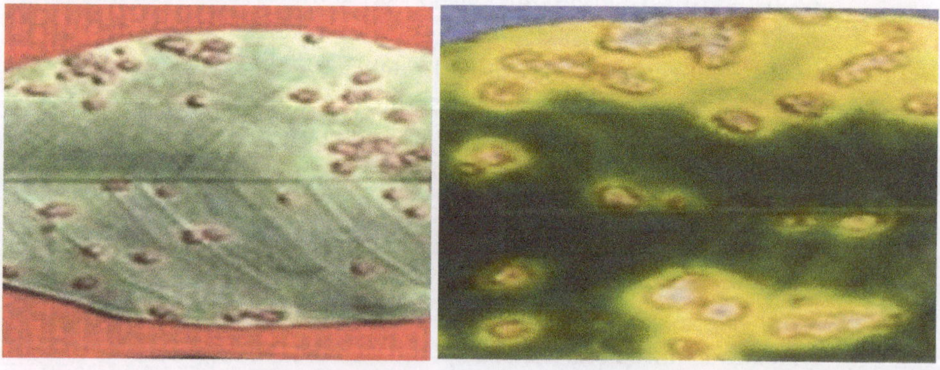

Figure 3. Cankerous Lesions on Leaves

On Twigs and Branches

Lesions on twigs and branches are quite similar. Branches of 50 to 75 mm in diameter are commonly infected. The canker growth often encircles the twigs causing the death of the portions above the infected area leading to die-back of shoots.

On Fruits

The infection advances to the fruits on which typical cankerous spots are formed without yellow halo. The cankers may be scattered all over the surface or several cankers may occur together forming an irregular scurfy mass. Gumming is sometimes associated with spots on fruits (Figure 4). Effect on pulp is higher and drastic reduction in the juice content is observed. The market value is very much reduced because of the canker spots on the fruits. Cankers paves the way for secondary parasite (rotting organism) infection. The affected plants are stunted and fruit yields are reduced considerably.

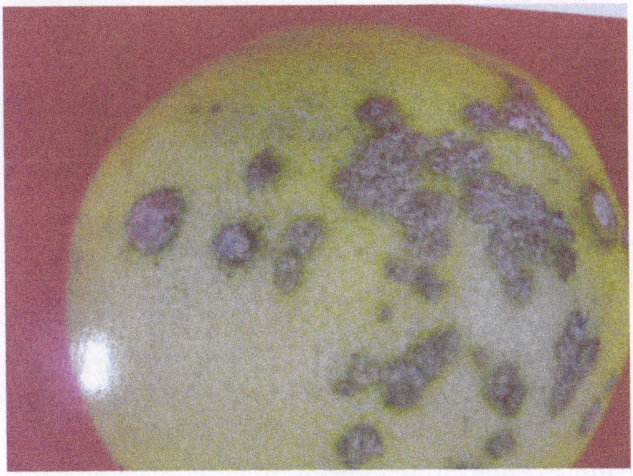

Figure 4. Cankerous Lesions on Fruits

Causal Agent

Xanthomonas citri pv. *citri* (Hasse) Vauterin *et al.* [Syn. *Xanthomonas campestris* pv. *citri* (Hasse) Dye.].The bacterium has a genome length approximately 5 Mbp. There are a number of types of citrus canker disease caused by different pathovars and variants of the bacterium. The Asiatic type of canker (Canker A), *X. citri* subsp. *citri*, caused by a group of strains originally found in Asia, is the most widespread and severe form of the disease. Cancrosis B, caused by a group of *X. axonopodis* pv. *aurantifolii* strains originally found in South America is a disease of lemons, key lime, bitter orange and pomelo. Cancrosis C, also caused by strains within *X. axonopodis* pv. *aurantifolii*, only infects key lime and bitter orange. "A" strains, discovered in Oman, Saudi Arabia, Iran and India, only infects key lime. Straight rods, 0.4-1.0x 1.5–4.0µm, motile with polar flagellum, growth on agar medium is usually yellow and most are slow growing. All spp. are plant pathogens found only in association with plants and plant materials.

Disease Cycle

The bacterium propagates in lesions in leaves, stems and fruit. When there is free moisture on the lesions, bacteria ooze out and can be dispersed to new growth and other plants. Rainwater collected from foliage with lesions contains between 10^5 to 10^8 cfu/ml. Wind-driven rain is the main natural dispersal agent and wind speeds 18 mph (8 m/s) aid in the penetration of bacteria through the stomatal pores or wounds made by thorns, insects such as the Asian leaf miner, Water congestion of leaf tissues can be seen following rainstorms with wind. Citrus foliage can hold 7 microliters/cm^2 of leaf area. Studies of inoculum associated with water congestion have demonstrated how as few as 1 to 2 bacterial cells forced through stomatal openings can lead to infection and lesion formation. Pruning causes severe wounding and can be a site for infection. The long distance dissemination takes place through diseased planting material.

Epidemiology

The disease is severe in acid lime, lemon and grape fruit. Rarely, it is found on sweet oranges and mandarins. Temperature range of 20 and 35°C with evenly distributed rains favours the disease. Presence of free moisture for 20 mm on the host surface is essential for successful infection. The bacterium spreads mainly through wind-splashed rains and long distance disseminations take place through diseased planting material. It survives in leaves, twigs and branches. It survives in the infected leaves for five months and infected twigs up to 70 months. Injury is caused by leaf miner (*Phytocnistis citrella*) paves the way for the entry of pathogenic bacterium. Different pathogenic strains of the bacteria such as A, B and C are well established. Among these, strain A (also called Asiatic strain) is most aggressive in India. The bacterium survives in tissues of infected stem, in cankerous lesions on leaves and crop debris. Bacterial cells enter through stomata, lenticels and wounds. Pruning is a major means of dispersal through citrus orchards. The disease is more severe under wet cloudy climates.

Management

1. Pruning of infected twigs to be done before monsoon and spraying with Bordeaux mixture (1 per cent) are found effective to control the disease (Ramakrishnan, 1954; Prasad, 1959). Pruning of infected twigs every year during November-December and 3 to 4 sprays of Bordeaux mixture (1 per cent) in a year can reduce the disease. Two prunings along with four sprays or 5000ppm copper oxychloride or 1.0 per cent Bordeaux mixture was reported to be effective against the disease (Kishun and Chand, 1987).
2. Streptomycin sulphate @ 500-1000 ppm was effective when sprayed along with glycerine (1 per cent) on acid lime. Six sprays of 100 ppm

streptomycin sulphate along with two prunings reduced the canker in acid lime (Balaraman and Purusotman, 1981). Ravikumar *et al.* (2001) reported that pruning of infected twigs before monsoon and spraying with streptomycin sulphate 500ppm + copper oxychloride 2000 ppm, three sprays at 20 days intervals were found effective to control the disease.

3. Biological control of citrus canker are in a preliminary stage. A strain of *Pseudomonas syringae* is reported to show antagonism to the citrus canker bacterium and also prevent enlargement of canker lesions as well as subsequent defoliation of infected leaves. The antagonist probably stimulates phytoalexin (citrinins) synthesis in the tissues. *Pseudomonas fluorescens* is a strong antagonist of *X. citri* pv. *citri*. *B. subtilis* when sprayed on leaves in high concentration (100 million cfu) reduced canker incidence by 61.9 per cent. A strain of *A. terreus* also reduced disease incidence by (47.5 per cent). *Pantoea agglomerans (Erwinia herbicola)*, a common phylloplane micro flora, grows more rapidly than the canker bacterium both *in vitro* and *in vivo* and eventually causes quick decline of the pathogen population. However, this bacterium grows only in the area where heterotrophic cells are established, but never in the front boundaries at which the pathogen attacks healthy tissues inducing development of hypertropic cells.
4. Use of resistant varieties would be the most desirable method. In India, seed less lime is reported resistant against citrus canker (Kishun and Chand, 1987). Variation in susceptibility/resistance to citrus canker is reported in different citrus species. Hybridization between these resistant types might result in the production of canker resistant varieties. Root stocks do not appear to impart scion resistance to the pathogen (Kishun and Chand, 1987).

Important Tips for Controlling Bacterial Canker

- ☆ After pruning, the secateurs should be disinfected with sodium hypochlorite (1 per cent) solution.
- ☆ Burning of the pruned twigs should be immediately.
- ☆ New flush of the tree should be sprayed properly.
- ☆ Sanitary measures like cutting of grasses, clean cultivation should be undertaken.
- ☆ Additional spray should be done during unseasonal, intermittent rains.
- ☆ Movement of planting material from canker infested to non-infested areas should be restricted.
- ☆ Harvesting of fruits should not be allowed until the foliage is completely dry.

Citrus Greening

Citrus greening, otherwise known as huanglongbing (HLB), is a common disease in many citrus growing countries of the world. The disease has been known since 1927 in South Africa as "Geeltak" ans as "die- back" in India. In India, the disease was first reported by Fraser in 1966. The disease has been reported from Assam, Andhra Pradesh, Bihar, West Bengal, Gujarat, Himachal Pradesh, Jammu and Kashmir, Karnataka, Kerala, Maharashtra, Madhya Pradesh, Odisha, Punjab, Rajsthna, Sikim, Tamil Nadu, Uttarakhand and Uttar Pradesh. It is particularly serious in northern states, where it considered more dangerous than tristerza virus. The greening bacterium causes disease in genera of Rutaceae with species affected by huanglongbing are: most severe symptoms are found on sweet orange (*C. sinensis*), mandarin (*C. reticulata*) and tangelo (*C. reticulata* x *C. paradisi*). Somewhat less severe symptoms are found on lemon (*C. limon*), grapefruit (*C. paradisi*), Rangpur lime (*C. limonia*), Palestinian sweet lime (*C. limettioides*), rough lemon (*C. jambhiri*), kumquat (*Fortunella* spp.) and citron (*C. medica*). Symptoms are even weaker on lime (*C. aurantiifolia*) and pummelo (*C. grandis*).

Symptoms

Leaf chlorosis is the major symptom of greening, which resembles zinc deficiency symptoms. Characteristically, green dots or islands appear against yellow background of leaves. A characteristic feature of greening is that the yellow areas are surrounded on one side by the mid-rib and on the other side by lateral veins. The yellowing expands towards the margins and size of the leaves is reduced drastically. The internodes of the branches are shortened, giving a bushy appearance to the branch. The diseased plants look stunted; they flower earlier and produce smaller fruits of poor quality.

Causal Agent

Citrus greening was earlier considered as a viral disease for a long time. Later, some pathologists claimed that MLOs are responsible for this disease, but later in 1996, it was finally accepted that it is a bacterial disease and *Candidatus* Libreobacter asiaticumwas identified as casual agent for Asian citrus greening and *Candidatus* Libreobacter africanum for African citrus greening. In 2005, a third species was found in Brazil and called *Ca.* Liberibacter americanus. They are all fastidious, phloem limited, Gram-negative bacteria with a peptidoglycan-containing cell wall. The African strain does not require as high a temperature for optimum expression as the Asiatic strain. Neither of them can be cultured on artificial media.

Disease Cycle

The pathogen is spread by vegetative propagation and two psyllid insects. The Asian strain is spread primarily by *Diaphorina citri* where as the primary vector for the African strain is *Trioza erytreae*, but both insect vectors can transmit either strain of the bacterium. The disease has also been transmitted from sweet orange to sweet orange by doddor (*Cuscuta reflex*).

Epidemiology

The three *Ca.* Liberibacter species present differences in their tolerance to temperature and transmission by psyllid vectors. *Ca.* Liberibacter africanus, transmitted by *T. erytreae* is heat-sensitive and disease symptoms do not develop in climates where temperatures above 30°C are reached several hours a day. *Ca.* Liberibacter asiaticus which is present in Asia and America, is heat-tolerant and withstands high temperatures, its optimum ranging from 24 to 32°C and is transmitted by *D. citri*. Natural transmission appears to be related to high vector populations and the extensiveness of the inoculum reservoirs. As psyllid migrations appear to be highest when host plants are flushing, natural spread is probably greatest in late spring and perhaps in other periods when new citrus flushes are available and psyllid populations are high. Vectors apparently spread the pathogen to either adjacent or nearby trees only a few spaces away. The secondary foci are at variable distance from the main cluster of disease and when vectors move, (naturally for feeding or when disturbed by cultural practices), they occasionally move at least 25 to 50 m.

Control Measures

Growers should adopt the under mentioned measures to control greening disease effectively:

1. Uproot and destroy the infected plants.
2. Raise new plants from disease free indexed stock or from nucellar seedlings.
3. Use of healthy plant material for replanting after eradication.
4. Citrus psylla should be kept under control by spraying systemic insecticides like rogor or metasystox (0.05 per cent).
5. Spray bavistin + ledermycin (500 ppm each) six times at 10 days interval.
6. Develope a certification program based on microshoot-tip grafting *in vitro* and produce plants in protected greenhouses to avoid the presence of vector.

4. Mango (*Mangifera indica*)

Bacterial Canker

It is the oldest disease reported on mango in 1881. The bacterial canker was first reported as bacterial leaf spot from Pune, Inida by Patel *et al.,* 1948. The disease is prevalent in Maharashtra, Tamil Nadu, Delhi, Haryana and Uttar Pradesh. It was probably existed since 1881 in Bihar (Kishun and Sohi, 1983). It is widely spread throughout the country but particularly severe in northern states of India. The disease causes loss with the range of 10-80 per cent depending on stage of infection and severity of the disease. The diseased fruits are highly vulnerable to insect and fungal invasion through the cankers. The market value of infected fruits is reduced. Most commercial cultivars are highly susceptible to canker and infections can result in drastic reductions of crop yield because of premature fruit drop, loss of fruit quality and induction of severe defoliation, especially following storms or hurricanes events (Gagnevin and Pruvost, 2001).

Symptoms

The pathogen affects leaves, stem, branches, petioles and fruits. On leaves water soaked lesion appear at the tip of leaves surrounded by yellow hallow. As the disease increases, these spots turn brown to black in colour. These spots coalesce and gives blighted appearance. Thus, it is called as bacterial blight. As the infection increases the infected leaves drop off. The pathogen also infects petiole and fruits. On fruits, cracks are found with black spots and dropping of fruits prematurely is most common. Thus, it is called as "Bacterial canker". The canker stage also invites the other saprophytic fungi infection leading to secondary infection. There will be attraction of insects in some cases.

Causal Agent

The bacterial canker is caused by *Xanthomonas citri* pv. *mangiferaeindicae* (Patel *et al.*, 1948) Robbs, Ribeiro and Kimura (Syn.: *X. campestris* pv. *mangiferae indicae*). The constatin bacterium is rod shaped, mono flagellated and Gram -ve measuring 0.36-0.54 x 1.44 µm. The pathogen has shown significant diversity with 10 different pathotypes in southern India and 3-4 pathotypes in northern India.

Disease Cycle

The disease is non-systemic and pathogen survives one fruiting season to next in localized cankers on twigs and leaves. Survival of *X. c.* pv. *mangiferae indicae* is very efficient in leaf lesions. Bacterial populations decrease relatively slowly in leaf lesions on trees maintained under controlled conditions. The presence of free water allows the release and redistribution of bacteria from the ruptured epidermis. Bacterium is also found pathogenic on *Spondias mangifera, Anacardium occidentale, Acanthosperma, hispidum, Lantana, camara, Ficus glomerata, Solanum tarbum, Psophocarpus fellagonolobus* and *Caesalpinia minosoldes.* Role of mango stones in survival of pathogen has been established by Kishun (1988). Pathogen has also been found to survive on weed hosts (Kishun and Chand, 1988). Disease spread is rapid during rains. In new area, disease spreads through infected planting materials, Myllocerus discolor var. variegate, *Orthaga evadrusalis* and *Canthecondita furcellate* are found associated in mechanical transmission of the disease. The pathogen enters intact leaves through stomata, in fruit through stomata and lenticels and in twig through lenticels. It is an exclusively intercellular parenchymatous pathogen and causes chloronemea, separation and collapse of the cells to form bacterial pockets.

Epidemiology

Bacterium survives in resident phase on different parts of host plant such as twigs, smaller branches, seeds of fruit, phylloplane micro flora on weeds and long distance spread is through use of infected planting material. In the garden, the disease development and spread is generally favoured by high RH (90 per cent) and temperature of 25-30°C. Rainfall during flowering/berry development causes fruit infection. The disease is also known to be transmitted by ashy weevil, leaf webber and bugs as non-persistent bacterium. Three month-old leaf lesions on a susceptible cultivar may contain up to 10^7 CFU and up to 10^5 CFU were detected in 18-monthold lesions. Incidence of diseased fruit is highly correlated with leaf spot severity observed 6 months previously. Twig cankers also serve as a source of inoculum, but their relative importance compared with bacterial inoculum in leaves is difficult to assess. Bacterial populations on mango surfaces can average as high as 10^6 CFU per leaf or fruit. Surface populations vary depending on climatic conditions but are correlated to the number of leaf lesions.

Management

- ✰ Field sanitation: Removal of weed hosts and several other infected debris helps to reduce the disease spread.
- ✰ Although all the cultivated varieties are susceptible, some genotypes showed moderately resistant type of reaction. The proper screening methodology is to be worked out to identify the genotype's reaction to various pathotypes prevailing in the country.
- ✰ Two sprays with streptocycline @ 0.5 per cent and copper oxychloride chloride (2g/lt) is recommended for disease management.
- ✰ Dipping the fruits in agrimycin- 100 @ 250ppm proved effective to control the disease. Two spray of streptocycline (250- 300 ppm) at 20 days interval reduced fruit infection (Bose and Singh, 1980). The disease can also be reduced with monthly spraying of 1000 ppm bavistin (Kishun and Sohi, 1984).
- ✰ Resistant varieties like, Bombay Green, Fazli and Suvarnrekha (Shekhawat and Patel, 1975, Kishun and Sohi, 1983) are reported to be resistant.
- ✰ *Bacillus coagulans*, a bacterium antagonist, provides effective biocontrol measure against *X. citri* pv. m*angiferaeindicae*.

5. Banana (*Musa* spp.)

i. Moko Disease/Bacterial Wilt

Bacterial wilt of banana was first recorded in Guyana in 1840. This disease is observed in many countries where banana is grown extensively. In India, the disease was first reported from West Bengal in 1968. Losses are difficult to assess, but decline in yield of 74 per cent in plantain and 64 per cent in banana were attributed to Moko disease in Guyana from 1963-73 and the export banana industry in Trinidad was virtually eliminated in the early 1960s. It is reported to occur on cv. *Robusta* and *Poovan* in Tamil Nadu causing 75 to 100 per cent losses.

Symptoms

Initially, the bacterial wilt is distinguished by the development of a yellow discoloration of the inner leaf lamina close to the petiole that is followed by wilting of the inner leaves, with the exception of the unrolled heartleaf. In *Cavendish* banana, the lower leaves become yellow, which progresses upwards. Leaves later exhibit dingy whitish yellow, become dry and flaccid and readily droop. The petiole breaks at its junction with the lamina or pseudo stem. Then the affected leaves break down in quick succession and droop around the pseudo stem. If diseased suckers are planted, the terminal leaf becomes necrotic and the plant dies.

Young sword suckers from affected plants show wilting and blackening. Infection may be localized in one section of the rhizome. Such plants may die later. When a diseased pseudostem is cut transversely the vascular discoloration ranging from yellow to dark brown or bluish black could be seen. The reddish tinge present in typical cases of Panama disease is usually absent, although in some advanced cases a reddish brown ring of discolored vascular tissue may be found. Moko wilt, the discoloration tends to be central while in Panama disease it tends to be more in the peripheral region. In some cases, no external symptoms are seen until the bunch is produced. But the individual fingers are distorted. The presence of yellow fingers in otherwise green stem often indicates the occurrence of Moko disease. Internally, a firm dry rot is found in the fruit. The pulp exhibits a characteristic dark brown discoloration. The production of bacterial ooze, which is grayish brown in colour is seen when the pseudostemof the affected plant is cut transversely. Care should be taken not to confuse the normal healthy latex ooze with bacterial ooze. Fruit rot and fruit stalk vascular discoloration, wilted or blackened regrowth of suckers, blackened and dead male flower buds and peduncles with vascular discoloration moving up into the fruit stalk and fruit are the other symptoms of Moko disease.

Causal Agent

Ralstonia solanacearum (Smith) Yabucchi *et al.* 1995 (Syn. *Pseudomonas solanacearum* E.F. Smith) race 2 is an aerobic, non-spore forming, rod shaped, Gram-negative bacterium with single polar flagellum. Colony on TTC medium for 48 h at 30°C is fluidal, > 5mm, white or pink, EPS producing colonies, grow at 37°C, it does no grow at 41°C, NaCl tolerance up to < 2.0 per cent, change the nitrite to nitrate, gas form nitrate, hypersensitive reaction show on tobacco plants. *Ralstonia solanacearum,* the bacterial wilt pathogen, is a heterogeneous species, as Substantiated by its large host range, pathogenic specialization and cultural and Physiological properties, as well as its phylogeny. The most important monocotyledonous crops susceptible to *R. solanacearum* are banana, plantain and edible ginger (Hayward, 1994). *R. solanacearum* has been classified into 5 races based on host range at the plant species level (Buddenhagen *et al.*, 1962) and 5 biovars based on the utilization of disaccharides and sugar alcohols. Fegan and Prior (2005), introduced another classification called phylotype, based on the geographical origin of the strains. Accordingly four phylotypes were defined, of which phylotype I, II and III are composed of strains mainly from Asia, America and Africa and surrounding islands, while phylotype IV is composed of strains from Indonesia and some isolates from Japan, Australia and the Philippines. Phylotype IV consists of strains including *R. solanacearum*, *R. syzygii* and the blood disease bacterium (BDB). Due to the intense diversity of *R. solanacearum,* this group of organisms is now commonly called as *R. solanacearum* species complex (RSSC). Within phylotype, different sequevar or sequence variants were identified which are defined as a group of strains with high sequence conservation with in the area sequenced (Fegan and Prior, 2005).

Ralstonia solanacearum has been recovered from *Heliconia* species from virgin forest in the Coto valley, southwest Costa Rica, suggesting that Moko was originally endemic in the rainforests of the Caribbean area (Sequeira and Averre, 1961). *Heliconia* species in the larger LAC, the Philippines and Indonesia, have also been known to harbor *R. solanacearum* (Elphinstone, 2005). *R. solanacearum* strains affecting *Musa* spp. A redefined by their symptom expression in the plant, cultural characteristics and whether the pathogen is spread mostly by insects or by mechanical and soil transmission (Sequeira, 1998). The "SFR" (small, fluidal, round) and "A" (Amazon basin) strains are known to be transmitted by insect whereas the "B" (banana) strain is transmitted through root contact and contaminated planting equipment (Sequeira, 1998). Until 1960, Moko was present in Trinidad (where it was first reported in the 1890s (Stover, 1972) and described by Rorer (1911), Guyana, Venezuela, Panama, Costa Rica and Honduras. Subsequently, the "SFR" strain transmitted by insects was reported in Guatemala, El Salvador, Nicaragua, Costa Rica and Mexico (Buddenhagen and Elsasser, 1962). The occurrence of Moko in Mexico was first reported in Chiapas in 1960

The name "Bugtok" is used when wilt symptoms, caused by *R. solanacearum,* appear on ABB cooking type bananas in the Philippines (Molina, 2006). It was a very serious disease of Saba (Musa BBB) in the early 1990s. The major transmission of this pathogen on Saba is by insects (Molina, 2006). Tool transmission seems less frequent most likely as a result of less intensive management. Through simple early male bud removal, the disease has been managed to acceptable levelsin both medium-scale farms (for banana chips production) and subsistence farms. Early de-budding has now become a cultural practice among farmers not only to control the disease but also to use the male bud as a vegetable. The same pathogen causes "Moko" in large-scale Cavendish plantations (Molina, 2006). The causal agent of Bugtok in the Philippines has been shown to be identical to the "B" strain of *R. solanacearum* from Honduras, as both strains are indistinguishable by numerical taxonomy or genetic analyses (Ilagan *et al.*, 2003; Raymundo *et al.,* 2005). Furthermore, the low genetic diversity among *R. solanacearum* strains isolated in the Philippines from *Musa* spp. (Raymundo *et al.,* 2005), suggests that a single genotype of the pathogen was introduced from Central America.

Disease Cycle and Epidemiology

The pathogen is soil borne. The banana pathotype of *R. solanacearum* survives in a wholesome manner in the presence of a susceptible host. Bacteria survive in host plants and, for several months at least, in soil. Bacteria enter roots through wounds, reach the xylem vessels and multiply and move through them. Bacteria are transmitted through infected banana rhizomes and through contaminated tools and equipment.

The pathogen is soil-borne. It survives in susceptible hosts like banana and *Heliconia* spp. High temperatures and high soil moisture generally favors disease development. The bacterium is soil borne and can survive in soil for 12-68 months. Infected suckers are the major source of inoculum. The bacterium is transmitted through tools. Insect pests help to disseminate the bacteria oozing out through the wounds. Similarly, the pollinating insects disseminate the oozing out through male flowers. The bacterium has a very wide host range infecting a large number of cultivated plants and weed hosts. *Heliconias* are susceptible to the banana strain of the pathogen and many of the members can serve as symptomless carriers. All varieties of commercially cultivated bananas are susceptible to the disease.

Management

- ✰ Strict plant quarantine and phytosanitary measures are necessary to prevent the spread of Moko disease. Fruits should not be transported from one country to another and the export and import materials should be carefully screened.
- ✰ Pruning wounds are the common sites of infection and effective way of preventing this infection is through disinfecting the tools used. Formalin, diluted with water (1:3) is a suitable disinfectant.
- ✰ Better drainage, lengthened pruning cycles and frequent disease surveys can reduce the incidence of Moko disease. Flower-visiting insects are major transmission agents. Hence removal of the male buds when the hands begin to turn upwards can reduce the incidence.
- ✰ *Heliconia* and several weed species serve as alternative hosts of the disease and should be removed from the plantations. Six to twelve month fallow is needed depending upon the strains. During the fallow period the area should be kept weed free. Growing cover crops like kudzu (*Pueraria phaseoloides*) and sorghum is helpful in reducing the soil inoculum. An expensive but very effective way of treating an infected plant is by fumigation with methyl bromide.
- ✰ Replanting can be done after two months of treatment. Herbicides like 2,4-D can be used to kill infected plants in *situ.*

- ☆ Insecticide like dieldrin may be sprayed on to the infected plants to prevent insect transmission of the disease to healthy plants.
- ☆ None of the commercial bananas is truly resistant to Moko, although in Panama, *Lacatan, Giant Cavendish, Poyo* and *Velery* are claimed to show moderate resistance.
- ☆ Moko wilt has been found effectively control by bacterization with *Pseudomonas fluorescens. Bacillus subtilis* strain SW116b has the highest activity against *R. solanocearum,* to suppress the development of Ralstonia wilt disease in banana.

ii. Tip Over and Soft Rot Disease

Disease is more severe in Haveri, Bagalkot, Gokak, Raibagh and Vijayapur districts of Karnataka and severity ranges- 10-25 per cent. The disease incidence ranges from 30-35 per cent in the districts of Kolarof Karnataka state and Bangalore.

Symptoms

As the bacteria affects the underground stem of young plants leads to extensive rotting and emitting of odour. The older plants show rotting at the collar region and leaf bases. Infected part will become so weak that it will snap off the corm easily even for a normal wind force. In severe cases, pseudostem base becomes swollen and splits (Fig. 5). Affected plants looks yellowish brown with a definite brown margin.

Causal Agent

It is caused by *Erwinia carotovora* subsp. *carotovora* (Syn: *Pectobacterium caratovora*). Bacteria are Gram –ve and rod shaped.

Disease Cycle and Epidemiology

This disease is caused by soil borne pathogen that enters the young banana plants through wounds that may be sustained at roots at the time of planting. During hot summer periods tender roots of tissue cultured plants may get damaged (burns/scorching) due to excessive heat build up of the surface soil resulting in decaying of the affected root portion. Disease also spreads to the daughter suckers of the clump. The bacteria spread in contaminated water.

Observed commonly during hot summer months and rainy season after 3 to 5 months of planting in the field. The disease comes more severe under hot wet weather.

Management

- ☆ Bacterial attack on tissue culture plants can be effectively controlled by maintaining sufficient moisture around the root zone of the plant during initial months (hot summer) to prevent heat injury to tender roots.
- ☆ Drip irrigation may be done preferably during early morning or late evening hours and avoid irrigating the plant during hot noon.

Fig. 5. Bacterial soft rot disease symptoms

iii. Bacterial Leaf Spot of Banana

The disease was first reported on *Musa paradisiacal* L. (AAB group) from Tamil Nadu by Rangaswami and Rangarajan (1965). The disease is widespread on cvs. *Monthan, Peyan, Poovan and Rasthali*. Most of the cultivars of *M. paradisiaca* (Monthan, Poovan, Rasthali and Peyan) are susceptible.

Symptoms

The disease affects only leaf blades. In the initial stage, the disease appears as chlorotic linear streaks along veins throughout the leaf blade. Later, the streaks spread and coalesce to form large, irregular, chlorotic patches. Brownish linear streaks also appear along the veins and they turn dark brown and finally black with a chlorotic halo and coalesce to from linear to irregular dark brown patches. In severe infection, the leaves may roll and dry up in the affected portions, presenting a scorched appearance. The affected plants are stunted and size of bunches is reduced.

Causal Agent

The disease is caused by *Xanthomonas musicola.* The bacterium is a Gram-negative, rod shaped with 1.5-2.0 x 0.5-0.75 µm in size. Bacterium forms chains and capsules and is motile by one polar flagellum. No spores are formed and they aerobic. Colonies of bacteria on nutrient agar are circular, straw yellow, slightly raised and glistening.

iv. Rhizome/Pseudostem Soft Rot

Symptoms

Soft rot of rhizome can cause severe losses in particular localities. A massive soft odorous rot of the centre or a portion of a rhizome characterizes the disease. It progresses upward the pseudostem and destroys the growing point, causing internal decay and discolouration of the vascular tissues. Leaves show yellowing, which wilt very soon. Sometimes, it is confused with fausarium wilt.

Causal Agent

It is caused by *Dickeya paradisiaca* formerly *Erwinia chrysanthemi* var. *paradisiaca*. *Dickeya paradisiaca* belongs to Enterobacteriaceae (class Gamma-proteobacteria) and is an aerobic, Gram negative rod, with peritrichous flagella, that appears single or in pairs that do not form spores. It is protopectinase, amylase, nitrate reductase, lecitinase and catalase positive; amylase, urease, oxidase and gelatinase negative and produce gas from glucose. It does not grow at 5 per cent NaCl but can grow at 40°C). Colonies in nutrient agar after 48 h are white to light gray, have irregular borders, fine granular growth,and after 4 days show a well-defined rising center. In YDC (Yeast extract-Dextrose-Calcium Carbonate) medium, some isolates may produce a blue non-diffusible pigment. On King B media, it produces a brown diffusible pigment.

Disease Cycle and Epidemiology

Dickeya paradisiaca infects the plants through open entries and wounds produced during sanitation of senescent leaves attacked by black Sigatoka and pruning of suckers (Thwaites *et al.*, 2000). Cultivars of plantain (AAB) and cooking bananas (ABB) are more susceptible to pseudostem wet rot than Cavendish cultivars. These conditions associated to poor sanitations practices enable severe symptom development including plant toppling.

Management

- ✰ Uproot and destroy the affected plants.
- ✰ Soil and plant drenching with bleaching powder at an interval of 10-15 days in quite effective for controlling the disease.

6. Apple (*Malus domesticus* L.)

i. Fire Blight

It was first noticed in the Hudson valley in the USA in 1780 and the bacterial nature of the disease was described by T. J. Burrill in 1882. In India Singh (1943) reported the possible occurrence of Fire blight (*Erwinia amylovora* (Burrill) Winslow *et al.*) on apple in some parts Jammu and Kashmir, Himachal Pradesh and Uttar Pradesh.

Symptoms

The first symptoms appears on the flower, which later become water soaked, then shrivel rapidly, turn brownish to black in colour and may fall off or remain hanging in the tree. Soon the symptoms proliferates to the leaves on the same spur or on nearby twigs, starting as brown to black blotches along the midrib and main veins or along the margins and between the veins. As the blackening spreads, the leaves curl and shrivel, hang downward and usually cling to the curled, blighted twigs. Terminal twigs and water sprouts (suckers) are usually infected directly and wilt from the tip downward. Bark turns brownish black in color and is soft at first but later shrinks and hardens. The tip of the twig is hooked and the leaves turn black in color and cling to the twig. From fruit spurs and terminals the symptoms extends down to the supporting branches, to form cankers. The bark of the branch enlarges and encircles the branch, the part of the branch which is infected dies. Fruit infection usually occurs through the pedicels. Direct infection is also common. The fruit becomes water-soaked, turns brown in color, shrivels, shape is mummified and finally turns black in color. Vegetative shoots exhibit a characteristic shepherd's crook in which the shoot tip bends to about 180°. Shepherd's crook is often associated with gray-green discoloration of the stem, attached leaves and ooze. Dead fruit may also hold on to the tree for several months after infection. Under humid conditions, droplets of a milky-coloured, sticky ooze may appear on the surface of any recently infected part. The ooze usually turns brown in color soon after exposure to the air. The droplets coalesce to form large drops which may run off and form a layer on the surface of plant parts.

Causal Agent

The disease is caused by *Erwinia amylovora* (Burrill) Winslow *et al.,* The bacterium is rod shaped and 1.0 to 2.0 × 0.8 to 1.2µm in size. The bacterium is motile by peritrichous flagella. Bacterium occurs usually singly but pairs or chains of 3 to 4 bacteria also exist. Each bacterial cell is enclosed in a capsule.

Disease Cycle and Epidemiology

The bacterium over winters in the margin of cankers formed during previous season. Bacteria survive most often in large branches and seldom in twigs less than 1 cm in dia. Flies, wasps, honey bees and rain splashes helps in spreading the bacteria. Sucking insects like aphids and leaf hoppers act as inoculating agents because they insert bacteria into freshly wounded tissues inside the leaf. Young tender twigs are infected by bacteria through wounds made by various agents, through their lenticels and through insects carrying bacteria and feeding on the twigs. The principal and most susceptible hosts come from the species in the sub-family Pomoideae of the family Rosaceae. The most susceptible species of all is pears (*Pyrus* spp.). Other important susceptible species are service berries (*Amelanchier* spp.), Japanese quinces (*Chaenomeles* spp.), rock sprays (*Cotoneaster* spp.), hawthorns (*Crataegus* spp.), quinces (*Cydonia oblonga*), loquat (*Eriobotrya japonica*), medlars (*Mespilus germanica*), photinia (*Photinia* spp.), firethorns (*Pyracantha* spp.), Japanese roses (*Rosa rugosa*), blackberries (*Rubus* spp.) and Japanese plums (*Prunus salicina*).

The pathogen grows optimally at 27°C, but can grow anywhere from 5 to 31°C. During times of temperature extremes (5°C or below, or 31°C and above) disease development and lesion extension slows down due to these unfavorable conditions. Temperature above 24 °C and heavy rain favours the infection and rapid spread of the disease.

Management

- ☆ During winter season all branches, blighted twigs and cankers should be cut out about 10 cm below the last point of visible infection and burnt. Source of inoculum is reduced by cutting of blighted twigs, suckers and root sprouts in the summer and prevent production of large cankers on the branches. The tools should be disinfested after each cut in mercuric chloride solution.
- ☆ Best insect management programme should be maintained to reduce or eliminate spread of bacteria by insects.
- ☆ Resistant varieties should be planted in new areas.
- ☆ Sprinkler system of irrigation should be avoided in vicinities having problem of fire blight.
- ☆ Excessive use of nitrogenous fertilizes should be avoided.
- ☆ Pruning should not be heavy, because it favours vigorous growth, which is susceptible to fire blight attack.
- ☆ Bordeaux mixture (2:6:100) or streptomycin 50 to 100 ppm are the effective blossom sprays. Use bordeaux mixture during quick drying conditions to avoid rosetting of fruit. Four sprayings of streptomycin should be applied either when maximum temperatures are above 15 °C or during the night.
- ☆ Spray streptomycin (100 ppm) or oxytetracycline (100 ppm) during flowering. It should be repeated after 4-5 days.
- ☆ Cell suspension sprays of *Pantoea agglomerans*, (Syn. *Erwinia herbicola)*, *Pseudomonas fluorescens* antagonistic bacteria are quite effective. Non-pathogenic, microbial epiphytes sprayed onto flowers can preemptively suppress fire blight by colonizing the niche (stigmatic surface) used by *E. amylovora* to increase its epiphytic population size. The bacterium *Pseudomonas fluorescens* strain A506, is registered and sold commercially for this purpose (BlightBan A506) as is the yeast, *Aureobasidium pullulans* (Blossom Protect).

ii. Crown Gall Disease

Agrobacterium tumefaciens was originally named *Bacterium tumefaciens* by Cavara in 1897. In 1907, the name was changed to *Phytomonas tumefaciens* (Smith and Townsend) Bergey *et al.*) by E. F. Smith, which was subsequently changed to *A. tumefaciens* and in 2001,Young *et al.* changed the name as *Rhizobium radiobacter*, the taxon name of current use. This organism was first isolated from grapevines in 1897 by Fridiano Cavara at the Laboratorio di Botanica del Recherci Instituto Forestale di Vallom Drosa in Naples, Italy. In 1907, E. F. Smith and C. O. Townsend in the United States isolated the bacterium from chrysanthemum galls. The disease occurs on apple, peach, apricot almond, pear, cherry and plum.

Symptoms

Small out growths are visible on the stem and roots. At early stages, the galls are soft, spherical in shape, white or flesh coloured. The galls vary in size from 7 to 10 mm in diameter. The galls are hard and corky on woody stems. They are generally knobby and knotty and become more cleft as they grow older. The affected plants are stunted with chlorotic leaves.

Causal Agent

The disease is caused by *Rhizobium radiobacter* (Beijerinck and van Delden) (Syn: *Agrobacterium tumefaciens* (Smith and Townsend) Conn. It is a Gram negative, rod and 0.7 to 0.8 × 2.5 to 3.0 μm. It is a biotrophic pathogen that can alter the physiology and morphology of its host plant without killing it, resulting in tumour like structures or galls. The bacterium can grow on relatively simple medium but its isolation from soil requires selective media. It consists of a heterogenous group of strains.

Disease Cycle and Epidemiology

The bacterium is soil borne. Tumour inducting principle transforms the normal plant cell to tumour cells. The bacterium causes abnormal growths or galls on roots, twigs and branches of 93 member families of plants including shrubs primarily in the rose family (apple, pear, cherry, apricot, almond, walnut) and *Chrysanthemum* spp. as well as on vines and canes such as grapevines and raspberries. Marguerite daisies, chrysanthemums and grapevines can become infected systemically. Field crops such as cotton, sugar beets, tomatoes, beans and alfalfa are also infected by the bacterium, but in these crops, the disease does not impact such crops economically. The pathogenic bacterium naturally resides on the rhizoplane of these hosts. Presence of bacteria in soils originates from galls that were broken or sloughed off from infected plants during cultivation practices or disseminated as infected plant material. Irrigation water helps in further dissemination of the *R. radiobacter* bacterial cells. *R. radiobacter* is also spread by infected and infested planting stocks originating as nursery stock from uncertified sources. Secondary spread then occurs through pruning and cultivation equipment, particularly when galls are removed manually with the same cutting tools used in pruning. Tilling equipment can be contaminated by cutting through galls at or near the base of trunks of infected trees. Rogueing (removal) of infected trees and replanting in the same spot where the infected tree had grown is poor practice because sloughed off galls serve as sources of abundant populations of *R. radiobacter*. *R. radiobacter* enter a plant wound, disease will be more severe and a large gall will be seen rapidly growing at the infection site. Plants that are systemically infected (such as grapevines) will harbor *Allorhizobium vitis* [syn. *Rhizobium vitis* (Ti), *Agrobacterium vitis* (Ti), *A. tumefaciens* biovar 3] for extended periods of time in the absence of overt symptoms. Tissue injuries induced by accidental wounds made during cultivation practices or by frost will elicit new infections with the appearance of numerous galls along the vascular system of the host.

Management

- ☆ Removal of infected root stocks by visual examination of the disease has been the conventional primary screen for diseased material. The method is limited for complete disease control because *A. tumefaciens* can reside on the rhizoplane and systemically in certain host plants such as grapevines, chrysanthemums and marguerite daisy.
- ☆ Fields that have grown cereal crops for a long period are favored as crown gall-free sites. Fields previously used for growing fruit and nut crops can remain infested with *A. tumefaciens*. Certain weeds such as morning glory (*Ipomoea leptphylla*) can serve as natural hosts of the bacterial pathogen.
- ☆ A crop rotation with cereal crops followed by green manuring helps reduce the population size of *A. tumefaciens*.
- ☆ Creosote-based compounds, copper-based solutions and strong oxidants such as sodium hypochlorite are transiently effective to eradicate the disease. The chemical eradicant application procedure is labor intensive and therefore costly both monetarily and to the environment. The superficial treatments are ineffective against systemically infected plants. Generally, chemicals are rarely used for control of crown gall.
- ☆ Certain strains of *A. tumefaciens* are sensitive to the antibiotic agrocin produced by *A. radiobacter*, a closely related soil-borne bacterium that does not infect plants. An example of an antibiotic produced is Agrocin-84, which is an analog of the opine agrocinopine A. Agrocinopine A is produced in crown gall tumors induced by *A. tumefaciens* strains whose Ti plasmid encodes for nopaline and agrocinopine A. Agrocin-84 mimics agrocinopine A and therefore is taken up by the same transport system used by *A. tumefaciens* to utilize agrocinopine A. Inside the *A. tumefaciens* cell, the antibiotic Agrocin-84 inhibits DNA replication and cellular growth. Plants are protected against Agrocin-84 sensitive strains of the pathogen by dipping the root system into a suspension of *A. radiobacter* K84 before planting in infested fields.
- ☆ Transgenic crop plants harboring one or more unique genes tailored to protect the plant from crown gall have been developed. Genes encoding products that degrade or inactivate the T-DNA strand complex when it enters the host cell, that prevent the expression of T-DNA genes encoding indoleacetic acid and cytokinin biosynthesis and that prevent *A. tumefaciens* attachment to its target are some examples currently being tested. Biotechnology companies, such as DNA Plant Technology (Oakland, CA), are applying sense strand messenger RNA or small-interfering RNAs to develop crown gall resistant fruit and nut crops.

iii. Hairy Root

In certain areas, hairy root is also a serious problem in temperate fruits and many other crops caused by *Rhizobium rhizogenes* (Syn. *Agrobacterium rhizogenes*). The disease was first identified as a pathogen of economic importance on apples in the early 20th Century (Riker *et al.,* 1930). The plant families Solanaceae, Rosaceae, Fabaceae, Crassulaceae, Caesalpinaceae, Brassicaceae, Polygonaceae and Asteraceae are susceptible to hairy root caused by *R. rhizogenes*. Hosts which are phenol accumulators tend to be more sensitive to infection by *R. rhizogenes. R. rhizogenes* has a world-wide distribution but it cannot grow above 30°C in culture.

Symptoms

The bacterium invades the roots through wounds and cause excessive growth of adventitious roots at the site of infection. It appears like a broom. Symptoms of hairy root disease may also include wrinkled leaves. Hairy root disease symptoms have been divided into 2 categories, 'simple or fibrous hairy-root' and 'woolly-knot form of hairy root.' Simple hairy root disease is identified as stated above where as the woolly-knot form is characterized by the initial formation of a smooth tumor on the plant stem followed by an outgrowth of adventitious roots from the tumor surface. The affected plants show stunted appearance. It mainly affects the plants in the nursery.

Causal Agent

The disease is caused by *Rhizobium rhizogenes* (Riker *et al.*) Young *et al.* (2001) (Syn.: *Agrobacterium rhizogenes*) is soil inhabitant, Gram-negative, rod shaped, size 0.6–1.0 µm × 1.5–3.0 µm, non-spore forming and motile by 1–6 peritrichous flagella. Colony morphology shows convex, circular and smooth characteristics, while its colour becomes non-pigmented to light beige on a nutrient agar plate. *R. rhizogenes* is an aerobic bacterium that uses oxygen for its growth needs. The root-inducing plasmid (*Ri-plasmid*) gives the bacterium the ability to induce the production and degradation of opine complex acids such as mannopine and agropinic acids as sole sources of nitrogen and carbon. The favourable conditions for growth of bacterium is a between 25 - 28°C temperature and 5–9 pH level and biotin as a growth factor (Brenner *et al.,* 2005).

Disease Cycle and Epidemiology

The plants suffer from wounds by soil pathogens or other sources. *R. rhizogenes* multiply through the process of binary fission, where a bacterial cell (called a parent cell) makes a copy of its DNA and grows larger by doubling its cellular content. Generally, bacteria reproduce either via binary fission or budding. However, on both cases, the DNA found in parent cells and the offspring is exactly the same. The bacterium can adapt and survive amid changing surrounding environment and ensure that its host is prepared to tolerate changes as well. Root plant cells would release compounds, which are sensed by the bacterium in the soil and which triggers it to be attracted to that area. The bacterium then transfers DNA from the plasmid into the host cell and integrates the T-DNA (Ri- plasmid) into the plant cell gene structure. After integration, the plant undergoes root architectural changes and produces an abundance of opines, which are beneficial for the growth of *R. rhizogenes*.

Management

Growers should follow the under mentioned measures to control this disease effectively:

- Follow quarantine measures strictly to avoid the entry of the bacterium into new locations.
- Mechanical injury to the plants should be avoided at every cost.
- Root chewing insects should be controlled in time.
- Infested fields should be planted with maize or other grain crops for several years.
- The crops like tomato and marigold should not be planted near the vicinity of orchards, having infestation of crown gall.
- Infected nursery stock should be destroyed.
- Biocontrol of hairy root with the use of *Agrobacterium radiobacter* has been very successful in many countries of world.

7. Peach (*Prunus persica* L.)

i. Bacterial Canker and Gummosis

The bacterial canker disease caused by *Pseudomonas syringae* pv. *morsprunorum* (Wormald) Young *et al.* (1978) (Syn. *Pseudomonas morprunorum* Wormald; *Bacterium pruni* E. F. Smith) was originally described in 1902 by Erwin F. Smith is a serious disease of the Japanese plum (*Prunus salicina*) in Michigan (Smith, 1983). The pathogen also causes canker and gummosis in sweet cherry, almond, cherry plum, sour cherry, European plum, damson flowering almond Japanese plum and however, it was capable of producing a serious disease of the peach, apricot and nectarine. Losses from *P. syringae* pv. *morsprunorum* are less severe than those from *P. syringae* pv. *syringae*, in part because pv. *morsprunorum* has a more restrictive host range and geographic distribution than pathovar syringae. The killing of buds, fruiting spurs and branches and infection of fruits reduces yields. Infection of young trees may cause tree mortality or adversely affect tree form. The cankers and dieback caused by *P. syringae* pv. *morsprunorum* on sweet cherry clones grown in forest plantations can reduce the quality of the timber production.

Symptoms

Small circular green spots occur on the leaves which later they become angular, deep purple to black. The spotted area falls and shot holes are seen. Severe infection leads to defoliation which devitalizes the tree. Circular spots are noticed on fruits which are water soaked initially and become black. On immature fruit, lesions are brown with a margin of wet or water-soaked tissue. The affected tissues collapse, leaving deep, black depressions in the flesh, with margins becoming yellow to red as lesions and fruit, age. On fruit stems, lesions are elliptical and brown with water-soaked margins. Due to intense spotting on the fruits, pitting or cracking occurs in the vicinity of spots. Dormant leaf and flower buds with internal infections may fail to open in spring resulting in a dead bud symptom. On spurs, blossom blight develops as infection spreads quickly from diseased flowers throughout the blossom truss and into the spur. Whole blossoms become brown to black. Cankers at the base of infected buds and spurs on trunks, limbs and branches are slightly sunken and darker in colour than the surrounding healthy bark and are often accompanied by gum exudation during late spring and summer. Girdling canker causes the terminal portions of limbs and branches to dieback. Occasionally, even large scaffold limbs are killed.

Causal Agent

The bacterial canker disease in peaches and other stone fruit caused by *Pseudomonas syringae* pv. *morsprunorum* (Wormald) Young *et al.,* 1978. The bacteria form greyish-white colonies of the levan type, 2-3 mm in diameter, which were shiny, smooth and mucous on the NA medium enriched with sucrose. They could be seen after 2-3 days of growth at the temperature of 24-26 °C. On the King's B medium, formed colonies were white, 2 mm in diameter and fluorescent under UV light. The bacterium is gram-negative, metabolized glucose only in aerobic conditions, created fluorescent pigment on the King's B medium. They formed levan and production of oxidase, pectolytic enzymes and arginine dehydrolase are negative. The bacterium shows gelatin and aesculin tests negative, while tyrosinase and tartrate positive.

Pseudomonas syringae pv. *persicae* Young *et al.* causes "bacterial decline" on peach and nectarine. The bacterium is Gram negative, obligately aerobic, short rods, motile by 3 - 6 polar flagella. Colonies are either non-fluorescent or weakly blue-fluorescent, occasionally producing a fuscous pigment, initially produced small (less than 1 mm diam.), grey, translucent, flat colonies on King's B medium. However, after 4 - 5 days incubation, some of these colonies gave rise to large (2 - 3 mm diam.), flat, creamy, opaque colonies. The *P. syringae* pv. *persicae* produces levan on sucrose, oxidase negative, do not rot potato, arginine dihydrolase test negative and produces a hypersensitive reaction in tobacco.

Management

- Mashobra paste (Stearic acid 25.5 g, morpholine 9.0 ml, water 340 ml, lanolin 136 g and non-sterile dihydro streptomycin sulphate 3.1 g) is effective.
- Fair Howen, Fleming and Worlds Earliest are tolerant peach varieties.
- Oxytetracycline is generally considered the most effective compound and the least prone to phytotoxicity problems (damage to plant tissue), but also perhaps most likely to lose effectiveness with the appearance of bacterial strains with resistance. There is some evidence that the

hydrochloride form of oxytetracycline (Fireline) is more effective than the calcium form Mycoshield.

- ☆ Spraying with zinc sulphate-lime solution is effective.

8. Plum (*Prunus domestica* subsp. *insititia* L.) Schneid.

i. Bacterial Canker/Bacterial Gummosis

Bacterial canker is a common and sometimes lethal disease of trees in the genus *Prunus* including cherry, plum and peach. Bacterial canker is sometimes also referred to as *gummosis, blossom blast, dieback, spur blight* and *twig blight.* The number of causal agents identified for this disease has increased dramatically in the past few decades, with majority being pathovars of *Pseudomonas syringae* or recently described *Pseudomonas* species. These include *P. syringae* pv. *syringae* (*Prunus* spp.), *P. syringae* pv. *avii* (*P. avium*), *P. syringae* pv. *morprunorum* (*P. avium* and *P. domestica*), *P. syringae* pv. *persicae* (*Prunus* spp.) and *Pseudomonas cerasi* (*P. avium*) (Bophela *et al.,* 2020). Bophela *et al.* (2020) reported that *P. viridiflava* causes bacterial canker in plum trees. *Pseudomonas syringae sensulato* is broadly defined as a bacterial group of species that cannot be distinguished phenotypically from *P. syringae* and these group bacteria represent the majority of the fluorescent phytopathogenic pseudomonads. Currently, >50 pathovars are defined in this bacterial group.

Symptoms

Disease affects all above ground parts of the tree. Small purple lesions are more commonly seen on leaves of plum and apricot. The bark and outer sap wood may show circular to elongated, water soaked lesions. Bark becomes brown and gummy. The girdled branches fail to develop and may die. The blossom blight has been observed in plums, cherries and pears. Fruits of apricot and cherry become spotted with sunken black lesions.

Causal Agent

The disease is caused by *Pseudomonas syringae* pv. *syringae* VanHall. It is a Gram negative rod, motile with lophotrichous flagella and measures 0.7-1.2 × 1.5-3.0 µm. The colonies are white and non-slimy on SPA, creamcolored and non-slimy on YDC and round, light cream-colored and slimy on SNA media. All the strains were gram negative and obligate aerobic. In LOPAT tests, levan production and hypersensitive reaction on tobacco leave are positive but oxidase, potato soft rot and arginine dihydrolase are negative. The bacteria are negative for starch hydrolysis but positive for fluorescent pigment production on KB medium after 24–48 h of incubation. Hydrolysis of casein and tween 80, motility, catalase and Hydrolysis of gelatin are positive. Tyrosinase activity is negative and hydrolysis of aesculin and utilization of tartrate acid are varied. The bacteria can not grow at 4°C and 41°C and 7 per cent NaCl.

Disease Cycle

The pathogen survives in infected plant parts and spreads through wind splashed rain and infected planting material.

Management

1. Sprays of Bordeaux mixture have been inconsistent in the control of the disease in field trials (Agarwala and Sharma, 1971). Mashobra paste (stearic acid 25.5 g, morpholin 9.0 ml, tap water 350 ml, lanolin 136g and non-sterile dihydro streptomycin sulphate 3.1 g) formulated by Agarwala (1964) has been reported effective against the disease on peach, plum and apricot. Agarwala and Sharma (1971) reported that mashobra paste having streptochlor as active ingredient was more effective and had longer storage life.
2. The root stock has little influence on the degree of susceptibility of scion. The varieties Black Champa, Barbank, Elephant Heart and mariposa were resistant to the disease (Agarwala and Sharma, 1970).

ii. Bacterial Leaf Spot

Bacterial spot disease of stone fruits is caused by *Xanthomonas arboricola* pv. *pruni* that was described for the first time in the USA (Michigan) in 1903 on *Prunus salicina* (Japanese plum). The main hosts of economic importance are *P. persica* (peach), *P. persica* var. *nucipersica* (nectarine),

P. domestica (plum), *P. armeniaca* (apricot) and *P. dulcis* (almond). The pathogen also infects Sino-Japanese group (*P. salicina*), *P. avium* and *P. cerasus* (sweet and sour cherries), *P. mume* (Japanese apricot), *P. davidiana* (Chinese wild peach), *P. buergeriana*, *P. crassipes* and *P. donarium*. Ornamental species such as *P. laurocerasus* (cherry laurel) are also affected by the bacteria (EPPO, 2021). In India, it was recorded in plum nursery during 1969 at Mashobra in Himachal Pradesh.

Symptoms

The disease appears as angular dark brown spots of 0.5 to 3.0 mm dia. The lesions are bordered by light yellowish green halo. The necrotic tissues fall off having shot holes on leaves. Severely affected leaves turn yellow and drop. On *P. persica* (peach and nectarine), symptoms on leaves can show a typical three-band colour (green-yellow-brown), followed by heavy defoliation. Lesions are also seen on twigs. On *P. persica* (peach and nectarine) and *P. armeniaca* (apricot) fruits, small circular brown spots appear on the surface. Lesions become sunken, the margins are frequently water-soaked and often light-green halos give a mottled appearance to the fruit. As a result of natural enlargement of the fruit, pitting and cracking occur in the vicinity of the spots.

Causal Agent

The disease is caused by *Xanthomonas arboricola* pv. *pruni* (Smith, 1903) Vauterin, Hoste, Kersters and Swing, 1995 (Syn. *Xanthomonas campestris* pv. *pruni* (Smith) Dye. *X. arboricola* pv. *pruni* colonies are domed, smooth, mucoid and glistening: colour is bright, creamy yellow, with a tendency to darken a little, turning yellow-orange with age or depending on the media. After incubation at approximately 28°C for 3–5 days, *X. arboricola* pv. *pruni* colonies on mXCP1 appear yellow and mucoid and the hydrolysis of Tween 80 and soluble potato starch can be observed as a light halo around them. *X. arboricola* pv. *pruni* is a Gram-negative bacterium, rod-shaped, motile by one flagellum, measuring 0.2–0.4 × 0.8–1.0 µm. It is a strict aerobe with an optimum growth temperature range of 24–29°C.

Management

The disease can be managed effectively by spraying with Zinc sulphate-lime solution.

9. Walnut (*Juglans regia* L.)

Bacterial Blight

Bacterial walnut blight caused by *Xanthomonas arboricola* pv. *juglandis* (Xaj) is one of the most serious and economically important diseases of walnut, occurring in all major walnuts-growing areas of the world (Adhikari *et al.*, 1988; CABI-EPPO, 2001). There are no effective control methods, thus bacterial walnut blight may cause considerable yield losses (up to 70 per cent) under favourable conditions for the disease(Lang and Evans, 2010).

Symptoms

Disease symptoms appear all parts including leaf, stem and fruits. Yellow halos around the brown irregular spots on leaves appear and later the lesions spread on the entire leaf. On the fruits, the lesions become sunken as they enlarge. Then the lesions tissue become soft in dark brown to black color and when they become wet, they expand and become burst, resulting in oozing out the bacterial infection from the fruits. Early infection causes nut drop while late infection results in nut shriveling and discolouration. Early blooming cultivars suffer more damage.

Causal Agent

The disease is caused by *Xanthomonas arboricola* pv. *juglandis* (Pierce) Vauterin *et al.* [Syn. *Xanthomonas campestris* pv. *juglandis* (Pierce) Dye].The colonies of bacteria on YDC agar medium appear as pale yellow, slimy, circular, mucoid round (with a diameter of 2 to 4 mm) after 3 days of incubation at 28°C. When single colonies cultured on NA medium, they appear yellow and round. The bacterium is an anaerobic, Gram negative, rod-shaped. It is positive for starch, esculin hydrolyses and levan production; while negative for arginine hydrolysis and indole production.

Disease Cycle and Epidemiology

Pathogenic bacteria can survive either on the surface of leaves or inside leaves, corresponding to an epiphytic and endophytic phase, respectively (Lindow and Brandl, 2003). The bacterium overwinters primarily in symptomless dormant buds. Primary inoculum of the pathogen comes from buds and possibly old stem infections. The bacteria are spread by rain, insects and wind-blown pollen from infected staminate florets and other infested or infected plant parts. The pathogen enters green plant parts through stomata when free water is present. Wounds, such as those made by the walnut husk fly or codling moth and from broken trichomes, are important portals of entry. The force of raindrops may drive the pathogen into natural openings in plant tissues, as occurs with other foliar bacterial diseases. Flowers are usually infected soon after they appear in the spring, even before the stigmas are receptive to pollen. This is a disease of new growth and nuts. Disease development is favored by wet periods around flowering time.

Management

- ☆ Prune out diseased twigs and branches, if practical.
- ☆ Management of this disease depends on the application of protective sprays to buds, flowers and developing nuts. In orchards with histories of walnut blight damage, protective treatments at 7- to 10-day intervals during prolonged wet springs are necessary for adequate disease control. In areas or years with less intensive rainfall, spray intervals can be stretched and weather forecasts can help with spray timing.
- ☆ In years with high rainfall during catkin flowering, treatments may be applied when 30 per cent –40 per cent of the catkins emerge. In most years, the first application can be delayed and should be applied when 30 per cent –40 per cent of the buds reach the "prayer" stage (when terminal leaves of pistillate flower. A second spray should be done 7–10 days later to effectively treat the pistillate flowers that weren't sufficiently open during the initial application.
- ☆ Applications of copper spray in June and July, in periods of prolonged wetting, at 7- to 10-day intervals, can help reduce the incidence of walnut blight infection. As a precautionary note, resistance of bacterial blight to copper spray has become a serious problem in other countries growing Carpathian walnuts. Use copper sprays only when necessary to control infections.
- ☆ Spray at early prebloom (when catkins begin to enlarge), late prebloom (when shoots begin to expand) and early postbloom. In California, initial treatments are recommended when the terminal buds break. Resistance to copper products has been seen widely in California and may be a problem in the Pacific Northwest. However, copper-resistant bacteria do not seem to cause as much disease as ones that are copper-sensitive. Most spray adjuvants have not improved blight control; however, silicone-

based adjuvants show promise. Biological Control Serenade MAX at 2–4 lb/A. Active ingredient is a protein from *Bacillus subtilis*.

10. Mulberry (*Morus* spp. White mulberry - *M. alba* L.)

i. Bacterial Leaf Blight

Bacterial blight caused by *Pseudomonas syringae* pv. *mori* is a disease of much economic importance. The disease was recorded on mulberry by Smith (1910), but its occurrence was first reported in India by Sinha and Saxena (1966) from Uttar Pradesh and later by Prasad and Siddaramaiah (1978) from Karnataka. This is prevalent in almost all the mulberry cultivating states of India. In addition of causing heavy defoliation, the disease impairs the nutritive value of mulberry leaves (Sato and Takahashi, 1973).

Symptoms

Numerous tiny water soaked and irregular spots later turning brown with yellow haloes are produced particularly on young leaves. Occasionally dark spots occur on leaf veins. In severe infection, the young leaves become curled and distorted. On young twigs elongated dark lesions appear which later become cankerous. From the infected twigs bacterial ooze can be seen. The bacterium may survive in the lesions on the twigs and it spreads through rain splashes. Diseased plants should be removed and burnt.

Causal Agent

The disease is caused by *Pseudomonas syringae* pv. *mori* (Boyer and Lambert) Young *et al.* The bacterium produces green fluorescent pigment in King's B medium. The colonies are white in colour, with a smooth and flat surface and an entire edge which gradually becomes undulated and grows slowly. They are gram negative, straight or curved rods, 0.5-1.0 x 1.5-4.0 µm in size. Cells are motile by one to several flagella and catalase, hydrolysis of Tween 80, levan formation and geltine liqufection are positive by the bacterium and strict aerobes except for a few which denitrify. The bacria are negative in H_2S production, indole production, starch hydrolysis, oxidase reaction, methyl red and casein hydrolysis.

Disease Cycle and Epidemiology

The population of bacteria decreases in summer, increases in autumn and maintains a definite level in winter. The pathogen population decreases rapidly in non- sterilized soil compared to sterilized soil. The bacteria survive for > 100 days in dead leaves and > 150 days in roots of mulberry in the field. The bacterium overwinters in the dead leaves of mulberry in the fields, on wounds, in exudates or inside the infected tissues. Low temperature and dry weather prolonged survivability of bacteria while it is shortened under high temperature and humidity. The bacteria also present in winter bud of healthy shoot at low population.

Perennating bacteria are carried to the infection site (stomata, hydathodes, wounds, *etc.*) by rain splash, insects or by wind during wet weather. After landing on wet leaves, they multiply in a film of water or guttation drop and penetrate through stomata, hydathodes or wounds caused by insects or mechanical injuries. Inside the tissues, they multiply and spread intercellular at a rapid rate. Rain by its washing or spattering effect facilitates dissemination to other plants. Necrosis on leaves and twigs appears 20 days after inoculation.

Management

- ✰ Pruning of young trees in autumn followed by spraying of Bordeaux mixture to control the disease.
- ✰ Varieties like Bulgariun-3, Bulgarian-24, Thileseuri and Galicen from Romania shows resistance (Coteance, 1971) and Husung, China, Shine Ichinose, Gumji and Ichihie of *M. indica* and *M. alba* shows resistance to bacterial blight under field conditions. Under Indian Conditions, Berhampore, berhampore local, Channapatna local, Coonoor series C-6, C-10, C-11, C-14, C-15, C-18, CRRHakkikalu, Jakkur local, Kajali, Kanva- 2, MR2, Mysore local, S-36, S-41, S-54 and Yenneranganakaddi and exotic varities, like Goshoerami, Haraban, Italian, Kairyo- nezumigaeshi, Kosen, Kurangi, *M. australis, M. cathayana, M. lembang, M. multicaulis,* Okinawa2, Philippine, Popua, Tai-song-2 and -3 and Tsukasaguwa are totally free from the bacterial blight disease.
- ✰ Foliar application of phytobacteriomycin (0.05 to 0.1 per cent) containing glycerine controls the disease and stimulates the growth as well as the resistance of plants.
- ✰ *Azotobacter* and *Azospirillum* based biofertiizers reduce bacterial blight.

ii. Bacterial Leaf Spot

The bacterial leaf spot (blight) of mulberry was earlier reported by Srinivaschary *et al.* (1996).

Symptoms

Characteristic water soaked lesions are seen on leaf and stem leading to blight. The bud infection is the typical symptoms of the disease and it results in bud necrosis.

Causal Agent

The disease is caused by *Xanthomonas campestris* pv. *moricola*. The bacterium is gram negative, rod shaped and measuring 0.4-0.7 x 0.7-1.8μm with a single polar flagellum (monotrichous). They are aerobic, catalase positive and weak producer of acids from carbohydrates. Colonies of most species and pathovars are mycoid, convex and yellow on YDC agar. The optimum growth of the bacteria is 25-30°C temperature and 95-100 per cent relative humidity and 7.5 pH. They produce extra cellular polysaccharides, which is the source of xanthan gum' on media containing glucose. The unique yellow membrane –bound non-water soluble pigments produced by all but three species/pathovars, are brominated arylpolulene esters (Xanthomonadins) soluble in petroleum either, methanol and benzene. *Xanthomonas* species can be identified by the yellow xanthomonodin.

Disease Cycle and Epidemiology

The bacterial pathogen enters the host tissues through stomata, hydathodes and injured parts. It grows intercellularly in leaf tissue and stem. The spread of the disease occur through rain droplet and mechanical injury. Infection aroused from the bacteria exuding from the shoot lesions during rain, which washed them over, the surface to bring about further infection. During unfavourable concitions, the pathogen perennates in dead leaves and soil.

Management

- ✰ Minimize wounds to limbs and new shoots.
- ✰ Prune out and destroy infected shoots and branches during the late dormant season.
- ✰ Space plantings to provide good air circulation.
- ✰ Nearly immune reaction to BLS has been reported in Rotundiloba followed by MS-8, Kolitha-9, Kolitha-3 and Mandalaya against bacterial leaf spot disease (Maji *et al.,* 2012). Mulberry varities like S146, Punjab local, Phillipines are resistant to bacterial leaf spot disease.
- ✰ Streptomycin sulphate, tetracycline oxychloride, oxytetracycline, plantomycin, estamycin and fungicide copper oxychloride are quite effective to control the disease (Maji and Qadri, 1999).

11. Karonda (*Carissa carandas* L.)

Bacterial Canker

The disease was noticed for the first time during monsoon 1962 at Pune (Moniz *et al.,* 1964). Later, the disease was also recorded in Rajasthan, Punjab, Haryana, UP and Delhi.

Symptoms

The disease first appear on lower surface of leaves as small, round, water-soaked, translucent dots, gradually expanding, turning dark brown and surrounded with halo. Spots on under surface of the leaf enlarge and become raised and upper surface may be slightly raised. The central tissue of the spots finally collapse and form a depression. Twigs show similar symptoms. Flowers and fruits do not show any symptoms.

Causal Agent

The disease is incited by *Xanthomonas campestris* pv. *carissae* (Moniz *et al.*). The bacterium is a Gram negative rod, motile with single polar flagellum and measures 0.66 × 2.5 µm. The pathogen also infects Rat -ki - Rani (*Cestrum nocturnum*) by producing very small water soaked spots on the leaves and pili- Kaner (*Thevetia neriifolia*) with typical blisters or cankers.

Disease Cycle and Epidemiology

The pathogen survives in diseased plant debris. The bacterium enters the host mainly through stomata and wounds created by thorns. Generally, the disease becomes severe during rainy season. Disease is favored by a mean temperature of 24 °C and relative humidity of 56 per cent.

Management

- ☆ Removing of affected leaves and pruning of the shrub at a height of 45 cm from the ground level is effective to control the disease.
- ☆ Spraying with phytomycin 250 ppm controls is also found effective to control the disease.

12. Tamarind (*Tamarindus indica*)

Bacterial Leaf S

Small round spots, at first pole brown with yellow margins but later jet black appear on the leaves. The disease is caused by *Xanthomonas axonopodis* pv. *tamarindii* Vauterin *et al.* (1995) [Syn. *Xanthomonas tamarindii* (Patel *et al.*)]. The bacterium is Gram negative rod shaped with 1.5-2.0 x 0.5-0.75 µm in size single polar flagellum. The bacterium forms chains and capsules and is motile by one polar flagellum. No spores are formed and they aerobic. Colonies of bacteria on nutrient agar are circular, straw yellow, slightly raised and glistening.

Chapter 10

Diseases of Vegetable Crops

1. Tomato (*Solanum lycopersicum* L.)

i. Bacterial Leaf Spot

Disease was first observed on Tomato in South Africa and was described by M. Doidge as tomato canker in 1914. Doidge identified causal organism as *Bacterium vesicatoria.* In 1918, for first time, Sherbakoff reported bacterial leaf spot on pepper. The disease has been reported from various parts of world including Egypt, Greece, Hungary, Israel, Italy, Russia, Australia, China, Korea, Pakistan, Philippines, Thailand, Turkey, Kenya, Ethiopia, Niger, India, Canada, Mexico, USA California and Argentina (Mbega *et al.,* 2012). In India, the disease was first reported from Agricultural Farm Puna, Maharastra in 1948 (Patel *et al.,* 1950). Later, the disease was reported from Kerala, Karnataka, Uttar Pradesh, Himachal Pradesh, Uttarakhand and Maharashtra. This disease was sever in warmer parts of world and causes more damage when infection is early. Losses due to bacterial leaf spot caused by *Xanthomonas* species complex presents a serious problem leading to significant crop losses up to 23–44 per cent especially in regions with warm and humid climate.

Symptoms

Small translucent and water- soaked spots are formed on the leaves. Soon they enlarge into circular dark brown to black in color, greasy spot surrounded with yellowish halo. Severe spotting on young leaves may cause distortion and death of the affected leaf lets. Defoliation starts from the lower leaves (Figure 1a). The lesions are large black in color and canker- like on the stem and petiole. On fruit, the spot appears as circular tan coloured upto 8-9 mm in dia (Figure 1b). Slightly sunken with a central scab and irregular lobed margins are visible on the fruits.

(a) (b)

Figure 1. Bacterial Leaf Spot Symptoms on Leaves (a) and Fruits (b).

Causal Agent

Bacterial leaf spot of tomato is caused by four species of *Xanthomonas i.e. Xanthomonas vesicatoria* (Ex Doidge) Vauterin *et al., X. euvesicatoria, X. perforans* and *X. gardneri*[Syn. *Xanthomonas campestris* pv. *vesicatoria* (Doidge) Dye]. Based on DNA± DNA hybridization studies and accordingly bacteriological code, species level elevation of all four groups were name group A as *Xanthomonas euvesicatoria,* group B as *X. vesicatoria,* group C *X. perforans* and group D as *X. gardneri* (Vauterin *et al.*, 1995; Jones *et al.,* 2004). Presently, bacterial leaf spot causing xanthomonands are placed in four different species, namely *Xanthomonas euvesicatoria, X. vesicatoria, X. perforans* and *X. gardneri*. The main host plants of *X. vesicatoria, X. euvesicatoria* and *X. gardneri* are tomato and pepper, *X. perforans* affects mainly tomato but there is one report on affecting the pepper. Some races infect both pepper and tomato (P/T strains), whereas other races are specific for tomato (T races) or pepper (P races). On pepper, 11 races have been reported based on the pathogenicity reaction on pepper differentials having five resistance genes in peppers *i.e., Bs1, Bs2, Bs3, Bs4* (Stall *et al.,* 2009).

The bacterium is Gram negative, short rod-shaped and has a single, polar flagellum and measures size of 1.7 X 0.9 μm. The colonies are dull white, mucoid, glistening slimy, convex odourless and round on nutrient agar. The bacterial spot-causing xanthomonads also have been classified into races, some of which infect only tomato or pepper and some of which can infect both crops.

Disease Cycle and Epidemiology

The pathogen remains alive in the volunteer plants and diseased plant debris. It is seed-borne. The major sources of infection for these bacteria are thought to be seed and infected crop debris. The bacterium enters through stomata or lenticels and injuries. The bacteria are spread primarily by splashing water and wind-driven rain or mists produced during storms. In the field, spread by equipment or workers is probably of lesser importance than it is in the greenhouse, unless wounds are being opened up at the same time, as in a pruning operation or when plants are injured by a cultivator. They also may be present on volunteer tomato plants and on the surfaces of contaminated equipment (farm machinery, racks, greenhouse structures, tools). It attacks, *Capsicum annuum, Nicandra physuloides, Nicotiana rushica, Physalis peruviana* and *Solanum* spp. Kumar and Khan (1996) reported that the bacterium was isolated for a period of 60 days from *Cyperus rotundas* and *Cynodon dactylon*, which could survive only for 15 days on *Marsilea quadrifieda* and *Hemilia sanchfolia*.

The disease is favoured by high relative humidity (80-90 per cent) low temperature (24- 30°C) and intermutant rainfall for further spread of the disease.

Management

- Seeds should be treated with hot water at 50°C for 25 min or immersed in 0.1 per cent mercury chloride for 5 to 7 min.
- Amending the field soil with the compost did not add to the level of control achieved by foliar sprays.
- Aqueous extract of farmyard manure or composted cowdung manure gives as good control of the disease as Actigard (Acibenzolar-S-methyl).
- Three sprays with streptomycin sulphate (500 ppm) + copper oxychloride (2000 ppm) were found to be very effective in reducing the disease (Ravikumar and Khan, 2001).
- The plant activator, acibenzolar-S-methyle (Actigard) inducses SAR in tomoto against many diseases including bacterial leaf spot and bacterial speck (*Pseudomonas syringe* pv. *tomato*). It reduces the number and diameter of bacterial spots and suppresses the growth of the bacteria in the plant. It can be integrated in a viable alternatives to copper-based bactericides for field management of these diseases. Cymoxanil and famoxadone in combination with copper hydroxide were found to be very effective against copper insensitive and sensitive strains of bacteria (Fayette *et al.*, 2012). Acibenzolar-S-methyl and aluminum Tris O-ethyl phosphate can reduce bacterial populations significantly (Ma, 2015). Plasma activated water (PAW) enhanced the tomato plants defences, it was effective in reducing the disease severity by giving relative protections up to 38 per cent 9 (Perez *et al.*, 2019).
- Bacteriophages have been used as biocontrol agents against bacterial spot of tomato. Efficacy of the phage is greatly reduced due to its short residual activity on the plant foliage. However some formulated products have given satisfactory control of the disease.

☆ Abbasi *et al.* (2003) have reported that weekly sprays of aqueous suspension of neem (Margosa) oil (0.5 per cent, v/v) reduced bacterial spot on tomato and pepper leaves and increases yields of fruits.

ii. Bacterial Canker

This disease of tomato was observed first during 1909 in the USA. In India disease was observed on tomato fruits in Nagpur market during 1949 (Jain, 1951). The disease is of less economic importance in India. A yield reduction of 70 per cent has been reported from UAS due to this disease. It causes serious losses to both greenhouse and field tomato crops either by killing young plants or reducing marketable yield. The disease occurs in many countries on tomato and pepper particularly in the regions characterized by rainy humid summers. As a seed borne pathogen, *Clavibacter michiganensis* subsp. *michiganensis* is included in the A2 (high risk) list of quarantine pathogens by European and Mediterranean Plant Protection Organization. Hence, it is under strict quarantine control and zero tolerance. Bacterial pathogen infects wild plants, *Solanum douglasii* and *S. nigrum*.

Symptoms

In the field it causes a general chlorosis, drying of the leaves and wilting. Both wilting and desications may occur together. Long, brownish stripes appear on the shoots, stem and petioles and these may split to show creamy white, yellow or reddish brown coloured cavities in the pith and exude bacteria. Infected fruit develops characteristic bird's eye spots, *i.e.* small, round, raised white spots. These later develop light brown roughened centres surrounded by a white flat halo. The spots are 1 to 6 mm in dia and occur singly or in groups. Heavy infection results in browning and shriveling of sepals and lessens the fruits attachment to the plant.

Causal Agent

The disease is caused by *Clavibacter michiganensis* subsp. *michiganensis* (Smith) Davis *et al.*, [Syn. *Corynebacterium michiganense* (Smith) Jensen]. Pathogen produces yellow colonies on nutrient agar. The colonies reach 3- 4 mm in diameter after 5 days of incubation. The bacteria can also be isolated on specific medium SCM medium. Bacterial cells can be pleomorphic in culture, but rod shaped when isolated from plants. Optimum growth can occur at 24 -27°C. It is a non motile rod, Gram positive, 0.7 to 1.2 X 0.6 to 0.7 µm and produces capsules in culture.

Disease Cycle and Epidemiology

Bacteria can occur on the seed coat as well as within the embryo. The bacteria are seed borne and its inoculums may serve as one source of the bacterium. The pathogen remains alive in the soil, in diseased plant debris and can also survive in soil covered crevices of wood stakes and posts where tomatoes are staked and trellised. In soil it survives for 1 to 4 years. The disease is transmitted from infected seeds to seedlings and mechanically from plant to plant during seedling production, grafting, pruning and harvesting. It enters the plant through wounds on the stem or roots. Spots on the fruit are probably caused by rain splash or overhead irrigation dripping from centres on stems and leaves. Long distance distribution is through seed and seedlings. The disease is influenced by various soil and air factors. It develops at air and soil temperatures as low as 10°C and as high as 36°C. The optimum temperature is about 28°C. It survives in *solanaceous* weed, *Solanum nigram*. The presence of root knot nematodes may increase the severity of disease.

Management

- ☆ Early detection of infected asymptomatic plants could prevent the dissemination of tomato bacterial canker throughout the crop and, thus, reduce associated economic losses. Three indices, of the shortwave-infrared zone, allowed the detection of bacterial canker inoculated plants in a fast and non-destructive way, up to one week before symptoms were visible: Normalized Difference Water Index (NDWI), Simple Ratio of Water Index (SRWI) and Water Index 1 180 (WI_{1180}). These indices arise as a promising tool for the early detection of vascular plant diseases, that could be integrated into sustainable management plans (Cordon *et al.*, 2021).
- ☆ Seeds should be treated with 5 per cent hydrochloric acid for 5 hr or hot water at 56°C for 30 min. Soil sterilization with methyl bromide and three year crop rotation with non- hosts are recommended. Application of 2 to 4 sprays of streptomycin to the bed followed by weekly field sprays of copper sulphate controls the disease. Treating the seeds in 1.05 per cent sodium hypochlorite for 40 min. followed by four weekly applications of streptomycin to the crop is effective.
- ☆ Sonicated extracts from cyanobacteria and microalgae of the genera *Leptolyngbya*-1267 and *Scenedesmus*-677 strains were effective against the disease. The biopesticidal effect demonstrated could be closely related to their ability to promote plant growth as well as root development of tomato plants (Toribio *et al.*, 2021).
- ☆ Sprays of copper fungicides and antibiotics do not suppress the disease but prolong survival of the transplanted seedlings in the field. *C. michiganensis* is susceptible to an antibiotic fragarin isolated and purified from soluble fraction of strawberry leaf extract. This pre-formed compound in the plant acts at the membrane level of the bacterium increasing cell wall permeability (Fillippone *et al.*, 2001).
- ☆ Pre-treatment of tomato seedlings with the defence activator acibenzolar-S-methyl suppresses the bacterial growth by 68.2 per cent and disease severity by 75 per cent (Soylu *et al.*, 2003) by induction of systemic resistance. The treatment causes progressive and significant increase in the defence enzymes peroxidase and glutathione peroxidase when challenged by the pathogen. Chemicals that have been described as activating plant resistance are salicylic acid (SA), jasmonic acid (JA), DL-

β-aminobutyric acid (BABA), potassium salts, 2,6-dichloroisonicotinic acid (INA), acibenzolar-S-methyl and specific volatiles such as nitric oxide and ethylene were found to be affective against canker pathogen.

- Resistance to *Clavibacter michiganensis* sub sp. *michiganensis* has been identified in wild relatives of cultivated tomato. These are *Lycopersicon peruvianum* and *L. hirsutum*. Both are cross compatible with *Lycopersicon esculentum* (Francis *et al.*, 2001).

iii. Bacterial Wilt

It is also called as southern bacterial wilt and is widespread world wide in tropical, subtropical and warm temperature regions. In India, loss may be upto 90.6 per cent and it is a threat to tomato in Karnataka, Madhya Pradesh, Maharashtra and West Bengal (Singh *et al.*, 2010).

Symptoms

Young plant shows a rapid wilting and yellowing affected plants are stunted and downward curling of leaf lets occurs, some times excessive adventitious roots are produced along the stem. Plants collapse and die. In some cases instead of collapsing, the xylem may blacken at the Junction between the leaf and the stem. Further down the stem the whole vascular system may be blackened and when cut cream coloured bacterial slime oozes from it.

Causal Agent

The disease is caused by *Ralstonia solanacearum* (Smith) Yabuuchi *et al.*, [Syn. *Psuedomonas solanacearum* (Smith)]. The bacterium is Gram negative, rod- shaped, often occur in pairs, motile with 1 to 4 polar flagella and size 0.5 to 0.7 X 1.5 to 2.5 µm. The optimum temperature for growth is 30 to 37°C. The organism grows aerobically and does not form endospores. *R. solanacearum* is catalase positive, oxidase positive and reduces nitrates. Bacterium does not hydrolyze starch and does not degrade gelatin. In broth culture, the organism is inhibited when the concentrations of NaCl is greater than 2 per cent.

Based on host range the pathogen species is subdivided into races. Currently, polymerase chain reaction (PCR) is the main means of definitive identification of pathogen race. Presently five races *i.e.* races 1, 2, 3, 4 and 5 have been categorized based pathogenic reaction of the different host like, tobacco, potato and tomato. Races 1, 2 and 3 are the most economically important races (Table 1). Before PCR, a tobacco hypersensitivity test, developed by Lozano and Sequeira (1970), was used to distinguish between the strains of bacteria. The species is also subdivided into six biovars (biovars 1, 2, 2T, 3, 4 and 5 based on the utilization of the disaccharides cellobiose, lactose and maltose and oxidation of the hexose alcohols dulcitol, mannitol and sorbitol as mentioned in Table 2.

Table 1. Races and Biovars of *Ralstonia solanacearum* (Adapted from Daughtrey, 2003; Schaad *et al.*, 2001)

Race	*Host Range*	*Geographic Distribution*	*Biovar*	*RFLP Division*
1	Wide	Asia, Australia	3, 4	I
		Americas	1	II
2	Banana, other Musa spp	Caribbean, Brazil, Worldwide	1	II
3	Potato, some other Solanaceae, Geranium; few other species	Worldwide except US and Canada	2	II
4	Ginger	Asia	3, 4	I
5	Mulberry	China	5	I

Table 2. Classification of *Ralstonia solanacearum* into Biovars (Aley and Elphinstone, 1995; Schaad *et al.*, 2001)

Physiological Tests	*Biovars*					
	1	*2*	*2-T*	*3*	*4*	*5*
Utilization of disaccharides						
Cellobiose	-	+	+	+	-	+
Lactose	-	+	+	+	-	+
Maltose	-	+	+	+	-	+
Trehalose	+	-	+	+	+	+
Oxidation of alcohols						
Dulcitol	-	-	-	+	+	-
Mannitol	-	-	-	+	+	+
Sorbitol	-	-	-	+	+	-

Recently, it has been suggested that *Ralstonia solanacearum* should be considered a "species complex." Also, some consider the current race and biovar system inadequate since it is based on phenotypic measures and propose switching to a phylogenetically-based hierarchical system. This scheme is based upon phylogenetic analysis of sequence information using RFLP typing, 16S rDNA phylogeny and hrpB PCR-RFLP data. Currently, over 20 sequevars have been identified based on the partial endoglucanase gene (egl) sequence. The system divides the species complex into four phylotypes (phylogenetic grouping of strains) phylotype I (Asian), Phylotype II (American), Phylotype III (Afican) and Phylotype IV (Indonesian) (Table 3).

Table 3. *Ralstonia solanacearum* Current Infra Subspecific Classification (Source: Garcia *et al.,* 2019)

Phylotype	*Sequevar*	*Biovar*	*Ecotype*	*Races*	*Species*
I	12, 14, 16, 18	3, 4, 5	BWa of Solanaceae, ginger, mulberry	1, 4, 5	*R. pseudosolanacearum*
IIA,	1, 2, 3, 4, 5, 6, 7, CIP10, CIP223, NCPPB3987	2-T, 1, 2,	BW of Solanaceae, Musa spp	1, 2, 3	*R. solanacearum*
IIB	1, 2, 3, 4, 5, 6, 7, CIP10, CIP223, NCPPB3987	2-T, 1, 2	Moko disease, NPB,b brown rot of potato, BW of tomato and geranium	1, 2, 3	*R. solanacearum*
III	19, 20, 21, 22, 23	2-T, 1	BW of Solanaceae	1	*R. pseudosolanacearum*
IV	8, 9, 10, 11	2-T, 1, 2	BW of Solanaceae, BDB of Musa spp., c Sumatra disease of clove	1	*R. syzygii*

a: BW = bacterial wilt. b: Nonpathogenic to banana (NPB) are new pathogenic variants highly virulent to Anthurium, Heliconia and cucurbits but not to bananas. c: BDB = Blood disease of banana.

Disease Cycle and Epidemiology

The bacterium remain alive in soil and they spread by transplanting of infected seedlings and through irrigation water. The bacterium survives for three years in fallow and for a unlimited period in cultivated land. Chilli, egg plant, ground nut, potato and tobacco are alternative hosts which help it to survive between tomato crops, pathogen enters the plant through the root and stem wounds or through stomata.

The disease is favoured by warm temperature, *i.e.* 24°C to 27°C and high soil moisture together with periods of wet weather or rainy seasons. Bacterial wilt is found in a severe form during monsoon (July-October) than during summer (March-June) and winter (March-February) seasons. It is serous on sandy loam, clay and peat soils. Root-knot nematodes increase the severity of the disease.

Management

- ✩ Use disease free seeds and seedlings should be used for transplanting.
- ✩ Grow bacterial wilt resistant varieties of tomato: Palam Pink, Palam Pride and F1 7711, Arka Rakshak, Arka Abha, Swarna and Sampada.
- ✩ In India, CRA 66 rootstocks were used in grafting of tomato to reduce bacterial wilt in tomatoes. Several Hawaiian lines (Hawaii 7996-7998)

have been identified as suitable candidates for resistance to bacterial wilt. This technique could be a very valuable tool for eliminating bacterial wilt in tomato, pepper and eggplant production systems.

- ✰ High grafting should be done at a higher position (rootstock length, 10–12 cm) than the usual position (above the cotyledon; rootstock length, 3–4 cm). The mechanisms which limit bacterial movement and multiplication are at work in the additional stem of resistant rootstocks. This could further restrict bacterial movement into susceptible scions and high-grafted tomato plants show no symptoms or delay onset of wilting generally, plants of rootstock cultivars have a latent infection and suggest that tomato rootstocks are tolerant rather than immune to *R. solanacearum* infection (Nakaho,2021).
- ✰ Disinfecting the cutting knives and other tools used for with suitable bactericide.
- ✰ Avoidance of surface water for irrigation and maintain proper drainage.
- ✰ Crop rotation is not effective as the pathogen can survive for a long period (several years) in the soil and also attack a wide range of crops and solanaceous weeds. Crop rotation 2-3 years with wheat, *Crotolaria juncea* and lady's finger (*Ahelomoschus esculentum)*, cowpea (*Vigna unguiculata*) reduces wilt of tomato. Two seasons rotation sequence growing tomato after bean, maize and maize-bean resulted in about 29 per cent average final wilt incidence reduction. The population of bacterial pathogens was recorded in a declining trend but detectable in the rhizosphere soils and roots of presumable nonhost crops at 30, 45, 60, 90 and 120 days after inoculation (Ayana and Fininsa, 2016).
- ✰ Do not grow crops in soil where bacterial wilt has occurred.
- ✰ Rogue out wilted plants from the field to reduce spread of the disease from plant to plant
- ✰ Control root-knot nematodes, since they could facilitate infection and spread of bacterial wilt
- ✰ Where feasible, extended flooding (for at least 6 months) of the infested fields can reduce disease levels in the soil
- ✰ Soil amendments (organic manures) can suppress bacterial wilt pathogen in the soil Good results have been encountered by using biofumigation as soil treatment for bacterial wilt.
- ✰ Soil solarization combined with fumigation reduces the pathogen considerably.
- ✰ Destruction of weeds, collateral hosts and other off season hosts will reduce the inoculum potential.
- ✰ The incidence of wilt diseases in tomato was reduced by applying panchagavya which contains 20 ml of cow ghee, 50 ml each of cow milk

and curd, 40 ml of cow urine, 400g of cow dung, 20 g of common salt and 10g bakers yeast.

- ☆ Seed treatment, root dipping followed by soil application with antagonistic *P. fluorescens* strain controlled bacterial wilt of tomato. Bio-control agents such as *Pseudomonas fluorescens, P. glumae, Burkholderia cepacia, Bacillus* sp., *Bacillus licheniformis, B. cereus, B. subtilis, Erwinia* sp. *Streptomyces griseus, Stenotrophomonas maltophilia* and a Hrp-mutant of *R. solanacearum,Trichoderma harzianum, Candida ethanolica* and mycorrihiza are found to be relatively effective in the control of *R. solanacearum* populations under natural conditions.
- ☆ Biopriming plants with some plant growth promoting bacteria (PGPB) can also provide systemic resistance against a broad spectrum of plant pathogens. Certain bacteria trigger a phenomenon known as induce systemic resistance (ISR) phenotypically similar to systemic acquired resistance (SAR). SAR develops when plants successfully activate their defense mechanism in response to primary infection by a pathogen, notably when the latter induces a hypersensitive reaction through which it becomes limited in a local necrotic lesion of brown, desiccated tissue. As SAR, ISR is effective against different types of pathogens but differs from SAR in that the inducing PGPB does not cause visible symptoms on the host plant. Most reports of PGPB-mediated ISR involve free-living rhizobacterial strains, but endophytic bacteria have also been observed to have ISR activity.
- ☆ Application of bleaching powder 12-15 kg/ha in the furrow during transplanting in the field.
- ☆ Seeds should be treated with 5 per cent hydrochloric acid for 5 hr or hot water at 56°C for 30 min. Soil sterilization with methyl bromide and three year crop rotation with non- hosts are recommended. Application of 2 to 4 sprays of streptomycin to the bed followed by weekly field sprays of copper sulphate 53 per cent controls the disease. Treating the seeds in 1.05 per cent sodium hypochlorite for 40 min. followed by four weekly applications of streptomycin to the crop is effective.
- ☆ Seedling dip with streptocycline avoids early invasion and infection by the pathogen through wounds formed during transplanting. Bacterinol-100 can be used as dry seed dresser, for nursery spray and field application. Streptomycin sulphate or oxytetracycline when sprayed at 200 ppm at 7 days interval provides good control.
- ☆ Chemical soil fumigation (with chloropicrin or D-D), soil solarization and anaerobic soil disinfestation (ASD) with with molasses are effective for controlling bacterial wilt and nematodes because the pathogens persist in deeper soil layers (30–60 cm depth) (Inoue and Nakaho, 2019).
- ☆ The exogenous application of L-histidine, inhibited wilt disease in tomato without exhibiting any antibacterial activity by inducing systemic

resistance (Nakaho,2021). Black Soldier Fly Pupal Exuviae (BSF) pupal shells are an attractive renewable raw material for the recovery of valuable products (chitin and chitosan) with promising ability as a new type of eco-friendly control measure against bacterial wilt caused by *R. solanacearum*. Disease severity was reduced by 22.57 per cent and 23.66 per cent, when inoculated tomato plants were subjected to soil amended with Black Soldier Fly Pupal Exuviae (BSF) chitin and chitosan, respectively (Kemboi *et al.*,2022).

- ☆ Combined application of potassium phosphite (K-Phite), an environmentally friendly agricultural chemical and the biocontrol agent *B. amyloliquefaciens* QPF8 (strain F8) significantly reduced the wilt incidence of tomato (Su *et al.*, 2022). Injecting *Ralstonia pickettii* QL-A6 suspension on stem of tomato resulted in 73 per cent biological control efficiency of wilt.
- ☆ Actigard (acibenzolar-s-methyl) has shown to be effective, but only when inoculum densities are low.
- ☆ In integrated management strategies, resistant varieties (Arka Abha), biocontrol agaent (*P. fluorescens, B. subtilis*) and chemical (bleaching powder) was found effective to control the disease. The disease incidence is reduced in Arka Abha cultivar of tomato treated with bleaching powder (0.01 per cent) + *P. fluorescens* followed by bleaching powder + *B. subtilis* (Singh *et al.*, 2017).

iv. Bacterial Speck

Black speck was first reported from Taiwan in 1933. In India, it was reported from Tamil Nadu and Uttar Pradesh. Bacterial speck is a common disease of tomato that occurs worldwide wherever tomatoes are grown. The disease can substantially reduce yield when it severely affects leaves early in the growing season. The disease can have an even greater impact on quality (and marketability for commercial tomato producers) when symptoms occur on tomato fruit.

Symptoms

The disease is characterised by small, round shape, water-soaked spots, which gradually become yellowish brown and finally blackish brown in colour with sunken ashy centre. Pale yellowish green halo becoming irregularly circular on maturity and measuring 1.0 mm in dia. distributed over the entire surface and in acute cases forming large blotches. The spots on petioles are more or less oval in shape. Margins of leaves can be brown, sometimes extending down into the leaf forming a wedge of brown tissue. Severely affected leaves often turn yellow and then die.

Causal Agent

The disease is caused by *Pseudomonas syringae* pv. *tomato* (Okabe) young *et al.* The bacterium is Gram negative, rod shaped with rounded ends arranged singly, in pairs and occasionally in chains, motile with 1 to 3 polar flagella, measures 1.5 to 6.5 X 0.5 to 0.9 µm non-capsulated, non- acid fast and non- spore forming. The bacterium grows best at 23 to 25°C with the range of 18 - 24°C in culture.

Disease Cycle and Epidemiology

Initial sources of the pathogen include infested seed, infected transplants, wind-driven rain from affected plants in another planting nearby and contaminated tools or hands from exposure at another planting. After bacterial speck has occurred in a field, the bacteria can survive in plant debris and seeds in the soil, on stakes used to support plants and in weeds (especially those in the nightshade family). Bacteria can also survive on greenhouse structures and planting materials after developing on seedling. These are all potential sources of the pathogen for subsequent crops. This pathogen can become established in an area, surviving on the surface of non-host plants. Penetration takes place through wounds at the base of broken hairs or following injury from wind blown sand particles.

Management

- Practice good sanitation (*e.g.* Use clean equipment and tools, avoid working in fields when they are wet, locate cull piles away from production fields).
- Use disease-free seeds or treating seed to eliminate the bacterial pathogen.
- Use certified disease- free transplants.
- Avoid overhead irrigation
- Practice crop rotation (at least 2 year rotation) and scout and remove symptomatic plants.
- Actigard is the most effective fungicide to use in commercial plantings, but applications need to start before infection to obtain the most benefit as this is a plant activator.
- The bark extract of *Albizia lebbeck* effectively inhibited biofilm development of *P. syringae* (94.2 ± 4.9 per cent) at 50 µg/mL. The bark extract has demonstrated antibacterial activity on phytopathogens and could be used against speck disease in tomato plant that causes economic loss in mature plant or at seedling stage in organic farming. Phytochemicals from natural sources are safe and most of the synthetic pesticides induced genetic mutations and antibiotics induced drug resistance (Arasu *et al.*, 2022).
- Applications of bacteriophages consistently reduced disease severity of bacterial speck. There was a disease suppression upto 50 per cent on treating with bacteriophages (Cement *et al.*, 2018).
- Foliar application of Para-aminobenzoic (PABA) applications on tomato seedlings reduced lesion incidence (Trueman *et al.*, 2021).

- ✰ Pesticide sprays (*e.g.* Copper compounds, combination products of copper mancozeb and maneb.)

2. Eggplant/Brinjal (*Solanum melongena* L.)

Bacterial Wilt

In India, loss in yield due to this disease in brinjal is upto 8.1 per cent.

Symptoms

The symptoms consist of sudden wilting of the affected plants and it is characterized by the drooping of the young top leaves and shoots of the plants. The plants die within 3 to 5 days after the appearance of first recognizable symptoms. Water soaked areas in the form of black streaks are often noticed on the stem. In severe cases these area rot completely and bacteria ooze out from those portions in the form of minute droplets. The infected region of the stem or root shows brown discolouration in the vascular region and bacteria ooze out in the form of milky white stream. The disease occurs severely in the monsoon when the fields are waterlogged.

Causal Agent

The disease is caused by *Ralstonia solancearum* (Smith) Yabuuchi *et al.* [Syn: *Pseudomonas solanacearum* Smith]. It is a short rod, 3.2 to 6.4 X 1.6 µm and remains in chains in 24 h old culture but become single later. The bacterium is non-acid fast, non-capsulated, non-spore farming and motility by a polar flagellum. The bacterium in dextrose produces acid but not the gas, sucrose, lactose and glycerol. Starch hydrolyzed with slight liquefaction of gelatin.

Disease Cycle and Epidemiology

Bacterium infects > 450 plant species and 54 botanical families mainly banana, chillies, *Dolichos lablab*, fennel, ginger, potato, radish, tomato, *etc.* Bacterium though a non-spore forming is found to be alive and viable for more than 16 months under laboratory conditions. The pathogen is found to survive in the infected plant debris for about 10 months. Presence of root knot nematode, *Meloidogyne javanica* increases the wilt incidence.

Management

- ✰ The varieties Arka Keshav, Azoud Kranti PBr 108 and S-1, Arka Neel Kanth, Arka Nidhi, Banaras Giant Green, Dingra multiple purple, Long Black 17-B, PBr- 61, PBr-129-1, PBr 129-2, PBr 129-5, PBr 129-6, Pusa purple cluster, pusa purple Round, Sinampiro, S-3, S-20, SM-56, SM-74, SMI -5, SMI-6, SMI-10, SMI-312, T-3, Vijay Hybrid and west coast Green Round X Pusa Granti are found resistant to bacterial wilt. Around Kranti, PBr. 108, Pusa purple long and S-1 are highly susceptible. Rotation of Brinjal- Frenchbean- finger millet reduces the disease.

- Genetic resources for bacterial wilt resistance in eggplant are available in the wild. *S. torvum, S. sisymbrifolium* and *S. khasianum* having high degree of bacterial wilt resistance due to natural selection for survival (Barik *et al.,* 2020).
- CARI-1 was found to be highly bacterial wilt resistant. F1 hybrids developed using CARI-1 as one of the parents have shown a varied level of resistance. Genetic analysis through six-generation mean analysis in the populations derived from two crosses *viz.,* Arka Neelachal Shyama x CARI-1 cross and Pusa Purple Long x CARI-1 indicated bacterial wilt is polygenically controlled and the existence of epistasis. High significant additive × additive type non-allelic interaction in Arka Neelachal Shyama x CARI-1 cross suggested for the delaying of selection in later generations for the accumulation of favorable alleles for improvement of bacterial wilt resistance. In the case of Pusa Purple Long x CARI-1 derived population, additive gene action is predominant indicating simple selection will be highly effective for bacterial wilt resistance (Barik *et al.,* 2021).
- Chemical soil fumigation (with chloropicrin or D-D), soil solarization and anaerobic soil disinfestation (ASD) with with molasses are effective for controlling bacterial wilt and nematodes because the pathogens persist in deeper soil layers (30–60 cm depth) (Inoue and Nakaho, 2019).
- *Pseudomonas fluorescens* inoculated to seed (@ 10 g kg^{-1} seed), seedling roots (2.5 kg of talc based formulation of *P. fluorescens* in 40 L water) and soil (2.5 kg of talc based formulation of *P. fluorescens* mixed in 50 kg of FYM $acre^{-1}$ soil) gave disease control of 48.91 per cent over control. Seedlings dip treatment with *Bacillus* sp. (RCh6) in antagonist suspension resulted in 81 per cent of disease reduction (Manickam *et al.,* 2021).
- *Bacillus amyloliquifaceans* strain (Bs_Abi) showed significant biocontrol efficacy (92.1 and 73.7 per cent, respectively) against brinjal bacterial wilt disease. The antimicrobial potential of Bs_Abi were further ascertained by presence of nine antimicrobial peptide (AMP) gene biosynthesis in PCR amplifications which confirmed the presence of peptide genes for six antibiotics (Iturin, bacillomycin D, mycosubtilin, surfactin, bacilysin and subtilin) (Sakthivel *et al.,* 2019).
- Chemicals that have been described as activating plant resistance are salicylic acid (SA), jasmonic acid (JA), DL-β-aminobutyric acid (BABA), potassium salts, 2,6-dichloroisonicotinic acid (INA), acibenzolar-S-methyl and specific volatiles such as nitric oxide and ethylene were found to be affective against wilt pathogen.

3. Potato (*Solanum tuberosum* L.)

i. Scab

Common scab occurs in most potato producing areas in Africa, Asia, Europe, North and South America. It has spread in some areas in Indo Gangetic plains due to cultivation of potato year after year in the same land. Although scab does not usually affect total yields, significant economic losses result from reduced marketability of the tubers. Economic losses are greatest when tubers intended for table stock are infected, since appearance is important for this market. While superficial scab lesions do not greatly affect the marketability of processing potatoes, deep-pitted lesions, however, do increase peeling losses and detract from the appearance of the processed product. Seed lots exceeding 5 per cent incidence is rejected by seed certification agencies (in India) causing huge loss to seed industry. The occurrence of scab and its severity varies by season and from field to field. Cropping history, soil moisture and soil texture are largely responsible for this variability. Potato scab lesions can be confused with powdery scab, a disease caused by an entirely different pathogen, the fungus *Spongospora subterranea*.

Symptoms

Potato scab lesions are quite variable and distinctions have been made between russet (superficial corky tissue), erumpent (a raised corky area) and pitted (a shallow-to-deep hole) scab. Individual scab lesions are circular but may coalesce into large scabby areas. Insects may be involved in creating deep pitted lesions. The term "common scab" generally refers to the response of the disease to soil pH.

Common scab is widespread and is caused by *Streptomyces scabies*. "Acid scab" seems to have a more limited distribution, but has been found in several states in the Northeast. This disease occurs in soils below pH 5.2, as well as at higher levels. The causal agent, *S. acidiscabies*, is closely related to the common scab pathogen and can grow in soils as low as pH 4.0. Acid scab lesions are similar, if not identical, to those caused by *S. scabies*. All of these can be caused by the same pathogen, *S. scabies*. However, the type of lesion probably is determined by host resistance, aggressiveness of the pathogen strain, time of infection and environmental conditions.

Causal Agent

At least 13 different *Streptomyces* spp. have been found to cause common scab on potato worldwide (Hao *et al.,* 2009). The prominent among them are *Streptomyces scabies* (Thaxter) Lambert and Loria, *S. acidiscabies* Bambert and Loria, *S. turgidscabis* Takeunchi, *S. collinus* Lindenbein; *S. griseus* (Krainsky) Waksman and Henria, *S. Longisporoflavus*, *S. cinereus*, *S. violanceoruber*, *S. alborgriseolus*, *S. griseoflavus*, *S. catenulae* and others. *Streptomces* are bacteria which resemble fungi due to formation of vegetative substrate mycelium that develop aerial filaments. However, the filaments are of smaller dimensions than the true fungi.

These filaments produce spores through fragmentation. *Streptomyces* spp. may be pathogenic or non pathogenic. The pathogenic species produce thaxtomins which are phytotoxins and cause hypertrophy and cell death. Considerable variation exists within the pathogen with respect to their pigment production in media, colour and shape of sporulating filaments and use of specific sugars. *S. scabies* form grey, spiral spore chains on several media and produce brown pigment where as *S. acidiscabies* produces peach coloured wavy chains of spores and brown pigment in the medium.

Disease Cycle and Epidemiology

The pathogen is both seed and soil borne. It can survive in soil for several years in plant debris and infested soil. Most if not all potato soils have a resident population of *S. scabies* which will increase with successive potato or other host crops. Scab-susceptible potato varieties appear to increase soil populations faster than scab-resistant varieties. Rotation with grains or other non-hosts eventually reduces but does not eliminate the *S. scabies* population. This pathogen is a good saprophyte and probably reproduces to some extent on organic material in the soil. Given the right environmental conditions and a scab-susceptible potato variety, scab can occur in afield that has been out of potatoes for several years.

Streptomyces scabies infects young developing tubers through the lenticels and occasionally through wounds. Initial infections result in superficial reddish-brown spots on the surface of tubers. As the tubers grow, lesions expand, becoming corky and necrotic. The pathogen sporulates in the lesion and some of these spores are shed into the soil or reinfest soil when cull potatoes are left in the field. The pathogen survives in lesions on tubers in storage, but the disease does not spread or increase in severity. Inoculum from infected seed tubers can produce disease on progeny tubers the next season. Soil moisture during tuberization has a dramatic effect on common scab infection. Light-textured soils and those with high levels of organic matter are favorable to scab infection. *Streptomyces* are generally involved in the decomposition of soil organic matter and therefore thought to be stimulated by its presence. Applying manure to potato fields can cause an increase in scab infection. Coarse-textured soils are conducive to scab, probably because of their moisture-holding capacity. Thus, gravelly or eroded areas of fields that tend to dry out rapidly are often sites of heavy scab infection. Infection of seedlings of many vegetables and fleshy roots of beet, cabbage, carrot, radish, spinach, turnips and other plants has been reported.

Management

- ✰ Use scab-free seed and seed treatments to prevent introduction of the pathogen into fields.
- ✰ Seed treatments do not eliminate the pathogen but will provide some suppression of disease.
- ✰ Rotation with small grains, as maize, cotton, grain sorghum, wheat, cabbage and onion or alfalfa appears to reduce disease in subsequent

potato crops. Red clover, however, stimulates problems with common scab and should not be used in fields where scab has been a problem.

- ☆ Using resistant varieties is an effective tool for management of scab. Resistant varieties are not immune, however and will become infected given high inoculum densities and favorable environmental conditions. Resistant varieties include Keuka Gold, Lehigh, Pike and Marcy. Varieties that are moderately resistant include Chieftain, Eva, Reba, Andover and Russet Burbank.
- ☆ Soil solarization with transparent polyethylene mulching during hot summer season is effective for control of russet scab.
- ☆ Common scab is controlled or greatly suppressed at soil pH levels of 5.2 or lower. Severity of common scab is significantly reduced in soils with pH levels of 5.2 and below, but losses can rapidly increase with small increases in pH above 5.2. *S. acidiscabies* ("acid scab") causes scab in low-pH soils. While low-pH soils provide good control of common scab, there are disadvantages associated with this management strategy. Maintain soil pH levels between 5.0 and 5.2 by using acid-producing fertilizers such as ammonium sulphate.
- ☆ Avoid or limit the use of such alkaline-producing amendments as lime and manure.
- ☆ Maintaining soil at moisture levels above -0.4 bars (near field capacity) during the 2 to 6 weeks following tuber initiation will inhibit infection by *S. scabies*. Bacteria that flourish at high soil moisture appear to out compete *S. scabies* on the tuber surface. Avoid moisture stress during the 2 to 6 weeks following tuberization.
- ☆ Bio-control of common scab using antagonists such as *Bacillus subtilis*, non pathogenic *Streptomyces* spp. and bio-pesticides such as *Geranium pretense* have shown promise for control of common scab.

ii. Bacterial Wilt/Brown Rot of Potato

Bacterial wilt or brown rot has a worldwide distribution. The common name of the bacterial wilt diseases is brown rot, Southern wilt, sore eye or jammy eye. It is a destructive disease of potato especially in tropical and subtropical parts of Asia, Africa, South and Central America. In India, the disease is endemic in west coast from Thiruvananthpuram in Kerala to Khera in Gujarat, Karnataka, Western Maharashtra, Madhya Pradesh, eastern plains of Assam, Himachal Pradesh, Jharkhand, Odisha, Uttarakhand, West Bengal, Chhota Nagpur plateau, Meghalya, Manipur, Tripura, Mizorum, Arunachal Pradesh and in Nilgiris, Annamalai and Palani hills of Tamil Nadu and Andaman and Nicobar Islands (Singh *et al.*, 2010; Sharma *et al.*, 2015). It has been estimated to affect potato crop in 3.75 million acres in approximately 80 countries with global damage estimates exceeding $950 million per year (Floyd,

2007). The disease is present and limiting potato production in some parts of India. Losses up to 75 per cent have been recorded under extreme conditions (Gadewar *et al.,* 1991).

Symptoms

The aboveground plant symptoms include wilting of one to two leaves on young plants during the heat of the day. Such plants tend to recover at night. On large-leafed plants, only the tissue on one side of the midvein may wilt. This is very characteristic for plants such as Nicotiana. Affected leaves turn yellow and remain wilted after a time. The area between leaf veins dies and turns brown. Usually, the main stem of the affected plants remains upright even though all the leaves may wilt and die. Internal symptoms include light tan to yellow-brown discoloration of the vascular tissue. Long sections of infected stems reveal dark brown to black streaking in the vascular tissue as the disease progresses. As invasion proceeds, the pith and cortex of the stem become dark brown. Typical symptoms include wilting, yellowing and some stunting of the plants, which finally die right back. Wilting is first seen as a drooping of the tip of some of the lower leaves similar to that caused by a temporary shortage of water. At first, only one branch in a hill may show wilting. Affected leaves later become permanently wilted and roll upward and inward from the margins. The wilting then extends to leaves further up the stem and is followed by a yellowing of the leaves. This yellowing, wilting and in-rolling of the leaves make diseased plants very obvious, especially when surrounded by healthy plants. The leaves finally turn brown and fall off, beginning at the base of the stem and continuing upward.

Symptoms on the tubers appear as brownish gray areas on the outside, especially near the point of attachment of the stolon. Cut tubers may show pockets of white-to-brown pus or browning of the vascular tissues which, if left standing, may exude dirty white globules of bacteria. As the disease progresses, bubbly globules of bacteria may exude through the eyes; soil will often adhere to the exuded bacteria, hence the name "sore eyes" or "jammy eyes."

Causal Agent

Bacterial wilt or brown rot is caused by *Ralstonia solanacearum* (Smith) Yabuuchi *et al.* (1995) (Syn: *Pseudomonas solanacearum).* The wilt pathogen has a very wide host range. It is a serious problem in many developing countries in the tropical and subtropical regions of the world. It is usually found between the latitudes 45°N and 45°S. The bacteria were first named *Bacillus solanacearum.* It is described as a nonspore-forming, Gram-negative, nitrate-reducing, ammonia forming, aerobic, rod-shaped (0.5–1.5 μm) bacteria with one polar flagellum. Populations within this genus and species can be further divided into races and biovars based on differing host ranges, biochemical properties, susceptibility to bacteriophages and serological reactions has been described in bacterial wilt tomato). However, Race 1 and 3 infect potato plants. Several Race 1 groups of the pathogen cause bacterial wilt of tomato and other crops, such as pepper, tobacco and rarely, potato. In contrast, R3b2 is highly pathogenic on potato and tomato (primary hosts) and can also infect and eventually kill other solanaceous plants (*e.g.* eggplant, nightshade weeds) as well as geraniums. The R3b2 is widely distributed in Asia, Africa and South and Central America and is found in some soils and waterways in Europe. Race 3 biovar 2 is listed as an organism under bioterrorism act in the United States of America.

Disease Cycle and Epidemiology

Infected tubers and plant debris in infested soil are two major primary source of inoculum. Symptom-less plants may harbour the bacterium and transmit it to progeny tubers as latent infection. This could lead to severe disease outbreaks when the tubers are grown at disease free sites. High soil moisture, temperature, oxygen stress and soil type affect the survival of the pathogen. The pathogen population decline gradually in soil devoid of host plants and their debris. The pathogen infects roots of healthy plants through wounds. Nematodes such as *Meloidogyne incognita* which affect potato roots and tubers increase wilt incidence. Transmission of *R. solanacearum* from one area to another occurs through infected seed, irrigation water and farm implements. Inoculum potential of about 10^7cfu/g soil favours infection which however is dependent on other predisposing factors. Mean soil temperature below 15 °C and above 35 °C do not favour disease development. The disease is generally favored by temperatures between 25°C and 35°C. It usually does not cause problems in areas where mean soil temperature is below 15°C. Under conditions of optimum temperature, infection is favored by wetness of soil. However, once infection has occurred, severity of symptoms is increased with hot and dry conditions, which facilitate wilting. The wilt bacterium is able to survive for periods up to 2–3 years in bare fallow soils and for longer periods in soils cropped to non-solanaceous crops.

Management

None of the method is self sufficient to manage this disease, so need based deployment of various components of IDM is recommended.

- ✰ Infected plants must be discarded as soon as possible.
- ✰ Adopt rotations with pastures, cereals and non-solanaceous crops for periods exceeding 5 years.
- ✰ Use certified seeds from reliable sources.
- ✰ Planting in areas where BW has not occurred previously.
- ✰ Control self-sown potatoes.
- ✰ Control weed hosts such as nightshade, thorn apple, Narrawa burr around dames, along channels and in the paddocks after cropping potatoes.
- ✰ Avoid deep plowing–the organisms survive in the deep, cool layers of soil.

- Irrigation water should never be allowed to run freely over or below the soil surface. It should never be allowed to return to the dam or stream from which it is pumped nor to any other irrigation source.
- Regularly inspect crops for disease symptoms and remove and destroy diseased plants, tubers and immediate neighbors.
- Use stock to clean up chats, discarded tubers and crop debris, but do not allow the stock back onto clean paddocks.
- Do not return potato waste, for example, oversized, misshapen and diseased tubers to paddocks, so as to minimize the spread.
- Machinery taken onto a diseased paddock should be left on the paddock while it is being worked.
- Machinery removed from the paddock should then be washed clean with a disinfectant solution in a dedicated area where equipment are washed.
- Use high-pressure wash to clean machinery, sheds and so on to remove soil adhering to any surfaces. Clothing and boots of people working in the paddock should be exchanged for clean items when leaving the paddock, or else boots should be washed in a suitable disinfectant.
- After harvest, all diseased and discarded tubers should be collected and buried at least 1 m underground.
- Choose transport routes that minimize travel through potato paddocks and regions.
- Incidence of bacterial wilt declines by application of bleaching powder @ 12 kg/h mixed with fertilizer or soil drenching after first earthing up together with the use of healthy seed (Gadewar *et al.,* 1991).
- Application of copra and pea nut meal has been reported to reduce the disease.
- Biocontrol of bacterial wilt by use of antagonists such as *P. flourescens, Bacillus* spp., avirulent *R. solanacearum* and actinomycetes have been found to be effective in some countries.

iii. Blackleg Disease

The bacterium causing the blackleg disease of potato is a tuber-borne pathogen. The blackleg disease can cause severe economic losses to the potato crop. However, the occurrence of blackleg depends very much on the growing conditions, particularly temperature and rainfall after planting.

Symptoms

On Foliage

Blackleg disease sometimes develops early in the growing season soon after the plants emerge. This is referred to as an early blackleg and is characterized by stunted, yellowish foliage that has a stiff, upright habit. The lower part of the below ground stem of such plants is dark brown to black in color and extensively decayed. The pith region of the stem is particularly susceptible to decay and in blackleg-infected plants, the decay may extend upward in the stem far beyond the tissue with externally visible symptoms. The typical blackening and decay of the lower stem portion is the origin of the "blackleg" designation for this disease. Young plants affected by blackleg fail to develop further and typically die. In more mature plants, blackleg appears as a black discoloration of previously healthy stems, accompanied by a rapid wilting and sometimes yellowing, of the leaves. Black discoloration of the stems always starts below ground and moves up the stem, often until the entire stem is black and wilted.

On Tubers

There are two ways by which the blackleg bacterium may reach the progeny tubers produced on the potato plant. One important route of tuber infection is via the stolon by which the tuber is attached to the plant. Tubers with blackleg disease generally first become decayed at the stolon attachment site where the tuber tissue becomes blackened and soft. As the disease progresses, the entire tuber may decay or the rot may remain partially restricted to the inner perimedullary (or parenchymal) tissue, that is, the tissue inside the vascular ring. An alternate route for the pathogen to attack progeny tubers is via the soil. As the blackleg disease causes the belowground stem and seed tuber to decay, the causal bacterium spreads from infected tissue into soil water and becomes distributed throughout the root zone in which the progeny tubers are growing. In a poorly managed potato storage environment, blackleg bacteria present on the surface of tubers can cause extensive decay. Sometimes when storage conditions are improved, decay lesions around tuber lenticels or mechanically damaged areas become arrested, resulting in a condition known as "hard rot." Hard rot is typified by slightly sunken, brownish-black, dry, necrotic lesions surrounding individual lenticels or damaged areas. Once decay of potato tubers is incited by the blackleg bacterium, growth of secondary bacteria often contributes to the decay process and certainly modifies symptomatology of the disease. Hence, a general bacterial soft rot develops from the initial blackleg infection in tubers. Bacterial soft rot is characterized by total maceration of tuber tissue and seepage of a putrid, dark-colored liquid

Causal Agent

The causal agent of blackleg is *Pectobacterium atrosepticum* (van Hall) Gardan *et al.* (2003) (Syn: *Erwinia carotovora* subsp. *atroseptica).* In another study, *P. atrosepticum, P. carotovorum* subsp. *brasiliense, P. carotovorum* subsp. *carotovorum* and *P. wasabiae* all were detected from blackleg-diseased potato plants in Canada (De Boer *et al.,* 2012). *Dickeya solani* has been isolated from black leg infected potato in the Netherlands and Israel (Toth *et al.,* 2011). It is a Gram-negative, rod-shaped bacterium. The most distinguishing feature of the blackleg bacterium is its pectolytic enzyme activity, but in contrast to many of the other pectolytic bacteria, it does not grow above 36°C. It is facultatively anaerobic, meaning that it can grow both with and without the presence of oxygen. It is motile with numerous peritrichous flagella. Most strains of *E. carotovora* subsp. *atroseptica* belong to serogroup I, although several other serogroups occur.

Disease Cycle and Epidemiology

The blackleg bacterium survives poorly in soil. Although other members of the pectolyic *Erwinia* survive in surface water and in the soil environment, all evidence suggests that the blackleg bacterium does not survive very well outside of association with host plant tissue. Hence, the seed tuber is the most important source of inoculum in the blackleg disease cycle. When a contaminated or infected seed potato is planted, one of three things may occur. Bacterial cells enter lenticels of the progeny tubers and either becomes inactive, or, when conditions are favorable, initiate decay.

Management

- ☆ Healthy seed tubers should be used for sowing the crop.
- ☆ Planting limited generation seed in well-drained soil after soil temperature has increased above 10°C is recommended for avoiding the development of blackleg.
- ☆ Roguing out blackleg-diseased plants including belowground portions reduces soil inoculum. It is required to avoid the contact of diseased tissue with other plants in the field.
- ☆ It is also necessary to avoid injury to potato tubers during harvest to minimize decay in storage.
- ☆ Removal of decayed potatoes before they spread their contents over grading lines and bin pilers avoids spreading the bacterium to other tubers.
- ☆ Wound healing is important in the early phase of potato storage to prevent development and spread of rots.
- ☆ During storage, however, the potatoes should be kept at a low temperature with adequate aeration to provide a dry environment and to prevent condensation of moisture on tuber surfaces.

iv. Bacterial Ring Rot

The disease is called "bacterial ring rot" because the rot appears in the vascular ring of the potato tuber. The disease was first time described by Spiekermann and Kotthoff in 1914 from Germany. It is an EPPO A2 quarantine pest. The disease has been reported from Austria, Belgium, Canada, China, Czech Republic, Denmark, Estonia, Finland, Greece, Germany, Japan, Kazakhstan, Korea DPR, Korea Republic, Latvia, The Netherlands, Norway, Poland, Russia, Sweden, Taiwan, Ukrain, USA and Uzbekistan. The disease was also occurred in France but it was eradicated. The disease causes damage particularly in temperate area where optimum temperature is 21°C. The damage of the crop was up to 50 per cent when the seed tuber was cut and used picker type planting machines. Natural infections of the pathogen have been found on potatoes only. In inoculation tests, many members of the *Solanaceae*, including tomatoes and aubergines are found to be susceptible.

Symptoms

In severely affected tubers, the vascular ring is brown to black in color, often with cheesy or creamy ooze and many hollow spaces where the flesh has disintegrated. Dry cracks can usually be found on the surface of the tubers. In milder cases, the vascular ring may show only broken, black lines or a yellowish discoloration. Leaves of infected plants may show interveinal yellowing, wilting, or no symptoms.

Causal Agent

The disease is caused by *Clavibacter michiganensis* subsp. *sepedonicus* (Spiekermann and Kotthoff, 1914) Davis *et al.*, 1984 (Synonyms: *Corynebacterium michiganensis* subsp. *sepedonicum* (Spieckermann and Kotthof, 1914) Carlson and Vidaver 1982; *Corynebacterium michiganensis* pv. *sepedonicum* (Spieckermann and Kotthof, 1914) Dye and Kemp 1977; *Corynebacterium sepedonicum* (Spieckermann and Kotthof, 1914) Skaptason and Burkholder 1942. *C. michiganensis* subsp. *sepedonicus* is a typical, small Gram-positive (occasionally giving a weak or negative reaction) coryneform bacterium. Cells are club shaped, mostly single, often in pairs described as 'elbows' because they are in a 'V' formation and occasionally in small irregular shaped groups. The pathogenic bacteria is negative in oxidase test, nitrate reduction, urease activity, H_2S production, indole production,citrate utilization,starch hydrolysis,growth at 37°C, growth in 7 per cent NaCl and gelatin hydrolysis and positive in catalase test.

Disease Cycle and Epidemiology

Once a crop or farm is infested with the bacterium, the disease will carry over from year to year and spread quickly. The bacteria can survive for 2–5 years in dried slime on the surface of machinery, crates, bins, or burlap sacking, even if frozen. After a diseased potato is planted, the bacteria multiply very rapidly and follow the vascular tissue into the stems and petioles. The bacteria reach the roots and maturing daughter tubers, sometimes within 8 weeks after planting. The bacteria do not overwinter in field soil but persist for years in dried slime on potato bags, boxes, machinery and other equipment. Volunteer potato plants and plant debris, including infected cull tubers also carry the overwintering bacteria. The bacteria can be spread in rain and irrigation water and by insects, but wounds are needed for infection. The most important means of infection is cutting seed potatoes with contaminated knives. The optimum temperature for the bacteria is rather low (21°C) which results in favorable climate in north, north-west and central Europe.

Management

- ☆ Prevention: Plant only certified seed potatoes.
- ☆ Avoid using table stock for seed or importing seed from other disease-prone areas for planting.
- ☆ Disinfect knives frequently while cutting seed pieces to avoid spreading any disease that might be present but unnoticed.
- ☆ Bury cull piles and control volunteer potatoes.
- ☆ Clean and disinfect storage bins between crops and pressure-wash equipment to avoid spreading ring rot.

v. Bacterial Soft Rot

Apart from potato soft rot bacteria cause rotting in Brassica, capsicum, carrot, edible aroids and other root crops, onion, tomato and many more as well as ornamentals.

Symptoms

Soft rot on tubers first appears as small, tannish, water-soaked spots on the surface. These spots rapidly enlarge and the tissue decomposes in a soft, blister-like area on the surface of the tuber. Often, a slimy or watery substance oozes from breaks in the blister. The blister margin is darker than the tuber skin. Soft rot often follows bruising and is first white to cream colored. After exposure to air, it becomes brown to black. The boundary between the disintegrated and the sound tissue is sharp. It is nearly odorless at the stage. As secondary rot occurs, the rot becomes very foul smelling. The rot typically progresses to the point of a chalky white, foul-smelling mass Soft rot symptoms on the foliage include weak, chlorotic (yellowed) plants with margins of leaflets curled upward. Stem lesions are usually light brown and can be colorless, but not black. Stems will rot and become very mushy. Tuber rot will occur as point of infections often on an eye but can be generalized on the tuber.

Causal Agent

Pectinolytic *Erwinia* spp. was reclassified into two species: *E. carotovorum* and *E. chrysanthemi* and now they have been incorporated into *Pectobacterium* and *Dickeya* genera, respectively (Hauben *et al.*, 1998; Samson *et al.*, 2005). *Pectobacterium carotovorum* subsp. *carotovorum, Dickeya chrysanthemi* (Syn: *Pectobacterium chrysanthemi), Pectobacterium atrosepticum* and *Dickeya* species. *Dickeya dianthicola* was detected in Australia for the first time in June 2017 in a Western Australian potato crop. *P. carotovorum* subsp. *carotovorum* is a facultatively anaerobic, peritrichously flagellated, gram-negative, rod-shaped bacterium. It is catalase positive, oxidase negative and can ferment glucose, but does not produce acid from sorbitol. This pathogen has pectolytic activity and produces deep pits on crystal violet pectate (CVP) medium. It should be emphasized that tubers also commonly harbour other pectolytic saprophytic bacteria (*Bacillus* spp., *Clostridium* spp., *Flavobacterium* spp. and *Pseudomonas* spp.) which, if given the opportunity, can also cause rotting.

Disease Cycle and Epidemiology

Potato soft rot bacterium is a common soil resident. The bacteria that cause soft rot can remain in both potato plants and tubers without any obvious symptoms. Bacteria enter lenticels, growth cracks, or any injury. The main cause of spread is by wounds or damage to the potato. These usually occur during harvesting and grading, allowing the bacteria to invade the tuber. When this is combined with water on the surface of the tuber, the bacteria can defeat the tuber's natural defences and start the tuber rot. Other factors include anaerobic conditions, enlarged lenticels and invasion by other pathogens. Infections in the field are favoured by high soil moisture and high temperatures. During and after harvest, soft rot is favored by immature tubers, adverse temperatures (pulp temperatures above 21°C at harvest), mechanical damage and free water on tuber surfaces. This bacterium can grow between the temperatures of 0°C and 32°C, with optimal growth between 21°C and 27°C. Tubers harvested at temperatures above 27°C can be predisposed to soft rot. High temperatures create ideal conditions as oxygen in the tuber is rapidly replaced by high levels of carbon dioxide, causing stress on the tuber. This will lead to soft rot infection. Exposure to sunlight also aids soft rot development by killing tuber cells. The presence of other diseases will lead to increased infection. Decay can be retarded by temperatures less than 10°C; the lower the temperature, the better. Immature tubers are susceptible to harvester related injury and bacterial infection. Suberizing seed and treatment with fungicide is a tactic to reduce the risk of other seed infections that could lead to soft rot breakdown of the seed.

Management

- Other diseases that produce tuber lesions need to be controlled.
- Delaying harvest until the skin has set reduces tuber injuries which reduce the entry points for the pathogen.
- Ensure good sanitary procedures to reduce spread of bacteria.
- Harvesting during wet, muddy conditions generally leads to an increase in bacterial soft rot in storage.
- Properly suberize potatoes to ensure wound healing and reduce the infection sites for the pathogen.
- Leave potatoes a minimum of 7 days after the vines are totally dead to encourage skin set and reduce bruising.
- Eliminate as much soil as possible before the tubers are stored, as soil can restrict air movement.

- ☆ Minimizing potato bruising, avoiding harvesting during wet conditions and placing the potatoes into a disinfected storage are three easy, cheap and effective control practices to reduce loss in storage from soft rot.
- ☆ Avoid washing seed or potatoes before storage unless absolutely necessary.
- ☆ Potatoes that have soft rot should be removed before storing and during packing.
- ☆ After washing, potatoes should be dry before shipping.
- ☆ Use high-quality seed.
- ☆ Harvest mature tubers with well-set skins and avoid mechanical injury.
- ☆ Avoid excessive soil moisture before harvest to reduce lenticel infection.
- ☆ Use clean water to wash potatoes and avoid water films on tuber surfaces during storage. Postharvest curing and storage temperatures can be critical components of soft rot management.

4. Lady's Finger/Bhendi/Okra (*Abelmoschus esculentus* L.)

Bacterial Leaf Spot

Bacterial leaf spot disease damaging winter crops of okra caused by *Xanthomonas campestris* pv. *esculenti.* The yield of the plant is reduced.

Symptoms

The disease is chiefly restricted to the leaf blade. Minute, water soaked lesions are seen on both the surfaces of leaf but more commonly on the under surface. The spots develop into angular or irregular spots, brown in colour and 2 to 5 mm in dia. The margin of the spot are dark brown. The bacterial ooze exudes from the spot and dries up leaving on yellow encrustation on the surface. The bacterium affects veins and vein lets and change them to yellow or arrange during rainy seasons. Severely infected plants become stunted.

Causal Agent

The disease is caused by *Xanthomonas campestris* pv. *esculenti* (Rangaswami and Easwaran) Dye. The bacterium is Gram negative, short rod and 2.1 - 0 8.9 X 0.7 - 1.4 μm and motile with single polar flagellum.

Management

- ☆ The disease can be controlled by proper application of fertilizers and proper plant spacing, use of resistant varieties and hot water treated seeds.
- ☆ Practicing of field sanitation is important to control the disease.
- ☆ Volunteer seedlings left after harvest can be destroyed to minimize the initial inoculum at the beginning of the season.

- ✰ Providing good drainage system can also manage this disease.
- ✰ Spray fresh cowdung water extract or lemon grass or mint extract at 20 per cent.
- ✰ Spraying streptomycin sulphate and tetracycline combination 500 ppm.
- ✰ Spraying of copper oxychloride @0.1 per cent is recommended for managing the disease.
- ✰ Cu and carbonate fungicide gave the best control against the leaf spot disease.
- ✰ immonomodulators like salicylic acid (SA), jasmonic acid (JA), DL-β-aminobutyric acid (BABA), potassium salts, 2,6-dichloroisonicotinic acid (INA), acibenzolar-S-methyl, can be recommended.

5. Pea (*Pisum Sativum*)

i. Soft Rot

A bacterial soft rot of green pea pods has been noticed for the first time in 1949 in the market of Delhi (Hingorani, 1951).

Symptoms

Pods show yellow discolouration while the seeds within are transformed into a mushy, maladorus bacterial mass.

Causal Agent

Bacterium is described as *Erwinia* sp. Bacterium infects potato, carrot, cucumber, tomato, chilli, radish, beetroot, egg plant, cauliflower and other vegetables under artificial inoculation.

ii. Bacterial Blight

The disease was reported in Colorado in 1915 by W.G. Sackett (Sackett, 1916). It occurs in countries where peas are grown intensively and humidity in high. These areas include Australia, England, Greece, Hungary, New Zealand and the United states. The bacterial blight can be sporadic and severe particularly during seasons with frost damage, high hymidity and wet conditions or when overhead sprinkler irrigation is used.

Symptoms

Disease infects all the aerial parts of plant. Spots on leaf lets are round, oval or irregular in shape, 2 to 5 mm in dia and reddish brown in color with translucent centre and a darker brown margin. Several such spots may coalesce to produce a blighted appearance. Chocolate brown linear streaks are observed on the petiole and stem. Later on, the whole stem turn chocolate brown and gets shriveled. Sometimes, severely infected plants are killed. Tender pods are chocolate brown in color, thin, twisted and shriveled. Lesions are large on older pods and they become thin, twisted and dry. Seeds are discoloured and shriveled. Dried bacterial ooze makes pod surface glossy.

Causal Agent

The disease caused by *Pseudomonas syringae* pv. *pisi* (Sackett) Young *et al.* They are gram negative, straight or curved rods, 0.5-1.0 x 1.5-4.0 mm in size, nonspore forming aerobic and produce fluorescent on KB medium. Cells are motile by are one to several flagella and to 6 + C content of the DNA, catalase and levan production positive and negative in oxidase test, potato soft rot, agrinine dehydrolase. Bacterial pathogen causes hypersensiti reaction in tobacco. At present, eight races of *P. syringae* pv. *pisi* have been described on the basis of pathogenicity to differential cultivars.

Disease cycle and Epidemiology

Bacterium over winters and are disseminated in or on plant debris and seed. From infected seed, the bacteria move in to young seedlings. Later, infections usually occur through leaf stomata. The bacterium can be carried through irrigation water and splashing rain. Mechanical injury can predispose a pea crop to bacterial blight infection through the creation of wounds due to hail, sandstorms, wind, rain or cultural practices in plants through which bacterium enters. It attacks cowpea and hyacinth bean, soybean, sweet pea, red clover hairy vetch, purple vetch. High relative humidity and free water facilitate disease development. The optimum temperature for bacterial growth is 28°C.

Management

- ☆ Clean seed offers the best control.
- ☆ Slurry seed treatment with streptomycin is effective.
- ☆ Good soil drainage, wide spacing, removal of weeds, surface rather than aerial irrigation and rotations of three years or longer helps to control bacterial blight.
- ☆ Late varieties generally are less susceptible.

6. Beans/French Beans (*Phaseolus vulgaris*)

i. Common Bacterial Blight and Fuscous Slight

Common blight was first reported by Smith in 1897 from the USA. In India, bacterial blight of French bean was reported from Pune during 1946. Later on, occurrence of common blight is reported from Delhi, Himachal Pradesh, Maharashtra and Uttar Pradesh. The common bacterial blight is one of the major diseases of bean and causes yield loss about 20- 40 per cent and some time, loss in yield rises upto 83.8 per cent. It is an EPP) A2 quarantine pest.

Besides *Phaesolus vulgaris* L. (French bean, principal host), bacterial pathogen infects other legumes like tepary bean (*P. acutifolius*) Jacq, soyabean (*Glycine max*

L.), *Dolichos lablab* L., *Lupinus polyphallus* Lindl., *Stizolobium deeringianum* Bort, *Vigna angularis* (Willd) Ohwi and H. Ohashi and cowpea (*Vigna unguiculata*) (L) Wilcz (Hall). It also infects *Vigna aconitifolia* (Jacq.) Marechal., *Vigna mungo* (L) Hepper and *Vigna radiata* (L.) Wilcz.

Symptoms

Disease is distinguished by irregular, sunken, reddish to brownish spots which later become brittle and surrounded with yellowish halo. Sever infection may cause defoliation. Lesions may also develop along the main vein, on the stems and pods causing brown canker. In advance cases, cracks are formed on the stems and petioles, infected seeds are discoloured and shriveled. When pods are less severely affected, seeds develop normally and show no sign of disease or become slightly wrinkled and especially on white seed, take on a polished yellowish discolouration. This yellowing may be in small to large blotches or follow the veins of the seed coat.

Causal Agent

The disease is caused by *Xanthomonas phaseoli* pv. *phaseoli* (Smith) Constantin *et al.*, 2016 [Syn. *Xanthomonas axonopodis* pv. *phaseoli* (Smith) Vouterin *et al.* 1995 and *Xanthomonas campestris* pv. *phaseoli* (Smith) Dye)] and *X. citri* pv. *fuscans* (= *X. fuscans* subsp. *fuscans*) (produce a brown, diffusible pigment). The colonies are yellow, mucoid, glistening and convex with entire margins on nutrient agar medium while on Yeast Dextrose Agar, the colonies are yellow, mucoid, convex and shining. On MXP, they are yellow mucoid, smooth, convex and surrounded by zones of starch hydrolysis. Bacterium is a Gram-negative rod, 1.9 X 0.8 µm in size, non-capsulated, strictly aerobic bacterium which does not reduce nitrates and motility by single polar flagellum. The bacterium is relatively intolerant to triphenyl tetrazolium chloride (TTC) and is inhibited by 0.02 per cent TTC. Eight races have been reported based on reaction in *P. acutifolius* (Opio and Male-Kigiwa, 1995).

Disease cycle and Epidemiology

The pathogen is seed borne and the disease spread through wind splashed rains from diseased to healthy plants. Disease spreads through infected seeds. Debris from diseased plants is considered a possible source for seasonal carryover of plant pathogenic bacteria and it can survive in dry leaves under laboratory conditions for at least six years (Gilbertson *et al.,* 1988). The pathogen is also dispersed by insects such as whitefly, root weevil, bean leaf beetle and Mexican bean beetle. Seed infect can occur internally (including the Embrio) and externally. Infection is through stomata and wounds. The bacteria prefer warm temperatures. Disease severity was more in the crop sown during May month. The bacterial pathogen is a warm temperature bacterium and causes greater damage to plants at 28-32°C than at temperature lower than 16°C. High temperature, rainfall and humidity favour rapid disease progress in the field. The pathogen is spread by windblown rain, soil and plant debris, contact between wet plant leaves, irrigation water, animals and insects like leaf miners and whiteflies.

Management

- ☆ Plant debris should be ploughed and buried.
- ☆ Two or three crop rotation should be followed.
- ☆ Equipments or people should not be moved from field to field without a clean-up between fields.
- ☆ *Phaseolus coccineus* and *P. acutifolius* show resistance against the disease. Biser, Green plentiful, Goliath waxy, Gribeck, Jule, Medal Rufugee, Ruse 6, Tara, Hawassa dumme and valley are resistant to common blight.
- ☆ Field operations should be avoided in bean fields when the plants are wet. Suspected seeds should be treated in streptomycin sulphate solution (500 ppm).
- ☆ Copper bactericides will reduce epiphytic populations of bacterial pathogens on bean foliage and also reduce disease severity when applied as a preventive.
- ☆ Seed treatment with streptomycin and foliar sprays of kocide 101 reduced severity significantly (Hailu and Tontosha, 2021). Seed can also be dressed with chemicals like quintozene to kill contaminant bacteria.
- ☆ Seed treatment with hot water at 52°C for 20 min or hot air treatment 60°C dry air at 45-55 per cent RH for 23-32 h.
- ☆ Immonomodulators like salicylic acid (SA), jasmonic acid (JA), DL-β-aminobutyric acid (BABA), potassium salts, 2,6-dichloroisonicotinic acid (INA), acibenzolar-S-methyl, can be recommended.
- ☆ Mixed applications with *Bacillus subtilis* str. QST 713 and copper hydroxide (CH) proven to be very effective against the disease and reduction in severity was upto 51 per cent (Belete *et al.*, 2021)

ii. Halo Blight

The disease is also known as grease spot. The halo blight was describe from USA by Burkholder in 1926. It has been reported world wide where moderate or cool temperatures prevail. Under these conditions epidemics substantial yield loss as high as 43 per cent may occur depending on percentage of infection and environmental factors. Pod infections reduce marketability of snap beans and seed production of dry beans. The main host of bacterial pathogen is French bean, snap beanstring bean and frizoles (*Phaseolus vulgaris*). The pathogen also infects adzuki bean, lima bean, mung bean, scarlet runner bean, soybean, tepary bean, *Macroptilium atropurpureum*, *P. polyanthus*, *P. polystachyus* and *Pueraria lobata*.

Symptoms

Initially, leaves appear as greasy, water-soaked spots about 1/16 inch in dia meter they are generally most visible on the under side of young leaf lets approximately one week after infection. One or more water - soaked spots are later surrounded by a greenish - yellow halo about 1/16 to ½ inch in dia meter. This halo is caused by a bacterial toxin. The size and development of this halo is different and may not be produced during high temperatures. During acute infection the disease may become systemic and cause yellowing and death of new leaves. Halo blight symptoms on pods appear as small circular, water - soaked spots or streaks. A light cream or silver coloured bacterial ooze is often associated with these spots. Pod infection causes discolouration, shriveling and bacterial contamination of seeds; however, some seed may appear to be healthy.

Causal Agent

Pseudomonas savastanoi pv. *phaseolicola* (Burkholder) Gardan *et al.,* [Syn. *Pseudomonas syringae* pv. *phaseolicola* (Burkholder) (Young *et al.*). On standard media, bacterial colonies are white to cream-colored. They are gram negative, straight or curved rods, 0.5-1.0 x 1.5-4.0 mm in size. Cells are motile by are one to several flagella and catalase positive and strict aerobes except for a few which denitrify. Maximum growth and production of the bacterial-induced toxin, phaseolotoxin is 21.1-22.2°C. Phaseolotoxin is responsible for systemic chlorosis. Nine races of *P. syringae* pv. *phaseolicola* have been identified and race 6 is predominant (Taylor *et al.,* 1996).

Disease Cycle and Epidemiology

The principal sources of halo blight bacteria are internally infected bean seeds, seed contaminated externally with the bacteria, bean plants infected in the field and debris from the last seasons infected bean plants. Bean seed is the most important inoculum source because it not only carries the bacteria from one region to another, but also it can over winter from one season to the next. Seed contamination is more prevalent in susceptible than in resistant varieties. Seedlings infected by seed borne bacteria serve as the source of inoculum responsible for further disease development in a field. Infected plants in the field are the major sources of inoculum in the local spread of the bacteria. Susceptible varieties appear to have a much greater potential to serve as a secondary inoculum source than do resistant varieties. Infection takes through wounds and stomata. The pathogen spread by infected seed and rain splash. Cool moist conditions favour spread of the bacteria. Sufficient free moisture from plant exudates and condensation of moisture in the microclimate of the plants facilitate spread.

Management

- Exclusion of beans from the rotation for at least two years is a good practice to keep initial inoculum low.
- Implement movement in bean fields should be delayed until free moisture has drifted from foliage.
- Fields to be planted with certified seed should be planted before others.

- ☆ Seeds should be cleaned thoroughly after threshing and later treated with streptomycin to remove surface organism. This treatment does not control bacteria inside the seed.
- ☆ Blue lake, California small while 59, clarendon wonder FM- 1, FM-51, GN Nebraska No. 1 selection 27, GN No. 16, Hawkes bury wonder, OSU-10183, PI-150414, PI-181954, PI-203958, Red Mexican, Richmond wonder, U.I-3, U.I-34, U.I-35 and Windsor long pod are resistant to halo blight. Red kloud and Red kote have good tolerance to halo blight.
- ☆ SaniDate and OxiDate are the antibacterial chemicals and also yield enhancers can be recommended (Harveson, 2019).
- ☆ Immunomodulators like DL-β-aminobutyric acid (BABA), potassium salts, 2,6-dichloroisonicotinic acid (INA), acibenzolar-S-methyl can be deployed to manage the disease.
- ☆ Spraying of bean plants with phage (phage 1, phage 2 and phage 1+2) before inoculation with bacterial pathogen showed significant reduction of disease severity either in individual or mixing treatment.
- ☆ Treating the plants with *Pseudomonas fluorescens* and *P. putida* before occurrence of the disease reduces disease index.

7. Cluster Bean (*Cyamopsis tetragonoloba*)

Bacterial Leaf Blight

In India, it was first observed at Patna, Bihar and Khopali, Maharashtra in 1953. It occurs in Bihar, Delhi, Haryana, Punjab, Rajasthan and Uttar Pradesh. It severely suffers from bacterial leaf blight causes 50–70 per cent yield loss under severe conditions, if appropriate control measures to combat this disease are not taken timely.

Symptoms

Small circular water soaked spots are formed on leaflets. They are oily and olive coloured on lower surface. Fully developed spots become jet black in color. They coalesce to form necrotic patches surrounded by chlorotic holo. The infection from blighted leaves spreads to stem through petiole. Diseased stems get cracked. The infected pods show heavy spotting.

Causal Agent

The disease is caused by *Xanthomonas axonopodis* pv. *cyamopsidis* (Patel *et al.*) Vauterin *et al.*, 1995 [Syn. *Xanthomonas campestris* pv. *cyamopsidis* (Patel *et al.*) Dye]. Bacterium is Gram- negative rod occurring singly or in pairs, 1.5 X 0.8µm, capsulated and motile by single polar flagellum. The bacterium is aerobic and positive for H_2S production, Esculin hydrolysis, oxidase test, Tobacco HR, indole production from tryptophan, nitrate reduction to nitrite and utilization of glucose, mannose, Trehalose, galactose and starch. Pathogen produces polygalacturonate transeliminase, macerating and cellulotytic enzyme.

Disease Cycle and Epidemiology

The bacterium is a seed-borne and it spreads to new area by infected seed. Disease increases with increase in amount of rainfall and number of rainy days and with an increase in the nitrogen level from 20 to 40 kg/ha. It is destructive to rainy season crop. Younger plants are more susceptible.

Management

- ✰ Seed treatment with hot water at 56°C for 10 min. eradicates seed-borne infection.
- ✰ Disease free seeds from summer grown irrigated crops should be used.
- ✰ Seed treatment with streptocycline 500 ppm + 2 sprays of copper hydroxide (0.2 per cent) at 15 days interval recorded the maximum yield and minimum disease intensity (Bagri *et al.,* 2019).
- ✰ Spraying with streptocycline (100 to 250 ppm) and agrimycin- 100(100 to 500 ppm) controls the disease.
- ✰ Streptocycline (100 ppm) followed by agrimycin (100 ppm) is also effective.
- ✰ Cultivars G 85, 102, 126 and 255 are reported as highly resistant varieties where as H.G 75, HG 182, HFG 119, HFG 340, G 28, 89, 98, 100, 119, 127, 139, 158, 166 and 172 are moderately resistant. Glabrous hairy grain type varieties are comparatively less susceptible.

8. Cowpea (*Vigna unguiculata*)

Bacterial Blight

TheDisease was first reported by Burkholder from the USA in 1944 and from India in 1950 by Patel and Diwan, 1950). The disease is also called cowpea canker, and bacterial pustule. Presently, the disease is prevalent in most of the cowpea growing areas in India. Loss due to disease is reported upto 90.24 per cent (Kishun, 1989). Cowpea is the main host however, hyacinth bean, Sudanese weed and kidney bean are laso infected by the bacterial pathogen.

Symptoms

Infected cotyledons become red in color and shriveled. In case of acute infection, primary leaves, cotyledons and growing points are killed in the initial stage. Occasionally wilting of plants is also observed, on the leaves, disease appears as light yellow in color, irregular to round spots on the lower surface which in 2-3 days get enlarged into reddish brown in color, angular to circular in shape, necrotic spots with chlorotic halo. Severely affected leaves appear blighted, turn brown and wither off. Dark green water- soaked spots 1.5-2 mm in diameter with varying number and shape are formed on pods. Cracks are found on stem and joint of pod shells.

Causal Agent

The disease is caused by *Xanthomonas citri* pv. *vignicola* (Burkholder) Constratin [Syn: *Xanthomonas axonopodis* pv. *vignicola* (Burkholder) Vauterin *et al.*, 1995; *Xanthomonas campestris* pv. *vignicola* (Burkholder) Dye]. The bacterium is a Gram-negative short rod in shape, occurring singly, rarely in pairs, measuring 1.2-2.4 X 0.6-1.1µm, capsulated, non spore forming and motile with single polar flagellum.

Disease Cycle and Epidemiology

Pathogen is seed-borne in nature. It can Bacterium survive in seed upto 250 days when stored at room temperature and volunteer plants in the field, the bacterium is dispersed by splash water and wind driven rain. Infection is through stomata and wounds at higher temperature *i.e.* 28- 32°C and high humidity. Disease spreads upto 62 per cent plants with an initial inoculum load of 1 per cent infested seed at 20-80 per cent, relative humidity and temperature range of 25-34°C. Growth regulator, (Rao and Hiremath, 1985) and intermittent rain have been found favourable for bacterial blight development. Kishun *et al.* (1987) have studied the spread of disease in relation to time and space in the plots having different levels of primary inoculum load. Cowpea plants are found most susceptible upto the age of 90 days (Kishun 1989).

Management

- ☆ Use healthy seeds, Use partial or resistant varities, 3 years crop rotation and intercropping are important to control the disease.
- ☆ Application of potassium and molybedenum as host nutrition reduced the bacterial blight disease incidence.
- ☆ Calcium and magnesium reduces the disease at higher doses.
- ☆ Disease can be managed upto some extent by manipulation in date of planting (Kishun and Chand 1987, 1989).
- ☆ Spraying with bavistin (1000 ppm) + agrimycin - 100 (250 ppm followed agrimycin – 100 (250 ppm) are effective.

9. Soybean

Bacterial Pustule

The disease was first reported from Japan by Nakano in 1919. The bacterial pathogen can also infect snap beans. Because this pathogen produces a pustule, it can easily be confused with soybean rust, a much more serious disease. Bacterial pustules will differ from those of soybean rust, however, because they will not have a natural opening in the pustule, or masses of spores like those of soybean rust.

Causal Agent

Bacterial pustule is a soybean disease caused by the bacterium *Xanthomonas citri* pv. *glycines* (Nakano 1919) Constantin *et al.*, 2016 (Syn.: *Xanthomonas axonopodis* pv. *glycines* (Nakano, 1919) Vauterin *et al.*, 1995; *Xanthomonas campestris* pv. *glycines* (Nakano, 1919) Dye; *X. glycines, X. phaseoli* f. sp. *sojense).* The bacterium was gram negative, rod shaped with round ends. The cells appeared as single, some in pairs and in clusters having single polar flagellum. The cells measured 0.4 -0.25 x 1.25- 3.00 µm in size. The biochemical features of the pathogen revealed that the bacterium liquified the gelatin, produced the H_2S, positive for starch hydrolysis and acid produced by carbon sources (sucrose,glucose and dextrose) but it was negative for acid production by nitrogen (peptone) sources. Circular, flattened, slightly raised, convex with yellow to bright yellow colour on SX and NA medium. Optimum temperature for the growth of pathogen is 30° C. pH of 7.0 is optimum for growth (Lambani and Jahagirdar, 2017).

Disease Cycle and Epidemiology

The bacteria overwinter on surface crop residue and on seeds. The bacterial cells are spread during the season by wind-blown rain, rain splashing up from old crop residue, mechanically during field work when the wet canopy and sometimes insects (*Carpocoris fuscispinus* in Russia. The weed hosts *Brunnichia cirrhosa* (red vine) in the USA and *Macrotyloma uniflorum* (*Dolicus biflores*, horse gram)in India, soybean volunteers and wheat roots can serve as a reservoir in the field. The bacteria enter the plant through natural openings (Stomata) and wounds in the leaves. Bacterial pustule develops best during very warm temperatures of 29.4-32.2° C, if conditions remain wet.

Management

- There are high levels of resistance to bacterial pustule in most soybean varieties.
- Since the disease can be seed borne, it is important to use disease-free and certified seed.
- If a field develops bacteria pustule, it should be rotated out of soybeans until all infested debris has decomposed. Tillage can shorten the length of rotation where soil erosion is not an issue.
- Avoid moving equipment through fields or scouting when the foliage is wet.
- Among chemicals, Bordeaux mixture at 1 per cent recorded maximum inhibitory zone.
- Seed treatment with carboxin can reduce disease incidence.
- Copper oxychloride (0.3 per cent) + streptocycline (750 ppm) is found very effective and bacterial pathogen under *in vitro* conditions.

- *Pseudomonas fluorescens* showed inhibition zone against *X. axonopodis* pv. *glycines,* followed by *Bacillus amyloliquefaciens* under *in vitro* conditions (Lambani *et al.,* 2017).
- Early spraying of with insecticides to prevent the disease by reducing the vector population.

10. Cucurbits

i. Bacterial Leaf spot

The disease was first reported in the USA in 1926, in India disease was observed on cucumber (*Cucumis sativus* L.) plants in Pusa (Bihar) during July 1930 by Prasad in 1931 and later from Sabour, Bihar by Sinha in 1989.

Symptoms

Minute water-soaked lesions appear on the lower surface with corresponding yellowing on upper surface of leave. Later on spots enlarge, become brown, surrounded with halo, but remain delimited by the veins. Spots may coalesce to form large dry patches. Sometimes linear lesions appear on tender stems and petioles.

Causal Agent

The disease is caused by *Xanthomonas cucurbitae* (ex Bryan 1926) Vauterin *et al.,* 1995 (Syn: *Xanthomonas campestris* pv. *cucurbitae* (Brayon) Dye).Yellow-pigmented, *Xanthomonas*-like colonies were observed after 48-h incubation at 28°C. The bacterium is a Gram-negative rods, measuring 0.75-1.5 X 0.5 -0.7µm, single or in pairs and motility by single polar flagellum. The bacteriaare, oxidase-negative, indole-negative, hydrolyzed starch and esculin and formed pits on crystal violet pectate and carboxymethyl cellulose media (Dutta *et al.,* 2013). The bacterial isolates do not produce nitrites from nitrates but produced hypersensitive reactions on tobacco. Pathogen attacks *cucurbita moxima, C. pepo, C. moschata* and *Citrullus lanatus* (Rangaswami and Rajgopalan, 1973) and watermelon (Dutta *et al.,* 2013). Temperature at 28°C and pH range 6.0 - 0 7.0 is optimum for the growth of the pathogen.

Disease Cycle and Epidemiology

The pathogen is seed-borne. Disease development is favored by wetness and high temperatures range of 25-30° C.

Management

- Removal and destruction of crop debris.
- Seed treatment with streptomycin sulphate + copper oxychloride can supress the disease.
- *C. okeechobeensis* subsp. *martinezii* showed greater degree of resistance *X. campestris* pv. *cucurbitae* which could use in resistance breeding programme (Sulley *et al.,* 2021).

- ✰ Extract of *Allium sativum*, *Gingiber officinale* are found to inhibit *X. campestris* pv. *cucurbitae* significantly (Lia *et al.*, 2018).
- ✰ Hot-water treatment of seed at 55°C for 15 min eradicates *X. cucurbitae* from the seeds without any significant adverse effect on either seed germination or seedling vigor (Babadoost *et al.*, 2018).
- ✰ Two strains of *Bacillus subtilis* (UMAF6614 and UMAF6639) showed high antibacterial activity against *X. campestris* pv. *cucurbitae*, under *in vitro* conditions and in melon leaves (Zeriouh *et al.*, 2011).
- ✰ Immunomodulators like DL-β-aminobutyric acid (BABA), potassium salts, 2,6-dichloroisonicotinic acid (INA), acibenzolar-S-methyl can be deployed to manage the disease.
- ✰ Spray of 300ppm antibiotic with 0.3 per cent copper oxychloride could be effective.

ii. Angular Leaf Spot

The disease was fired time reported by E. F. Smith and Bryan in 1915 from the USA. Disease has been observed for the first time in severe from at Durgapur, Jaipur on cucurbits. The favourable environmental condtions for the disease (temperature 23-28°C and humidity in the greenhouse causes substantial yield loss due to less fruits and less weight per fruit by reducing the size of the infected leaves.

Symptoms

Symptoms appear as water soaked lesions on the leaves and are confined by the veins. They turn grey to tan in color and form an exudate on the lower surface of leaf. Many of the spots loosen and fall out. Infected fruits show a brown, circular, superficial, firm rot. Which causes rapid deterioration. The rot may extend into the flesh.

Causal Agent

The disease is caused by *Pseudomonas syringe* pv. *lachrymans* (Smith and Bryan, 1915) Young *et al.*1978 [Syn. *Pseudomonas lachrymans* (Smith and Bryan) Carner]. The colonies on nutrient agar are circular, smooth, glistening, transparent and white in color. The bacterium is rod shaped, motile with 1 to 5 polar flagella and 0.8 to 1.2μ m in length. It forms capsule and a green fluorescent pigment in culture.

Disease Cycle and Epidemiology

Infected seeds may bear the bacterium. Bacteria survive in soil or debris from diseased plants for two years. They spread by irrigation water, wind blown sand, hands, insects (*Diabrotica undecimpunctata* and southern corn rootworm) and machines. The bacterium attacks cucumber, gherkin, muskmelon, pumpkin, squash and watermelon, optimum temperature for disease development is found as 24 to 26.5°C. Moisture is the most limiting factor to disease development. Angular leaf spot is promoted by wet conditions frequently associated with rainfall and sprinkler irrigation. Infection is promoted by water congestion of leaves, a condition associated with extended periods of rainfall, high relative humidity and a combination of cool air and warm moist soil at night following warm days. Two weeks of dry weather condition block disease development. Disease develops over a wide temperature range, but is favoured by 24 - 28°C. A high temperature of 37°C for 5 days does not arrest disease development. Excessive nitrogen levels in the plant result in increased levels of disease.

Management

- Use of disease free seed and crop rotations 1 or 2 years without cucurbits are major control measures. Three year old seeds can be assumed to be free of bacteria.
- Resistance or tolerance to angular leaf spot is reported in cucumber varieties/lines such as Cherokee, chipe, Dal'nevoslochryi-27, Dasher, Early Truimph, H-3559, monastyrski, MSU-61-116, MSU-3075, PI-169400 from Turkey, Poinsett, Premier, Raider, Slicemaster, Southernelt, Superset, sweet slice and Vladivostokskii-155. Resistance appears to be governed by multigenic factors.
- The chances of seedling infection can be reduced by soaking the seeds for 5 to 10 minutes in mercuric chloride 0.1 per cent solution, rinsing in water and drying quickly.
- Spraying the plants with 400 ppm solution of streptomycin sulphate or copper fungicide effectively controls the disease.
- The most effective means of eliminating inoculum in the seed is to grow crops for seed in arid regions.
- Exogenous application of pipecolic acid modulates plant defence responses in cucumber and supresses *P. syringae* pv. *lachrymans* (Pazarlar, *et al.,* 2021).
- Pretreatment (priming) cucumber plants with prohexadione-Ca (Pro-Ca) enhances the defense response of cucumber leaves to subsequent bacterial infection as shown by the suppression of disease development (Skodowska *et al.,* 2021).
- Biocontrol agent, *P. fluorescens* and *P. putida* can supress *P. syringe* pv. *lachrymans* significantly (Al-Fallooji *et al.,* 2021).
- Limiting the use of overhead irrigation and avoiding use of excessive levels of nitrogen controls the disease.
- Fruit should be picked when plants are dry if the planting is planned for multiple harvest or if fruit is to be in transit or storage for an extended period.
- To reduce post harvest fruit rot associated with bacterial leaf spot, minimize harvest wounds on slicing cucumber by cuttings.

iii. Stem and Soft Rot of Muskmelon

In India, the disease was first recorded in Gujarat. Melon stem soft rot like other muskmelon bacterial diseases is becoming quite common and serious disease in many muskmelon-planting areas of China (Lei *et al.,* 2022).

Symptoms

Bacterial soft rots are a group of diseases that cause more crop loss worldwide than any other bacterial disease. Soft rot bacteria degrade pectate molecules that bind plant cells together, causing plant structure to eventually fall apart. Soft rots commonly affect cucurbits (*e.g.,* cucumbers, melons, squash, pumpkins). These diseases can occur on crops in the field, as well as on harvested crops in storage. Soft rot found in many vegetables is common in cucurbits. It is mainly a fruit disease and occurs due to injury to fruits and poor transit and storage conditions. Rot can occur over a wide temperature range, particularly when oxygen is limited (Meng *et al.,* 2017). Initially, bacterial soft rots cause water-soaked spots. These spots enlarge over time and become sunken and soft. Interior tissues beneath the spots become mushy and discolored, with the discoloration ranging anywhere from cream to black. Seepage from affected areas is common. Soft rots are known for a strong, disagreeable odor that accompanies the breakdown of plant tissue.

Causal Agent

The disease is caused by *Erwinia carotovora* sub sp. *carotovora* (Jones) (Bergey *et al.*). The bacterium is Gram negative, rod shaped with peritrichous flagella. Thepathogen is rod shape, gram negative, capsulated, peritrichously flagellated, non-spore forming and non-acid fast. The bacterium shows positive for pectate degradation on CVP medium, growth at 36.8C and 39.8°C, catalase production and negative for indole production. The bacterium produces acid from various sugars such as lactose, trehalose, maltose and cellobiose. The virulence of soft-rot *Erwinia* species is dependent mainly upon secreted enzymes such as pectin lyases, pectinases and proteases that cause maceration of plant tissue. Bacteria cause the greatest harm in areas with rather high temperature and increased air humidity during the vegetative period. The most intensive rotting occurs at high temperature (optimum temperature for the pathogen is 25-27°C) and long damp weather (especially at intensive rains and relative humidity of air of 50 per cent or higher). The bacteriosis severity is especially high in storehouses if storing production is infected with the pathogen and in the field on seed shoots in the next vegetation period (Meng *et al.,* 2017)

Management

- ✰ Once soft rot bacteria have infected plant tissue, there are no but avoid to compost or burry.
- ✰ Do not bury or compost this material.

- ☆ Wet conditions should be avoided to manage the soft rot disease. Plant vegetables in well- drained soils and control watering times and amounts, making sure plants are watered adequately (but not excessively) and maintain wider spacing for more rapid drying of plants and soil.
- ☆ Avoid damaging vegetables when weeding and during harvest.
- ☆ Make sure that soil fertility (particularly soil calcium) is optimal for the vegetables that you are growing based on a soil nutrient test.
- ☆ Keep insects that can wound vegetables such as cabbage maggot.
- ☆ Harvest only during dry conditions.
- ☆ Closely inspect vegetables from infected gardens that will go into long-term storage and be sure not to store any diseased vegetables.
- ☆ Use soft rot-resistant vegetables in rotation with susceptible considered susceptible to soft rot. When growing broccoli, avoid varieties with flat/ concave heads that trap moisture and promote soft rot.
- ☆ Minimize any handling of soft-rotted plants, but if you must handle such plants (*e.g.*, to remove them from the garden), wash your hands afterwards with soap and water.
- ☆ Decontaminate garden tools before and after use by treating them for at least 30 seconds with 10 per cent bleach or preferably (because of its less corrosive properties), 70 per cent alcohol. Rubbing alcohol and many spray disinfectants typically contain approximately 70 per cent alcohol (Singh *et al.,* 2020)
- ☆ Spray of copper oxychloride @0.3 per cent and streptomycin @500ppm can reduce the disease significantly.

iv. Bacterial Wilt

Losses in cucumber may be upto 25 per cent. The bacterium attacks cucumber, pumpkin and many wild cucurbit species. It is less common and less severe in pumpkins, squash and watermelon (Sanogo *et al.,* 2011). *Erwinia tracheiphila* is found throughout the United States, although it is most severe in the East. It also occurs in central and northern Europe, South Africa and Japan.

Symptoms

Initially wilt symptoms appears on individual leaves, but soon they affect lateral shoots and entire plant. Sometimes initial leaf symptoms are associated with cucumber beetle chewing injury. Leaf tissue surrounding such an injury becomes dull green in color and wilts rapidly. A single infected leaf may wilt and die within 5 to 6 days after infection. Wilt of lateral shoots and plant development occurs two weeks after infection. Initially symptoms development occurs on young vigorous plants. The larger the plant at infection, the longer the period of time between initial wilting and plant collapse. When older cucumber plants are infected some times only one shoot wilts initially. At fruiting, entire plants may collapse. On squash fruit, small areas of water soaked tissue appear on the fruit surface and minute glistening exudations appear on cut surfaces. The most common symptom is the rapid wilting of individual runners or whole plants. Wilted runners will appear dark green at first before becoming necrotic as the wilting becomes irreversible. Symptoms are more severe early in the season when plants are growing more rapidly. Long strands of bacterial slime can be observed if wilted runners are cut and the cut ends are pressed together and then pulled apart. *E. tracheiphila* is known to invade squash and cause internal spoilage (a slime) during storage. The fruit rots internally with little or no outside appearance. After several months in storage the exterior surface may show dark spots or blotches which coalesce and enlarge. In addition, infected squash become prime targets for invasion by soft-rot microorganisms.

Causal Agent

The disease is caused by *Erwinia tracheiphila* (smith) Bergey *et al.* (1923). Colonies of bacteria are grayish white-to-cream, circular, smooth and glittering. The bacteria are gram negative, motile rod in shape with 4 to 8 peritrichous flagella and capsulated. Bacteria do not grow at 39°C and produce hydrogen sulfide gas from hydrolysis of cysteine and do not hydrolyze litmus milk and starch.

Disease Cycle and Epidemiology

The bacteria overwinter in cucumber beetles and they appear to multiply in the beetle. The bacterium is not soil borne or seed borne. Bacteria can survive in stems for one month. Beetles prefer to feed on plants which are infected by bacteria than on healthy plants. Beetle can remain infective for at least three weeks. Striped cucumber beetle (*Aculymma vittata* Fubricius) and the 12 spotted eucumber beetle, *Diabrotica undecimpunetata howardi* Barbar help in the spread of the bacterium. *E. tracheiphila* is reported to attack non-cucurbit hosts, *i.e.*, corn, but apparently with no significant losses. Golden rod (*Solidago nemoralis* Ait. and *A. altissima* L.) and johnson grass (*Sorghum halepense* L.) are believed to host *E. tracheiphila* over winter, though these hosts do not show symptoms of bacterial wilt.

Weather appears to have little effect on the incidence of the disease but may affect wilting rate. Wet weather and heavy rain, especially if the sun is bright favour the bacterial growths and movement. Bacterial wilt is favoured by low nitrogen and potassium levels in unbalanced nutrient situations. Bacterial wilt develops when daily temperatures average between 7 - 22 °C with an optimum of 11 to 15 °C. The bacterium grows best at temperatures of 25 - 30 °C and under conditions of high relative humidity. Bacterial isolates differ in their virulence on each species of cucurbit (Mohammadi *et al.,* 2015).

Management

- ☆ Roguing diseased plants reduce secondary inoculum. Sprays with methoxychlor, corbary and endosulfan control beetles.
- ☆ Rotate crops on a 3-year cycle. Avoid susceptible or highly susceptible cucurbits 2 out of 3 years during this rotation.

- Plastic mulches can reduce overwintering beetle populations by reducing larval survival.
- Manage wild cucurbit species (such as bur cucumber and wild gourd) that can attract vector beetles and serve as reservoir hosts.
- Provide season-long control of spotted and striped cucumber beetles and other insects, such as grasshoppers and squash bugs, with insecticides.
- Apply a recommended insecticide frequently after removing the covering
- Start applications as the plants start to crack the soil, before the leaves appear, even if no beetles are evident.
- Frequent applications at 4 to 5 day intervals are necessary at seedling stage, to keep the foliage free of beetle-feeding wounds.
- Early-season sprays or dusts are the most important step in controlling bacterial wilt.
- The use of a systemic, soil-applied insecticide will provide moderate control for 5 to 6 weeks.
- In small plantings, pull the wilted plants at the first sign of disease and destroy them after first spraying thoroughly or dusting with a suggested insecticide.
- Where feasible, grow adapted, partially resistant cucumber varieties.
- Trap cropping to attract beetles followed by insecticide control on the trap crop has been used to reduce beetle populations before a main crop is transplanted in an adjacent area.
- Streptomycin 500 ppm and terramycin 100 ppm when applied 11 times at 4 to 7 days interval to control the disease.

v. Bacterial Fruit Blotch

The documented history of bacterial fruit starts in 1965 with a report from Georgia, US, of an unidentified, seedborne bacterium isolated from necrotic watermelon cotyledons of different plant introductions from Turkey (Webb and Goth, 1965). At the time, the disease was restricted to seedlings and no symptoms were observed on fruits. The causal agent of this disease was subsequently classified as *Pseudomonas pseudoalcaligenes* subsp. *citrulli* by Schaad *et al.,* 1978. The bacterial fruit blotch has great potential to cause significant economic losses to cucurbit production and has been responsible for up to 90 per cent losses of marketable yield in some watermelon fields. Bacterial fruit blotch affects the foliage at all growth stages and fruit of a wide range of cucurbitaceous hosts. Fruit blotch has been reported in many other cucurbits, such as honeydew, citronmelon, cucumber, squash and pumpkin.

Symptoms

Symptoms can be initially observed on cucurbit seedlings, between five and eight days after planting; depending on the environmental conditions. For most cucurbits initial seedling symptoms include water-soaking on the undersides of cotyledons. Watersoaked lesions start as discrete spots but then coalesce and extend along the veins of cotyledons. Lesions can extend along stems to tissues of true leaves and in severe cases they can cause seedlings to collapse and die in a "damping-off fashion".

Causal Agent

The disease is caused by biotrophic Gram-negative *Acidovorax avenae* subsp. *citrulli* (Syn: *Pseudomonas pseudoalcaligenes* subsp. *citrulli* (Schaad *et al.*,1978). The bacterium rod-shaped with average dimensions of 0.5 x 1.7 µm and is motile by a single polar flagellum. The bacterium is non-fluorescent on King's B (KB) medium, negative for hypersensitive response (HR) on tobacco and unable to induce fruit rot in field tests.

Disease Cycle and Epidemiology

The disease also spread worldwide, likely via infected seeds and the primary source of inoculum is often infested seeds. In many cases, before transfer to the field, seeds are sown in transplant facilities, which are usually characterized by mild temperatures, high humidity and use of overhead irrigation. These conditions are ideal for the spread, proliferation and establishment of *A. avenae* subsp. *citrulli* in young seedlings. The healthy plants might serve as inoculum sources for maturing fruits, which are very susceptible to the pathogen. Fruits can be infected either externally, by physical contact of the pathogen with the fruit followed by pathogen penetration through natural openings or wounds, or internally, through infection of plant's female blossoms. It is well established that pathogen spread from infected to healthy plants and organs is achieved mainly through water splash, from rain or irrigation. The role of insects, including pollinators, as vectors of the disease in the field has not been established but a recent report suggests a possible role for honeybees in watermelon seed infestation through blossom inoculation (Fessehaie *et al.*, 2005).

Management

- ✰ The most important management strategy for bacterial fruit blotch is exclusion *A. avenae citrulli* (Dutta *et al.*, 2013).
- ✰ It is impossible to guarantee pathogen-free seed, but seed health testing reduces the risk of outbreaks. The standard seed test for bacterial fruit blotch is the seedling grows out assay in which 10,000-30,000 seed/lot are planted under conditions conducive for disease development.
- ✰ Seed fields are visually inspected and only disease free seed are used.
- ✰ There are no commercial cucurbit cultivars with immunity to BFB. While several watermelon plant introductions are resistant, all commercial cultivars appear to be susceptible to the disease.

- ☆ Three-year rotations with non-cucurbit hosts are routinely employed.
- ☆ Debris, including infected watermelon culls from fields, is deep ploughed at the end of each growing season.
- ☆ Subsequent cucurbit crops are established in new fields physically separate from the outbreak site.
- ☆ If possible, alternative hosts such as wild and for preventive management, biweekly applications of copper-based bactericides *e.g.*, cupric hydroxide, copper hydroxosulfate, or copper oxychloride should be employed at the recommended.
- ☆ Preventive sprays are initiated at or before anthesis (flowering) and continue until fruit are mature.
- ☆ If BFB symptoms develop, weekly applications of the full recommended rate of copper-based bactericides are employed.
- ☆ To prevent spread, work is not conducted in fields with disease when the foliage is wet and field equipment is decontaminated before moving between fields.

11. Cruciferous Vegetables

i. Peppery Leaf Spot of Cauliflower

The causal bacterium of peppery leaf spot has been disseminated with seed world wide wherever cauliflower is grown. It is common in cool coasted areas such as long Island, the tidewater area of Virginia, Ontario and the coastal valleys of california, Czechoslovakia, Denmark, Finland and Sweden. Losses occur from unsightly discolouration of leaves surrounding the head or from leaf drop.

Symptoms

Diseased cauliflower leaves are covered with numerous small, brown to purplish spots ranging from specks to lesions 3mm in dia or larger if they have a yellow halo. Dead areas unite and produce irregular bleaches upto 1.3 cm in size. Spots appear first and most abundantly on the lower side of leaves. Each spot originates at a stomatal opening. Infection on veins results in retarded growth and puckered foliage. Extensive infection causes leaves to drop off. The bacteria affect stems, leaf petiole and seed pods.

Causal Agent

The disease is caused by *Pseudomonas syringae* pv. *maculicola* (McCulloch) Young, Dye and Wilkie. Bacterium is straight bacilli, moving by one to five flagella, usually 0.9 × 1.5–3.0 μm in size and Gram-negative and aerobic, forming a fluorescing pigment. On potato agar, colonies are round, brilliant, smooth, transparent in transmitted light, opalescent; later they become dirty-white with dense and raised center. Few characteristics of this pathogen are as follows: they do not curdle and peptonize milk; do not restore nitrates; do not form H_2S; produce indole; hydrolyze starch poorly; form NH_3; and utilize glucose and saccharose with production of acid, but do not utilize lactose and maltose. The optimum temperature for growth is 24°C–25°C maximum is 29°C and minimum is lower than 0°C.

Disease Cycle and Epidemiology

Cauliflower, broccoli, cabbage, kohlrabi and turnip are the hosts infected. The bacteria survive between crops, in infected seed and in diseased cauliflower left in the field. They remain alive for at least one year in soil. Bacteria can be introduced from one country to another with seed and can be distributed to shorter distance in infected seedlings. The bacteria are disseminated by washing soil, running water, splashing rain, blowing leaves and by some insects.

Peppery leaf spot is most prevalent in wet cool weather and many disappear entirely when the temperature rises to 30° C. the optimum temperature for growth of the bacterium is 24 to 25° C. The infection/pathogen survives on vegetation residues and seeds. Secondary infection takes place into the plant through injuries. Increased humidity (90 per cent and higher) and temperature (17°C–20°C) play a significant role in bacteriosis development. Favorable weather conditions for pathogens of bacteriosis may provoke epiphytotics, which may destroy plantings of cauliflower. High severity of the disease is possible on seed shoots of late cauliflower planting, if the second half of vegetation period is excessively rainy

Management

- ✰ Important control measures include having the seed and seed bed free from the pathogen practicing crop rotations and hot water treatment of the seed.
- ✰ If the disease has occurred previously in the seed bed, the soil should be treated with steam or methane sodium.
- ✰ Use seeds that are certified free from the disease, carefully remove plant residues, separate seeds from shrunken grains, treat the seeds with pesticides before sowing and treat the plants with pesticides during vegetation period.
- ✰ The rotation should provide at least one year between susceptible crops.
- ✰ Cauliflower debris should be buried by ploughing.

ii. Black Rot

Disease was first observed in the UAS on cabbage in 1891, but the causal agent was identified during 1895. In India, disease was first described in 1928 on cabbage from pune, Maharashtra. Now disease is known to be prevalent in most of the crucifers growing areas of country. The disease occurs on broccoli, brussel sprout, cabbage, cauliflower, Indian mustard, kohlrabi, mustard, radish, turnip and other cruciferous plants it is believed that the causal organism was introduced into this country with the cruciferous crops from the western countries. The bacterium has wide range of hosts especially crucifers. Broccoli, radish, cabbage, cauliflower, kale, kohlrabi, mustard, *Lepidium sativum, Brassica juncea,* Chinease cabbage, Chinese sarson, rutabaga (*B. napus* var. *napobrassica*), savoy cabbage, turnip, (*B. rapa* sp. *oleifera, Raphanus sativus* var. *caudatus, B. juncea,* fodder cabbage, *Capsella bursa-pastoris* (shepherd's purse) and *Cantella asiatica.*

Symptoms

Young plants in the nursery are killed out right infected plants become stunted, often one sided. The cotyledons turn yellow to black in color, hang down and drop/off prematurely. Initially most of the lower leaves shed early. The remaining leaves turn yellow in color with blackened veins. In the field "V" shaped chlorotic to yellow lesions develop from the leaf margin. The involved veins and veinlets turn black in color. The leaf tissues become necrotic and brittle (Figure 2a and b). The lesions appear on midrib which becomes black in color. From midrib the systemic infection spreads into the root. The vascular bundles in these plant parts become black,followed by internal breakdown of fleshy tissue. The attack of soft rot organisms causes further deterioration. Complete black rotting of the internal tissue in the lower part of the stem is noticed. Due to local infection, black sunken spots develop on succulent seed stalks and siliqua. Cabbage heads and cauliflower curds are also infected and become discoloured.

Causal Agent

The disease is caused by *Xanthomonas campestris* pv. *campestris* (Pammel) Dowson. It is Gram negative, short rod with rounded ends, 1.2 to 2.1 x 0.5 to 0.1µm, non capsulated, occurring singly, rarely in parts and motile with single polar flagellum. Two strains of the pathogen, *i.e. X. campestries* pv. *armoraciae* (McCulloch) Dye. and *X. campestris* pv. *aberrans* (Knosel) Dye have been reported. The optimum temperature for the pathogen growth is from 26 to 30°C, the minimum is 5°C and the maximum is 36°C. First strain affects cabbage, cauliflower, French bean, radish and turnip and while second infects cauliflower.

Disease Cycle and Epidemiology

It affects beet root, broccoli, cabbage, carrot, cauliflower, knol khol, mustard, radish and turnip. The bacterium lives on or in seeds and in debris from diseased plants in the fieldupto 244 days. Seed contaminated with black rot bacteria is considered the most important source of the pathogen and significantly contributes to the spread of this disease worldwide. As few as 3 infected seeds per 10,000 (0.03 per cent infected seeds) can result in a black rot epidemic. Seeds should be tested and certified to be disease free with less than 1 in 30,000 infected seed. The bacterium survives in infected crop tissue left on the soil until the crop tissue rots. However, the bacteria do not survive very long in soil as unprotected free living organisms. The black rot bacteria can also infect and survive on many crucifer weeds including *Centella asiatica.* It can grow and multiply on host tissue without infecting or causing disease. Rain-splashed bacteria from contaminated plant residue left on the soil or from neighbouring diseased plants is the primary method of disease spread throughout a field. The bacterium enters the cotyledons through the marginal stomata. From the cotyledons the bacterium passés into the true leaves. In the leaves, the entry is through hydathodes located at the edges and tips of leaves. Hydathodes often produce a drop of water during periods of high humidity early in the morning. On entry into the host, bacterium moves intercellularly and infects vascular bundles and establishes systemic infection in the true leaves. The pathogen spreads very quickly when rain droplets contaminated with bacteria splash onto healthy leaves and enter the hydathodes. The bacteria move into the leaf veins through hydathodes and begin to multiply, rot and plug the veins. Contaminated water droplets that exude out of hydathodes of infected leaves can then be rain-splashed to other plants. Equipment, people, animals, insects and overhead irrigation can further spread the disease. Water in the form of rain or persistent dew is required for the development of the disease.

(a) (b)

Figure 2. Black Rot Disease Symptoms on Cabbage (a) and Cauliflower (b).

Cabbage planted in Sept-Oct and January and cauliflower in October, show higher disease incidence than cabbage planted in November and February and cauliflower in February. Infected seedlings grown in the greenhouse under cool conditions (< 15°C–18°C) frequently do not show any symptoms of the disease. When infected seedlings are transplanted to the field and temperatures rise to 25°C–35°C during periods of high relative humidity (80–100 per cent), they become stunted with dead spots on the cotyledons and may eventually wilt and die. In regions with temperate climates (where temperatures remain cool), disease symptoms on infected seedlings may not always be obvious or appear severe. Infected seedlings grown under cool conditions may ooze bacteria from pores and lesions, which then serve as a source of the pathogen for neighboring plants. The bacteria produce a sticky polysaccharide called xanthan that eventually plugs the vascular tissue inside the veins causing them to collapse and turn black. During hot humid environmental conditions, the bacteria can move from the leaf into the stem through the xylem. Once inside the stem, the bacteria can move up or down to other parts of the plant including the roots. Systemically infected plants may produce chlorotic areas anywhere on the leaf. Severely infected leafy cole crops such as kale and cauliflower tend to shed their leaves from the bottom up leaving only a tuft of distorted leaves separated from the root system by a scarred barren stem.

Disease Management

- ☆ Soaking seeds in hot water at 50°C for 25–30 min is the most effective treatment for seed-borne black rot control. Weak seed, seed stored for several years and seed of certain crucifer crops, such as cauliflower, kohlrabi, kale, rutabaga and summer turnip, may be damaged by hot-water treatment; soak for 15 min at 50°C only.
- ☆ Reducing seeding rates and densities to promote good air circulation, facilitating the quick drying of plants, timing irrigation when plants will dry quickly and restricting field activities until later in the day when fields are dry will help reduce disease spread.
- ☆ Wash and disinfect equipment before moving from one field to another.
- ☆ Crop rotation 2-3-years with non crucifer crops is recommended.
- ☆ Weed Control Black rot bacteria can infect and survive on many crucifer weeds including bird rape (*B. campestris*), Indian mustard (*B. juncea*), black mustard (*B. nigra*), shepherd's purse (*Capsella bursapastoris*), globe-podded hoary cress (*Cardaria pubescens*), pepper grass (*Lepidium densiflore*) and wild radish (*Raphanus raphanistrum*).
- ☆ Good weed control within fields will aid disease management; however, careful attention to weed control in ditches and along fence rows is also important.
- ☆ The crucifer flea beetle (*Phyllotreta cruciferae*) can transmit black rot bacteria from infected plants to healthy ones; however, their importance in the spread of the disease is limited. Wounds caused by insects provide an entry point for the disease to infect plants during heavy dews or periods of rain. Insect control will help reduce the spread and severity of disease.
- ☆ Drenching the nursery soil with form aldehyde 0.5 per cent helps in checking thc disease.
- ☆ Late cultivars of cauliflower KN-81, BR-2, SN-445, ACC-641, KK-104, RSK-1301 Avans, Iglory, Pant Shubhra MGS 2-3, Pusa Ice and SN 445 are found resistant to black rot. Cabbage-cultivar 294 bc, crosses 83- 6 x MR-1, Cabbage No 8 x Sel. 8, cabbage No. 44x Sel. 8 and 83-6 x Sel. 8 are found resistant against to the disease. Resistant cabbage varieties which include Cabaret, Defender, Gladiator and Guardian are useful in black rot endemic areas.
- ☆ Application of bleaching powder at 10.0 to 12.5 kg/ha controls black rot and soft rot diseases.

- ☆ Spraying with agrimycin 200 ppm is suggested for its control.
- ☆ Kumud and Shyam (2003) reported that the integrated treatment with streptocycline 0.01 per cent + mancozeb 0.25 per cent for 3 hr as seed treatment, streptocycline + mancozeb as foliar spray and *Eucalyptum hybrida* applied on cauliflower cv. PSBK- 1 in the field to control the disease.
- ☆ Potential antagonistic bacteria are *P. fluorescens*, *B. amyloliquefaciens* and *B. pumilus* (Wulff *et al.*, 2002; Singh *et al.*, 2010b).

iii. Zonate Spot of Cabbage

A bacterium isolated from translucent zonate lesions on fully developed heads of cabbage was identified as *Pseudomonas cichorii*, a new host record for India (Trivedi *et al.*, 1978).

Symptoms

The spots on cabbage heads are circular to oblong in shape, translucent, buff to brown in color, 7 to 12 mm dia., sunken and have typical zonations. Lesions on the inner leaves are irregular in shape, black in color and are localized near the margin of young leaves. The vascular bundles on stem become dark brown I color. Due to systemic infection, elongated and sunken lesions of 5 to 7 cm size develop on petioles and mid ribs of lower leaves.

Causal Agent

The disease is caused by *Pseudomonas cichorii* (Swingle) Stapp. The bacterium is a Gram negative rod and 1.5 to 2.5 x 0.5 to 0.7 μm. It frequently occurs in pairs and motile by 1 to 4 polar flagella.

Disease Cycle and Epidemiology

The casual organism, *P. cichorii*, enters cabbage through wounds. Lesions may occur on any part of the leaf but there is no evidence that the disease involves the vascular system. Best growth of the pathogen occurs at 23°C.

Management

- ☆ Hot-water treatment of the seed at 50°C for 30 min is found effective.
- ☆ Drenching the seed bed with 5 per cent commercial formalin may help in checking the disease.
- ☆ Plant pathogen-free seed.
- ☆ Avoid sprinkler irrigation in the seedbed once the crop has germinated and established.
- ☆ In the greenhouse, flow irrigation is preferred to overhead sprinkling. Shred and turn under diseased crop refuse promptly after harvest to hasten the breakdown of the infected plant material.

- ☆ Do not plant cole crops the following year if the field has a significant level of infection.
- ☆ Agrimycin 100 gave good control of the disease. Streptocycline was found to be effective in checking the *in vitro* growth.
- ☆ Sprays of copper containing bactericide may reduce the incidence of disease.

iv. Curd Rot/Soft Rot

Symptoms

Any plant part can be affected, but the most serious economic loss is from head infections. Plant tissue at first appears water soaked and rapidly breaks down into a soft, mushy rot. This bacterial infection commonly affects the leaves. The disease is more likely where warm and wet conditions prevail. Leaves become yellow starting from the edges and moving toward the center. The infection causes the leaf veins and plant stems to turn black. As the infection increases, the broccoli head develops brown sunken areas of rotting matter, which rapidly extend. Much defoliation takes place especially for the more mature plants. Proper growth is inhibited. Plants remain small or die prematurely.

Causal Agent

The disease is caused by *Pectobacterium carotovorum* subsp. *carotovorum* (syn. *Erwinia carotovora* subsp. *carotovora*).

Disease Cycle and Epidemiology

The soft-rot bacterium, *P. carotovorum* subsp. *carotovorum*, enters through natural openings or wounds caused by insects and equipment. Prolonged moisture from rain or irrigation and mild temperatures encourage disease development. Warm, wet conditions favor soft rot development. Bacteria survive in soil and plant debris.

Management

- ☆ Set out plant rows that will allow good air drainage.
- ☆ Cultivate carefully to minimize injuring plants.
- ☆ Control frequency and source of irrigation water.
- ☆ Avoid frequent irrigation during head development.
- ☆ Time irrigation to allow the head to dry rapidly.
- ☆ Use well water, which generally is free of bacteria.
- ☆ Avoid stagnant water sources.
- ☆ Keep storage house relative humidity 90-95 per cent and temperature 0-4°C.
- ☆ Crop every 2 years is helpful to control the disease.

- ✰ Avoid high nitrogen rates with susceptible varieties.
- ✰ Cut heads such that the stem stump is angled to permit water runoff. Water pooling on a flat stump will provide favourable conditions for soft rot. Bacteria in rotting stem stumps can be dispersed to heads not yet harvested.
- ✰ Sub-MICs of hexanal could significantly decrease AHLs production. Hexanal could inhibit activity of extracellular enzyme of the spoilage bacteria. Hexanal could inhibit soft rot of Chinese cabbage and lettuce leaf scorch (Zhang *et al.*, 2018).

12. Carrot (*Daucas carota*)

i. Bacterial Blight

Bacterial blight was first found in California in 1931 by J.B. Kendrick (Kendrick, 1934). The seed crop is heavily damaged by the disease compared to root vegetable crop. Since then it has been found in Australia, Canada, Russia, Spain, Korea, Mauritius and the U.S.A. Cultivated carrot is the primary host of *Xanthomonas hortorum* pv. *carotae*, although the pathogen has been reported to infect a number of other umbelliferous plant species (Gilbertson, 2002; Poplawsky *et al.*, 2004).

Symptoms

The bacterium cause irregular brown spot on leaves, dark brown streaks on petioles and a blighting of floral parts. Lesions on leaves begin as small yellow spots. Soon the centre of the spot they become dry in nature and brittle with an irregular yellow halo. The discolouration may be present only on the divided leaf segments or it may affect the entire leaflet. A stickly bacterial exudates flows downward on petioles.

Causal Agent

The disease is caused by bacterium *Xanthomonas hortorum* pv. *carotae* (Kendrick) Vauterin *et al.* (1995) (Syn: *Xanthomonas campertris* pv. *carotae* (Kendrick) Dye. Colonies of bacteria on nutrient agar are circular, straw yellow, slightly raised and glistening. The bacterium is a Gram-negative, rod shaped with 1.5-2.0 x 0.5-0.75 µm in size. Bacterium forms chains and capsules and is motile by one polar flagellum. No spores are formed and they aerobic. It is a positive for catalase and negative for oxidase test and also positive for mucoid growth on yeast extract-dextrose-calcium carbonate agar, growth at 35°C, hydrolysis of esculin, protein digestion, alkaline in litmus milk, acid production from arabitol and utilization of glycerol and melibiose. It is negative for growth on SX medium, hydrolysis of starch and ice nucleation.

Disease Cycle and Epidemiology

The bacterial pathogenis seed borne, survives on seed and is spread with carrot seed. The bacteria also survive in carrot debris but cannot survive in the soil in the absence of debris. Water is an essential ingredient needed for symptoms to develop with disease spread occurring most rapidly under warm (25-30°C) and wet conditions. Therefore sprinkler irrigated carrot crops and areas with high rainfall seem to have more disease outbreaks. Optimum temperatures are between 26 - 30°C and infection does not occur below 17°C. The pathogen is dispersed in splashing water. Plant-to-plant spread may occur under heavy dew conditions. In most carrot-growing areas, bacterial blight does not warrant control. Under rainy warm conditions, epidemics occur rapidly.

Management

- ✰ Use certified, disease-free seed or treat seed with hot water at 52°C for 25 min.
- ✰ Avoid overhead irrigation.
- ✰ Crop rotation for 2–3 years should be followed.
- ✰ Plow infected crop debris after harvest under the soil to hasten decomposition.
- ✰ Carrot cultivars like Walthan Hi-color, Danvers and PI 306588 are tolerant to the disease.
- ✰ Protect the crop by spraying early with copper oxychloride 0.2 per cent

ii. Bacterial Soft Rot

Bacterial soft rot has been described in many countries under various names. It occurs worldwide, where ever carrots are grown. The bacterium attacks artichoke, asparagus Brussel sprouts, cabbage, carrot, parsnip, pepper, potato, radish, rhubarb, rutabaga, tomato and turnip. Bacterial soft rot can be serious on carrots marketed with their tops, especially when storage temperature control is inadequate. Soft rot can start on foliage and then spread to roots. Soft rot develops during transit or in storage if temperature rises above 4.5° C.

Symptoms

Parenchymatous tissues attacked by soft rot bacteria decay very rapidly than the other tissues. Cells become water soaked, the middle lamella is destroyed and the cells collapse into a soft, watery slimy mass. The epidermis usually remains intact. The rotted tissues are grey to brown in color. The may be accompanied by a foul odour. The decay develops most rapidly along the core of the root. When roots are split, a distinct line of demarcation between healthy and diseased tissues *Erwinia carotovora* is evident. Under conditions of high soil temperature and high moisture, soft rot bacteria can cause pitting of carrot roots. Tissues partially decayed by *E. carotovora* subsp. *carotovora* are invaded rapidly by fungi and other bacteria. *Rhizopus arrhizus* and *R. nigricans* frequently are found in storage on partially rotted carrots by *E. carotovora* subsp. *carotovora*. In the field, *Botrytis cinerea* and *Sclerotinia sclerotiorum* often permeate these decay tissues.

Causal Agent

The disease is caused by *Erwinia carotovora* subsp. *carotovara* (Jones) Bergey *et al* and *Erwinia carotovara* subsp. *atroseptica* (Van Hall) Dye. Bacteria are relatively large, Gram-negative and motile with large peritrichous flagella.

Disease Cycle and Epidemiology

Soil is the principal source of primary inoculum for stored carrots. Soil that contains debris from plants that were diseased in the previous year is the most important sourceof inoculum. The pathogen lives and multiplies within the soil. If soft rot occurs on carrot roots in fields, the inoculum source can be traced back to carrot foliage from which it moves directly down to the roots. Harvest bruises, freezing injury, fungus invasion and insect wounds offers site of penetration. Flies like *Hylemus cillicrura platura* and *H. brassicae* carry the bacteria in the tracts of intestine. When they lay eggs near or on plants, the larvae emerge and become contaminated with the soft rot bacteria. When larvae bore into carrot roots, they introduce the bacteria into the carrots. There the bacteria multiply rapidly, produce an enzyme that softens the middle lamella between cells and results in tissue softening associated with separation of cells from one another. By products of the bacterial growth cause the cell contents to flow into the intercellular spaces where it serves as food for the bacteria. Such cell collapse results in the typical watery or slimy type of decay.

Management

- ✰ Carrot should follow alfalfa, bean, beets, clover; corn, small grains or grasses in rotation and it reduces the bacterial population from proceeding crops.
- ✰ For the earliest and latest seedlings of carrots, avoid fields subject to a high water table during wet conditions.
- ✰ In the field, maintain good drainage and avoid practices that could wound roots.
- ✰ Avoid prolonged irrigation of mature carrots during warm months of the year.
- ✰ Careful handling at harvest minimizes bruising of carrot roots and control of temperature and humidity during storage effectively controls the disease.
- ✰ When soil temperatures are high, carrots harvested for immediate sale should be washed and cooled promptly and rinsed with clean, chlorinated water before being placed in a refrigerated holding area.
- ✰ Harvest crops intended for long-term storage after soil and air temperatures drop.
- ✰ If carrots are washed before storage or transit, dipping in a solution of 1:500 of sodium hypochlorite (5.25 per cent) or drying in warm forced air should be made.

- ✰ Storage rooms dump tanks and boxes should be disinfected each season with formaldehyde or copper sulphate.
- ✰ Keep storage as close as possible to 0°C temperature and 85 per cent -90 per cent relative humidity.

13. Radish (*Raphanus sativus* L.)

Root Rot

The disease was reported from Bangalore, Karnataka, during 1983 and its incidence varied from 15 to 50 per cent. The bacterial pathogen *Erwinia rhapontici* lives in association with rhubarb (*Rheum rhaponticum*) and cause crown rot and also appear on other plants like wheat and pea (pink seed). It is also has pathogenic effect on human.

Symptoms

The characteristic symptoms of disease are rotting of pith tissues resulting in cavity formation and wilting of plants. Disease is more serious in seed crop.

Causal Agent

The causal organism, *Erwina rhapontici* (Millard) Burkholder is a Gram negative rod, 1.2-1.5 X 0.5-0.8 µm, in size and possesses 3-7 peritrichous flagella. *E. rhapontici* has been identified as a plant pathogen that produces a distinct diffusible pink pigment on sucrose-peptone agar and creates pink seeds in the hosts (Huang *et al.*, 2003)

Disease Cycle

The bacterium is soil-borne and disease spreads through implements, irrigation water, insects, *etc.*

Management

- ✰ Crop rotation with non- host crop such as beans, corn and soybean plants reduces the disease incidenceand rotation with cabbage and potatoes should be avoided.
- ✰ Harvest may be made during dry weather.
- ✰ Washing with 25 ppm of free chlorine eliminates the soft rot bacteria.
- ✰ Agrimycin 100 (100 ppm) is found best in reducing the disease when applied as root dip.
- ✰ Various commercial species of *Bacillus* and *Pseudomonas* are found to be effective against pathogen. Proper humidity and CO_2 management in storage.

14. Beet Root (*Beta vulgaris*)

i. Crown Gall

The disease occurs in India.

Symptoms

Crown gall is characterized by the production of galls or outgrowths on the root. The galls usually begin to appear during March-April and continue to increase in number and size until harvest. At that time all sizes of tumours may be present and galls start as slight swellings. Later on galls become round, furrowed with corky or warty surface and are usually attached to the side of the crown portion. Excessive development in the crown may lead to excessive top growth causing crack in the crown portion.

Causal Agent

The disease is caused by *Rhizobium radiobacter* (Syn: *Agrobacterium tumefaciens* (Smith and Townsend) Conn. The genus *Agrobacterium* belongs to the family Rhizobiaceae. The bacterium is aerobic, Gram negative, rod shaped, 0.6 - 1.0 x 1.5-0 µm, occur singly or in pairs, non endospore forming, motile by one to 6 peritrichous flagella. On PDA + $CaCO_3$, *Agrobacterium* strains grow well and their colonies are distinguishable from other species. They are convex, glistering, circular with an entire edge and white to tannish cream in colour.

Disease Cycle and Epidemiology

Agrobacterium tumefaciens naturally resides on the rhizoplane of woody and herbaceous weeds. Its presence in soils originates from galls that were broken or sloughed off from infected plants during cultivation practices or disseminated as infected plant material. Irrigation aids in further dissemination of the *A. tumefaciens* bacterial cells. *Agrobacterium tumefaciens* is also spread by infected and infested planting stocks originating as nursery stock from uncertified sources. Secondary spread then occurs through pruning and cultivation equipment, particularly when galls are removed manually with the same cutting tools used in pruning. Tilling equipment can be contaminated by cutting through galls at or near the base of trunks of infected trees. Rogueing (removal) of infected trees and replanting in the same spot where the infected tree had grown is poor practice because sloughed off galls serve as sources of abundant populations of *A. tumefaciens.*

The bacteria enter a plant wound, disease will be more severe and a large gall will be seen rapidly growing at the infection site. Tissue injuries induced by accidental wounds made during cultivation practices or by frost will elicit new infections with the appearance of numerous galls along the vascular system of the host.

Management

- Crop rotation of atleast three years with non- host crop plants reduces the disease incidence.
- Careful handling at harvest minimizes bruising roots and control of temperature and humidity during storage effectively controls the disease.
- Injuries by insects should be prevented.
- Seed treatment with galltroll gives effective control.
- Agrocinis also found to be very effective. Bio-agents like *Agrobacterium radiobacter* K-84 are very effective.
- Seeds for sowing should be obtained from healthy beets.
- Beet tops should not be returned to future beet field.
- Seed suspected of carrying bacteria should be treated with hot water at 50°C for 20 min before planting.
- *Pseudomonas japonica* strain NBRC 103040, *Bacillus megaterium* strain CtST3.5, *Pseudomonas* sp. strain Gamma-81, *P. tolaasii* strain ATCC 33618, *P. chlororaphis* strain Lzh-T5 and *P. mosselii* strain CV25. These HCN producers were able to inhibit growth of *Agrobacterium tumefaciens* (Abdel-Rahman *et al.*, 2019).

ii. Bacterial Vascular Necrosis and Rot

The disease first appeared in the U.S.A. as reported by Thomson and Scroth in 1972, where its incidence ranged from 30 to 40 per cent. The pathogen can infect tomato, potato and chrysanthemum. Besides causing quantitative losses, infected roots prevent efficient extraction of sugar. In addition to beets, *Pectobacterium carotovara* subsp. *betavasculorum* can also infect tomato, potato, carrots, sweet potato, radish, sunflower, artichokes, squash, cucumber and chrysanthemum.

Symptoms

The bacteria invade the vascular tissues of the roots and petiole and usually causes an extensive rot. Vascular bundles of infected roots are necrotic and areas surrounding them turn pink in color on exposure to air. Initially season infections reduce yield of tubers and sugar more than the later ones. It is usually not fatal to the plant, but in severe cases the beets will become hollowed and unmarketable.

Causal Agent

The disease is caused by *Pectobacterium carotovorum* subsp. *betavasculorum* (Thomson *et al.*; Hauben *et al.* (Syn.: *Erwinia. carotovora* subsp. *betavasculorum* Thomson *et al.* and *Pectobacterium betavasculorum*). Bacterium is a gram negative, rod bacteria with peritichous flagella.

Disease Cycle and Epidemiology

The pathogen survives between sugarbeet crops in and on unharvested beets, soil and weeds. The bacteria thrive in warm and wet conditions, but cannot survive long in fallow soil. However, it is able to persist for long periods of time in the rhizosphere of weeds and non-host crops. Several races of the pathogen occur and can also attack tomato, potato and chrysanthemum. Splashing water, insects, soil and contaminated equipment may spread the pathogen. Injury is necessary for infection and temperature of 25.2- 30.2°C favour rapid disease development.

Management

- ✰ Crop rotation of atleast 3 years with non- host crop plants reduces the disease incidence.
- ✰ Careful handling at harvest minimizes bruising roots and control of temperature and humidity during storage effectively controls the disease.
- ✰ Injuries by insects should be prevented.
- ✰ Seed treatment with streptomycin gives effective control.
- ✰ Seeds for sowing should be obtained from healthy beets.
- ✰ Seed suspected of carrying bacteria should be treated with hot water at 122 °F for 20 min before planting.

iii. Silvering Disease

It was first reported in United Kingdom in 1945. It causes upto 75 per cent loss of plants. It infects marigold also.

Symptoms

The disease occurs at every stages of growth of the plant, but are most obvious about mid life as one or more leaves on a plant become silvery at first along the veins. The whole leaf lamina is soon affected and the silvering appears on other leaves. Finally plant wilts and dies.

Causal Agent

The disease is caused by *Curtobacterium flaccumfaciens* pv. *betae* (Collins and Jones) (Syn.: *Corynebacterium betae* Keyworth *et al.*). The bacterium is gram-positive bacteria, small irregular rods, lateral flagella, aerobic and catalase-positive. The colonies of the pathogen produce a pale-yellow pigment. The optimum temperature for the bacteria to grow is 24 - 27°C. The bacteria could utilize a wide range of organic compounds, including hydrolyzed casein, starch, esculin and Tween 80 and released H_2S from cysteine, cystine and $Na_2S_2O_3 \cdot 5H_2O$, but could not produce urease, oxidase, or indole. The cell wall peptidogly can was based on ornithine (type B2β). The predominant menaquinone was MK-9. Polar lipids contained several glycosyldiacylglycerols. The DNA G+C content of a type strain of the new pathovar, T30T, was 67.5 per cent.

Disease Cycle

The bacteria spread from such plants in the seedling bed during the autumn and winter, probably entering other plants through their leaf edges. The disease is readily transmitted on knives which nave cut infected roots.

Management

- ☆ Seed treatment with streptomycin or erythromycin gives effective control.
- ☆ Seeds for sowing should be obtained from healthy beets. At least a three year crop rotation should be practiced.
- ☆ Beet tops should not be returned to future beet field.
- ☆ Seed suspected of carrying bacteria should be treated with hot water at 50°C for 20 min before planting.

15. Cassava/Tapioca (*Manihot esculenta* Crantz.)

Bacterial Blight

The disease was originally discovered in Brazil in 1912 by G. Bondar, the disease has followed cultivation of cassava across the world. The disease has been reported from Asia, Africa and Latin America. It is the second most devastating disease after Cassava mosaic virus disease complex and may cause more damage to the crop than any other bacterial disease. The disease causes losses of fresh roots and also of planting material. Root yield reduction level may vary with the susceptibility of cassava cultivars, the climatic conditions and the inoculums pressure. The poor yield of storage roots due to severe outbreaks of the disease can affect the population as well as the livestock in areas where cassava is a major staple food.

Symptoms

Characteristic symptoms are leaf spotting and blight, wilting, die-back, Gum exudation and stem and root vascular necrosis. Primary symptoms, on planting material are wilting of the young germinated shoots followed by die-back. Symptoms of secondary infection are angular leaf spotting followed by blight, defoliation, wilting and die-back. At first the leaf spots are small, angular and water soaked later they enlarge and cover all or part of the leaf and turn brown or dark brown in color. The initial spots are clearly on the lowerside of leaves. In some varieties, a yellow halo is observed around the spots. The necrosis leaves roll and dry up. They remain attached to the stem for a short period and fall off. Gum is exuded from young. Infected stems, petioles and leaf spots collects as droplets. On leaves, these droplets occur mainly on the lower sides along veins or vein lets. On petioles and stems, the gum is exuded from cracks. The gum forms a yellowish glistening scab when dried. Vascular tissue of affected petioles and stems becomes necrotic and appears as brown strings. Leaves supplied by these vascular tissues wilt and young stem tissues rot. The young green stems rot faster than the older greenish brown ones which maintain a healthy appearance. The rotting of the young stems causes characteristic die-back symptoms. Roots have a dry vascular necrosis symptom. On young immature plants complete destruction of the aerial portions may be seen. New shoots are usually produced but they are infected, during the rainy season. The leaf blight is due to production of toxins by the bacterium, such as 3- methylthio) propionic acid (Perreaux *et al.*, 1986) tiglic acid, phenylactic acid and cyclopentane carboxylic acid (Ewbank, 1992).

Causal Agent

The disease is caused by *Xanthomonas phaseoli* pv. *manihotis* (Bondar, 1912) Constantin *et al.*, 2016 [Syn.: *Xanthomonas axonopodis* pv. *manihotis* (Berthet and Bondar, 1912) Vauterin *et al.*1995, *Xanthomonas campestris* pv. *manihotis* (Bondar) Dye]. The colonies of the bacteria on agar are yellow, mucoid, convex and round. It is an obligate aerobic bacterium and uses oxygen as terminal electron acceptor. It grows optimally between 25- 30°C. Its development is favoured between 6.5-7.2 pH. It is a Gram negative, slender rod in shape, motility by using single polar flagellum, non encapsulated and non spore forming. It is an aerobic, fast growing bacterium which forms no pigment on sugar containing media. Bacteria hydrolyse starch and gelatin and reduce litmus milk. It produces arginine dehydrolase, catalase and lipase but does not produce H_2S indole, urease, tyrosinase or phenyl alanine deaminase. The bacterium utilizes nitrate ad ammonium as sources of nitrogen.

Disease Cycle and Epidemiology

Primary infection is by use of infected planting material. The bacterium survives for one or two months. At the end of dry season, the rainfall starts,residual epiphytotics population of bacterial pathogen multiplies and penetrates the leaves tissues through epidermal wounds and stomata. Secondary infection results from gummy exudates by rain splash and movement of human beings, animals and insects (Grass hopper-*Zonocerus variegatus*) through the crop and the use of contaminated tools also spread the disease. Development of the disease is retarded during dry periods. Bacteria remain viable in the plant and are reactivated in later wet periods.

Management

- ✰ Resistant varieties can be grown.
- ✰ Only healthy planting sets should be used.
- ✰ Clean seed material can be obtained by rooting disease-free stem tips of apparently healthy plants.
- ✰ Fallowing for 6 months before planting helps to control the disease.
- ✰ Tool used for cutting the planting material should be regularly disinfected to avoid mechanical dissemination.
- ✰ Avoid fields subject to a high water table during wet conditions.
- ✰ Avoid prolonged irrigation of mature roots during warm months of the year.
- ✰ Spray of immunomodulators like salicilyc acid, BABA *etc.* can be recommended.
- ✰ Chemical spray with coc and streptocycline is recommended.

16. Yam (*Dioscorea* spp.)

Bacterial Rot

It is wide spread on *Dioscorea* spp. It is a problem during storage.

Symptoms

The first symptom appears as bacterial ooze comes out from infected tubers. Bacteria dissolve the walts causing a watery soft rot. The liquid rot appears off-white to brown in colour and emits a distinctive pungent odour.

Causal Agent

The disease is caused by *Erwinia carotovora* (Jones) Holland. Bacterial rot of Chinese yam caused by *Pseudomonas allii* has also been reported by Honda *et al.,* 2022. Colonies on yeast peptone (YP) agar plates are pale yellow to light cream in color, smooth and glistening. All tested strains were gram-negative and produced a fluorescent pigment on King's medium B. The bacterium is positive for levan formation, oxidase activity, potato soft rot, arginine dihydrolase activity and negative for tobacco hypersensitive reaction, indicating that they belonged to LOPAT group IVa.

Disease Cycle and Epidemiology

The disease is easily disseminated by splashing rain and the numerous insets that are drawn to the pungent odour. Yams which are not promptly removed from the storage area can serve as an inoculum source for yam tubers stacked beneath. Periods with high relative humidity and cooler temperatures favour disease development.

Management

- ✰ Crop rotation with beans, corn and soybean should be followed and rotation with cabbage and potatoes should be avoided.
- ✰ Injuries by insects should be prevented.
- ✰ Harvest may be made during dry weather.
- ✰ Washing with 25 ppm of free chlorine to eliminate the soft rot bacteria.
- ✰ Various commercial *Bacillus* and *Pseudomonas* spp. are found to be effective against pathogen.
- ✰ Proper humidity and CO_2 management during storage is required.

17. Taro/Coco yam/*Colacasia esculenta* Schott.

Bacterial Leaf Necrosis

Symptoms

The disease is found in India, Latin America Pacific Island and West Africa. Initial symptoms are marginal necrotic areas separated from apparently normal leaf tissue by a bright yellow chloratic border. The initial point of entry is believed to be via hydathods. Torn leaves also allow areas of entry. The bacteria multiply and enter the vascular system causing vascular necrosis. As the disease advances, it ramifies the entire leaf, which becomes dry and brittle. A slower, gradual necrosis of the petiole follows.

18. Elephant Foot Yam [*Amorphophallus campanulatus* (Rox b.) Blume ex Dene].

Bacterial Leaf Spot

It was observed in Bihar and Andhra Pradesh states in 1968 and 1970 respectively (Jindal *et al.,* 1972).

Symptoms

Circular toelongated spots with translucent or deep green water- soaked centre surrounded by a yellow border appear on the leaves. Further enlargement of the spots is restricted by the veins and thus linear stripes develop between the veins. These stripes often look yellow on the upper surface of the leaf and deep green on the lower surface. These ultimately turn brown and necrotic. Severely diseased leaves become yellow, flaccid and dry. The diseased tissue may often become papery which may fall down, leaving shot holes and may give ragged appearance to the leaf.

Causal Agent

Xanthomonas euvesicatoria pv. *amorphophalli* (Jindal *et al.*) Constantin *et al.,* 2016 (Syn.: *Xanthomonas campestris* pv. *amorphophalli* (Jindal *et al.*) Dye. The bacterium is a Gram-negative, rod shaped with 1.5-2.0 x 0.5-0.75 μm in size. Bacterium forms chains and capsules and is motile by one polar flagellum. No spores are formed and they aerobic. Colonies of bacteria on nutrient agar are circular, straw yellow, slightly raised and glistening.

Disease Cycle and Epidemiology

Under high humidity the long leaves develop rapidly translucent stripes. But under dry conditions, the spots do not enlarge but quickly turn necrotic.

Management

Hot water treatment of seed corm at 52°C for 30 min followed by dipping the corms in agrimycin 100 ppm for 30 min and three field sprays of agrimycin 50 ppm at 30, 45 and 60 days after planting resulted in 85 to 89.5 per cent control.

19. Amaranthus (*Amaranthus* spp.)

i. Bacterial Leaf Spot

It was first reported from New York. In India, it was reported from Pune, Maharashtra in 1952. It has been reported from Delhi, Karnataka, Rajasthan, Tamil Nadu and Uttar Pradesh.

Symptoms

Disease is noticed only on leaves in nature. It produces a few water soaked spots on the lower surface, become visible on the upper surface of leaf within 5-6 days as pale brown, round areas surrounded with yellowish halo. These water soaked spots look translucent against light and measure 0.5 mm in diameter. On the advancement of the disease, the spots become deep brown and depressed on the lower surface where as at upper surface spots are raised at margin and parched in centre.

Often these spots coalesce and form irregular lesions measuring 1.5-2.0mm. Spots invariably start from margin and some times bacterial ooze are present in the centre. Stem and petiole get infected under favourable environmental conditions when inoculated artificially. In advance stages of disease, side of the stem gets cracked becoming jet black where as the control portion remains grey. In case of petiole, cracked portion remain brownish.

Causal Agent

The disease is caused by *Xanthomonas campestris* pv. *amaranthicola* (Patel, Wanker and Kulkarni) Dye. Bacterium is Gram negative short rods with rounded ends, measuring 1.3 X 0.6 μm mostly single rarely in pairs, capsulated and motility by single polar flagellum.

Management

Hot water treatment of seed corm at 52°C for 30 min followed by dipping the corms in streptocycline 100 ppm for 30 min and three field sprays of agrimycin 50 ppm at 30, 45 and 60 days after planting resulted in better control.

ii. Bacterial Leaf Spot

The disease is caused by caused by *Burkholderia andropogonis* (Stapp) Gills *et al.*, A new bacterial disease of *A. cruentus* in Japan is described by Murai and Goto in 1996. Early symptoms include the presence of dark green water soaked spots on leaf blades, usually beginning on the lower leaves. These enlarge to form circular spots several mm in diam. light brown to reddish brown in colour.

Chapter 11

Bacterial Diseases of Field Crops

1. Rice (*Oryza sativa* L.)

i. Bacterial Blight

Bacterial blight of rice was reported in 1884–85 in Kyushu, Japan and the causal agent, the bacterium *Xanthomonas oryzae* pathovar *oryzae*, was identified in 1911, at that time having been named *Bacillus oryzae* and Philippines sixty years ago. The disease and causal organisms were described in detail from Japan by Uyeda and Ishiyama in 1922, who named the organism *Pseudomonas oryzae*. Dowson (1949) named it as *Xanthomonas oryzae.* Dye (1982) renamed it as *Xanthomonas campestris* pv. *oryzae.* Now it is called as *X. oryzae* pv. *oryzae*. Thriving in warm, humid environments, bacterial blight has been observed in rice-growing regions of Asia, the western coast of Africa, Australia, Latin America and the Caribbean.

The disease apparently remained a minor one until 1950 when it attracted attention in Japan. Bacterial blight was described in our country by Srinivasan *et al. (*1959) from Pune and Rangaswami, 1961, (unpublished) from Tanjore. But the disease was not taken seriously until 1962 when epidemics broke out in Bihar and other parts of north India. This sudden flave up was followed by the introduction of exotic rice varieties, especially Taichung Native -1 which is highly susceptible. Now the disease is considered a severe menace to rice production in India and occurs all over the country in the exotic and indigenous rice varieties. In recent years, disease is prevelant in non-traditional rice growing areas of Punjab and Haryana.

Symptoms

The bacterium induces either leaf blight or wilting of plants. Wilting syndrome is known as 'Kresek' and occurs sporadically in the fields causing serious damage. It commonly occurs within 3-4 weeks after transplanting of the crop. Kresek results either in wilting of only a few leaves or death of whole plant. Leaf blight phase is the most predominant phase of the disease occurring between tillering and heading stages of the crop. A third type of less evident symptom caused by the bacterium is yellowing of leaves, such leaves show blighted appearance.

Initial symptom of the blight phase is the appearance of dull greenish water soaked or yellowish spots 5-10 mm in length on the leaf towards the tip or margins, leading to tip and marginal drying. The infection soon expands along one or both margins, sometimes to the leaf sheath. As the disease progresses several lesions coalesce to form straw brown large lesions or blighted portion. The inner margin of the blighted patch in contact with the adjoining green portion of the leaf is ragged or wavy. Occasionally, the lesion may extend from the tip downward along the midrib itself and the leaf margins remaining green. Small droplets of bacterial ooze, pale amber in colour are found on the affected portions. When dried, these droplets form minute crusts and impart a rough touch to the leaves when passed between the fingers. The lesions are usually covered by saprophytic fungi, giving them a dark brown colour. In severe infections, all the leaves are attacked and premature drying is observed. When the affected leaves are cut and immersed in clear water in a test tube, a turbid ooze of the bacterium streaming from the vascular bundles, can be observed. The loss in grain yield due to this disease may range up to 60 per cent.

Causal Agent

The disease is caused by *Xanthomonas oryzae* pv. *oryzae* (Ishiyama) Swings *et al.* The bacterium is rod shaped, measuring 0.5-0.8 x 1-2µm, occurring single or in pairs, Gram-negative, aerobic, capsulating and non spore-forming, with a single polar flagellum. The mucous capsule is soluble in water, precipitated by acetone and is a polysaccharide called Xanthomonadin.

The virulence of the isolates varies on different cultivars. Isolates from India are more virulent than those from East Asia. Five different pathotypes have been identified in Japan based on the reaction of the isolates on a set of differential varieties. Indian studies using international set of differentials (IR8, IR20, DV85, BJ1 and Chemposelete) have so far revealed the existence of two distinctly different pathotypes. The isolates differ in their plasmid profiles and virulence of isolates is influenced by plasmids.

Disease Cycle and Epidemiology

Primary infection may result from the inoculum overwintering in the seed, husk as well as in the endosperm. It also survives in soil or plant stubbles and debris and the initial inoculum may be built upon the nursery seedlings. The bacterium is reported to infect some grasses, like rice cut-grass (*Leersia oryzoides*), *Cyperus rotundus,* Chinese sprangletop (*Leptochloa chinensis*) and common grasses and weeds which might play a role in the spread of the disease. It is disseminated through irrigation water and windborne rain splash, while hydathodes and wounds are its portals of entry. Once enters the host, the bacterium becomes systemic in the vascular bundles. The disease symptoms may not show up until few weeks after transplanting and become more severe at the time of flowering.

A combination of rainy weather, dull windy days and an atmospheric temperature of 22-26 °C are conducive for the development of the disease. Plants in shade and close planted crops supplied with high doses of nitrogen show more disease incidence.

Management

- ☆ Several bactericides are being used to control the disease, but none is very effective. However, since the disease is seed-borne, the seeds should be disinfected. Soaking seeds for 8 hours in agrimycin (0.025 per cent) followed by hot water treatment for 10 minutes at 52-54°C eradicates the bacterium from the seed.
- ☆ A mere soaking of seed for 8 h in streptocycline-crude agricultural preparation (3 gm in 11 liters of water) is also recommended.
- ☆ Spraying with copper fungicides alternately with streptocycline (250 ppm) is reported to be effective in controlling bacterial blight.
- ☆ Varieties like TKM6, IR42 and Chinsurah Boro II are tolerant to both kresek and leaf blight phases of the disease.
- ☆ The most-common method of defending against rice bacterial blight is the cultivation of rice varieties with genes that confer resistance to *Xoo* infection. Over 30 resistance genes, termed *Xa1* to *Xa33*, have been identified in rice plants and some, such as *Xa21*, have been integrated into the genomes of commercial rice strains. These resistant rice varieties have been largely successful, dramatically reducing yield losses in many rice-producing countries.
- ☆ Vidyashekharan *et al.* (2001) have reported induction of systemic resistance by *Pseudomonas fluorescens* Pf. 1. However, the induced resistance was only transient. Leaves were not protected when challenge inoculated beyond 60 days. Rangarajan *et al.* (2003) have reported that four strains of *Pseudomonas* tested by them suppressed bacterial leaf blight as well sheath blight (*R. solani*).
- ☆ Leaf extracts of *Adhatodavasica*, *Allium cepa*, neem (margosa) and turmeric are reported to significantly reduce blight incidence and increase yield (Madhiazhagan *et al.*, 2002).

ii. Bacterial Leaf Streak

Although less damaging than leaf blight, leaf streak is of some importance. This was first identified as a distinct disease in 1957, when the pathogen was described in China (Fang *et al.*, 1957). It was earlier confused with leaf blight. The disease occurs in India and it may be present in other tropical countries.

Symptoms

Bacterial leaf streak appears as fine, interveinal, long or short lines on the leaf blade which are water soaked and grayish. Yellowish beads of dried bacterial exudates are seen on closer scrutiny. The lesions enlarge and coalesce to form big patches and become yellow. Ultimately a large portion or often the whole leaf blade, becomes yellow or dirty white. At this late stage the symptoms are difficult to differentiate from those of bacterial blight.

Causal Agent

The disease is caused by the bacterium *Xanthomonas oryzae* pv. *oryzicola* (Ishiyama) Swings *et al.* (1990); Fang *et al.* (1957) described the morphology and physiology of the bacterium. They are rods, measuring 1.2 x 0.3-0.5 µm, single, occasionally in pairs but not in chains; no spores and no capsules; motility by a single polar flagellum. The bacterium is Gram negative, aerobic, grows preferably at 28°C. Colonies on nutrient agar are pale yellow in color, circular in shape, smooth, margin entire, convex or viscid.

Disease Cycle and Epidemiology

Shekhawat *et al.* (1969) reported that seed from an infected crop harbors the bacterium and is a potential source of the disease and its dissemination to distant places. The bacterium is found to be present both inter and intracellularly. It gains entry through the injured epidermal cells or directly through the stoma of the leaf, colonizing the substomatal cavity and mesophyll tissues (Parenchyma tissue). Slight discolouration of the invaded cells are noticed in the early stages of infection but in advanced stages of the disease the leaves turn completely yellow and degeneration of the protoplasts in the mesophyll occurs. The vascular bundles are not invaded by the bacterium. The bacterium seems to survive the off season under the glumes and when the seed is sown, contaminates the plumule during germination. Infection of coleoptile, leaf sheaths and first leaf occurs through the stomata. The bacterium is carried to the foliage by the first leaf. The bacterial exudates from the flag leaf causes seed infection in the panicles. Young leaves are more susceptible to the disease than the older ones. Continuous high humidity for 2-3 days or dew during early morning hours favours infection. Lesion development is high at moderate temperatures (26.0-20.3°C).

Management

- ✰ Seed treatment has been found to give partial control using a soak treatment in 0.5 per cent Thiram for three hours or 0.2 per cent Thiram for two hours.
- ✰ The use of resistant varieties will hopefully give good control. In India, 132 out of the 372 varieties tested at the National Rice Research Institute, Cuttack were very resistant and in the Philippines, most japonica varieties have been found to be highly resistant, although most of the indica varicties tested were highly susceptible.

iii. Culm and Foot Root Disease

Bacterial foot rot occurs in Japan, India, Bangladesh, Korea and the Philippines (Goto, 1979 a,b).

Symptoms

Leaf sheaths of affected plants typically exhibit a dark brown decay and attached leaves turn yellow in color and wilt. Infection often begins in the ligules. The nodes, culms and crown also become rotted and infected tillers can easily be detached from the crown. Infected culms and internodes turn black in color. Roots attached to infected nodes decay and fall off. Bacterial ooze may be present inside the culms and infected plants have an unpleasant odor. In some cases, the young leaves of tillers that show no sheath browning may wilt as a result of systemic infection of the crown alone.

Causal Agent

The disease is caused by *Dickyea zeae* (Syn: *Erwinia chrysanthemi*). The bacterium is a Gram-negative rod with four to six peritrichous flagella.

Disease Cycle and Epidemiology

The bacterium has a wide host range. In Japan, Iris plants, which are common around rice fields, were shown to be susceptible to the bacterium and capable of serving as an inoculums source. The bacterium can also be isolated from the rhizosphere of healthy rice plants and thought to be disseminated primarily by the irrigation water. The bacterium enters rice plants through wounds in the root.

Minor Diseases of Rice

i. Seedling Blight

Seedling blight is a relatively new disease that occurs in Japan and is associated with the production of seedling in nursery boxes. Seedlings for transplantion are raised in nursery boxes under conditions of high humidity and temperature that are favourable to infection. Initial symptoms are characterized by a basal chlorosis and withering of the second or third leaves. Infected seedlings later become reddish brown in color and desiccated but do not exhibit a soft rot symptom. Under severe infection, root growth is retarded and seedlings easily lodge.

The disease is caused by *Pseudomonas plantarii* Azegami *et al.,* 1987. The bacterium is a Gram-negative, nonspore forming, non-encapsulated rod, 0.7-1.0 x 1.41.9/μm, in pairs, or in short chains, motile with one to three polar flagella (Azegami *et al.*, 1987a). *P. plantarii* produces a compound called tropolone, which is responsible for the retardation of root growth and leaf chlorosis of infected rice seedlings (Azegami *et al.*, 1987b). Aerobic. Colonies are translucent, glistening, convex and round with entire edges and a smooth surface having a slight yellow tint. Colonies weakly produce water-soluble reddish brown pigment depending on conditions of the culture and media. Bacteria are positive in Gelatin liquefaction, Lecithinase, oxidase, denitrification and catalase test. It does not require an organic growth factor and accumulates poly-p-hydroxybutyrate and does not produce fluorescent pigments and produce levan from sucrose. Ithydrolyzes Tween 80, cotton seed oil and casein and does not hydrolyze starch, esculin, or arbutin, reduces nitrate to nitrite. It is negative in arginine dihydrolase and phenylalanine deaminase.

The application of iron compounds inhibits seedling blight, because the production of tropolone is supressed in the presence of iron. A thionine producing gene from oat, coding for antimicrobial activity, when transferred to rice by biotechnological intervention, conferred resistance against the pathogen.

ii. Bacterial Brown Stripe

Bacterial Brown Stripe also called as bacterial stripe, occurs in upland and wetland nurseries as well as in nursery boxes. Although it is widely distributed in the rice growing countries (Shakya *et al.,* 1985), it does not cause much damage.

The symptoms of the disease in the seedling stage were divided into four types, *viz,*.inhibition of germination, brown stripes occurring interveinally or along the midrib or leaf margins, curving of a sheath and abnormal elongation of mesocotyl. If the infected seedlings were transplanted to paddy fields, symptoms were masked. Natural occurrence of the disease was not seen after the tillering stage, except for the case where rice plants were submerged in the water by flood (Kadota and Ohuchi, 1983).

Bacterial brown stripe is caused by both *Pseudomonas avenae* and *P. syringae* pv. *panici. P. avenae* is a Gram-negative, non spore-forming, non-encapsulated rod. 0.92-2.4 x 0.5-0.7 μm, with one or two flagella. *P. syrinage* pv. *panici,* is also a Gram-negative, spore forming, non-encapsulated rod (Shakya *et al.,*1985). *P. avenae* is seedborne and *P. syrinage* pv. *panici* is likely seed borne. Natural infection of *Panicum miliaceum, Hordeum vulgare* and *Setaria italica* by *P. syringae* pv. *panici* has been reported.

Dry heat treatment at 65°C for 6 days can eliminate the pathogen from seeds. In nursery boxes, spraying of kasugamycin can control the pathogen.

iii. Halo Blight

In 1985, halo blight in rice caused by *Pseudomonas syringae* pv. *oryzae,* was first reported by Kuwata and is currently limited to Aomori Prefecture, Japan. The disease is characterized by circular in shape, pale green to yellowish brown lesions, 2-10 μm in diameter on leaf blades. The lesions are surrounded by a distinct halo and have a dark brown spot or stripe in the center. The lesions may coalesce to form large blotches. The disease does not cause serious damage to date (Kuwata, 1985).

iv. Leaf Sheath and Grain Rot

Sheath brown rot has been reported in Asia, Latin America, South America, Central Africa and Madagascar. The disease has been reported to be widespread in irrigated rice between 1300-2000m elevations in Madagascar.

On seedlings, a systemic streak of the leaf sheath occurs, which may extend to the midrib or veins of the leaves. In case of mature plants, symptoms typically occur on the flag leaf sheath from the booting to heading stage, become dry and the panicle withers. Glumes of panicles emerging from infected sheaths exhibit water soaked lesions that turn light brown in color. Grains of infected panicles are colorless, deformed or empty (Zeigler and Alvarez, 1987, 1990).

The disease is caused by *Pseudomonas fuscovaginae*, which is a Gram negative, non-spore forming rod in shape, 0.5-0.8 x 2.0 x 3.5 μm with one to four polar flagella (Tanii *et al.*, 1976). *P. fuscovaginae* survives on rice seed at a low level and as an epiphyte on *graminaceous* weeds in rice growing areas. Disease development on mature plants is favored by cool day time temperatures (17-23°C) that delay panicle emergence.

Use of healthy seed or seed treated with dry heat at 64°C for 6 days is important in management of sheath brown rot. Antibiotics such as streptopmycin, alone or in combination with oxytetracycline also, can affectively manage sheath brown rot if applied at or a few days after panicle emergence.

v. Sheath Rot of Rice

Sheath rot caused by *Pseudomonas syringae* pv. *syringae* (Syn.: *Psuedomonas oryzicola*) is identical to the symptoms of sheath brown rot caused by *P. fuscovaginae*. It is the only reported sheath rot pathogen of rice in Chile and has been reported from Asia, Australia and Hungary (Zeigler and Alvarez, 1990).

vi. Bacterial Panicle Blight and Grain Rot

Grain rot occurs in Japan, Korea and Taiwan. The disease is demonstrated as a grain rot of mature plants in the field and also as a seedling rot (Chien *et al.*, 1983). On seedlings, symptoms consist of a brown, water-soaked soft rot of leaf sheaths accompanied by wilting or soft rot of the leaves. On the panicle grains are shrunken and pale green, becoming dirty yellow to brown and dry. Between the infected and healthy parts of the grain a brown margin is a diagnostic feature of the disease. The flag leaf sheath collar or a mild rot of the flag leaf sheath occurs (Zeigler and Alvarez, 1990). Grain rot is caused by *Burkholderia glumae*. The bacterium is a Gram negative, rod shape, size 0.50.7 x 1.5-2.5 μm with one to three polar flagella. The bacterium is seed borne and invades the spaces between the cells in the outer epidermis and spongy parenchyma of the lemma. From seed, it may grow epiphytically at a low level until panicle emergence. The population of bacteria increases rapidly on the grain. The disease is favored by high temperatures (28°C) and high humidity.

The pathogen may be eradicated from small seed samples with dry heat treatment of 65°C for 6 days. Pre-treatment of rice seeds with a high concentration of (10^{10} cfu/ml) of the avirulent strain of *P.glumae* is the most effective method for reducing the incidence.

vii. Bacterial Palea Browning

The disease occurs in Japan and affects grain quality. Disease incidence was as high as 32 per cent and reduction in 1000 seed weight by as much as 15 per cent have been reported (Azegami *et al.*, 1983). Initially symptoms appear at early stages. First, light brown, water soaked lesions occur on the lemma or palea. The lesions then turn dark brown in color, The discoloration occurs mostly on the palea. Affected panicles have more immature grains and lighter grains at harvest and infected grains become brown after milling.

The bacterium responsible for the disease is *Erwnia herbicola*. It is Gram negative and fermentative, with peritrichous flagella. Disease cycle and epidemiology High epiphytic population of *E. herbicola* are common on rice. The disease occurs when heading corresponds with periods of rain and high temperature in the range of 30-35°C. Disease incidence is increased in fields with high levels of nitrogen fertilization, especially at heating

2. Wheat (*Triticum* spp.)

i. Tundu Disease or Yellow Ear Rot

This disease has been reported from Punjab and Haryana and some parts of Uttar Pradesh and Rajasthan. It was first recorded by Hutchinson in 1917. The same disease has also been reported from Egypt, parts of Australia and China. The disease is also called spike blight, yellow ear rot, yellow slime rot, disease in wheat and barley.

Symptoms

Yellow slime on the inflorescence and stem, which dries up to form sticky yellow layers and causes curling and twisting of the spikes. Most of the grains in the ear head are replaced by galls formed by the nematodes and these galls carry the bacterium. When the nematode is associated with the bacterium, the damage caused is often very severe.

Causal Agent

The disease is caused by a bacterium *Rathayibacter tritici* (Carlson and Vidaver) Zgurskaya *et al.*, 1993 with association of nematode *Anguina tritici* which produces seed galls (ear cockles) in certain plant varieties and facilitates spread of infection. The bacterium can be differentiated from other corynefrorm group of bacteria by the cell wall as peptoglycan of this genus including plant pathogens is based on diamino butyric acid (DAB) and absence of mycolic acid. Predominated menaquinones contain 9 and 10 carbon chains (MK-9 and MK-10). The v+c ratio are

67-68 mol per cent. The bacterium is gram positive, coryneform rods, non-motile with size 1.1–1.6 µm x 0.5–0.6 µm. Colonies appear on media (nutrient dextrose agar) for sub culture only, yeast peptone glucose agar, nutrient yeast dextrose agar after 20 days at 21°C incubation, pin point, cream and domed within 10 days due to slow growing. Typical colonies are-creamy white or ivory, round, smooth, raised convex (domed), mucoid - fluidal with entire edges and usually 1 - 3 mm in diameter. On an aerobic NBY agar plate, *R. tritici* formed round colonies and produced yellow pigment within 96 h at 25°C.

Disease Cycle and Epidemiology

Disease may spread through infected seeds along with seed-borne nematode galls of *Anguina tritici.* Since both 'ear cockle' and 'tundu disease' occur on wheat frequently in association, the bacteria are carried along with the galls developed due to ear cockle disease caused by the nematode. Such galls get mixed with the seed lot and the pathogen perennates therein. In addition, the nematode galls containing bacteria may survive in soil for 20 years or more. It has been studied that the bacteria can survive inside the nematode galls for a fairly long lime, probably for at least 5 years. The bacteria survive in the nematode which serve as vector. When seeds contaminated with bacteria and nematode galls are sown in the field, the galls absorb moisture from the soil and release out the nematode larvae and bacteria. If galls are already perpetuating in soil, they also release nematode larvae and bacteria in the same manner. However, the nematode larvae shortly start climbing up on the young wheat plants carrying bacteria with them. After the plants flower, the nematodes enter the floral parts wherein the bacteria establish primary infection causing tundu disease. Simultaneously, the nematodes may cause "ear cockle" disease and the two diseases may appear in association on wheat plants. Chances of secondary infection during the same growing season are least as the disease appears when the crop is reaching maturity. But, however, if most favourable conditions prevail in the field for a considerable time period, the bacteria produced in primary infection and released with yellow slimy ooze can cause secondary infections on healthy plants.

Under humid conditions, oozing of the slime is visible from affected plant tissue. Under dry conditions, the slime becomes hard and turns to dark yellow or brown in colour. Low temperature of 10-15° C with high humidity of 70-100 per cent the bacterium multiplies and expresses itself in the form of yellow slimes on the leaves of young plants. The disease cycle is repeated when seeds with galls are used for sowing.

Management

- ☆ The sowing of gall-free seeds in non-infected soil will help to reduce the incidence of disease.
- ☆ The seeds can be freed from the galls by being floated in brine, at 160 g of sodium chloride per litre of water.

- Wheat, barley, oat or other susceptible crops should not be sown in infected soil.
- Solar heat during May and June in the northern wheat belt of India can be utilized for destroying the galls and nematodes.
- Infected plant should be carefully taken out and burnt.
- Seed for sowing purpose should be taken from disease free areas only.

3. Sorghum/(Jowar) (*Sorghum bicolor* L.)

i. Bacterial Leaf Streak/Blotch

The disease has been reported worldwide but not reported on sorghum from India. It is a minor importance. The bacterial pathogen infects *Panicum miliaceum* (millet), *Setaria italica* (foxtail millet), *Sorghum halepense* (Aleppo grass), *Sorghum sudanense* (sudan grass) and *Sorghum bicolor* (common sorghum) (Williams *et al.,* 1978).

Symptoms

It is found on plants of all ages the lesions are elliptic, orange brown to deep red coloured spots, which coalesce to form large patches, occur on the blade producing an orange red pigment which diffuses and turns the leaf blade into bright red. The affected plants are stunted and the younger leaves appear to be chlorotic. Ears rarely emerge and if they do they are small and chaffy. Abundant bacterial exudates are produced as light-yellow droplets, which dry to thin white or cream scales.

Causal Agent

The disease is caused by *Xanthomonas vasicola* pv. *holcicola*(Elliott) Vauterin *et al.,* 1995, [*Xanthomonas campestris* pv. *holcicola* (Elliott) Dye].The bacterium is a rod with a single polar flagellum, measuring 1.0 - 2.0 X 0.5-1.9 μm.

ii. Leaf Stripe

In general, the disease is of minor importance on sorghum. Bacterial leaf stripe is considered to be a low to intermediate priority disease of sorghum in eastern Africa. The pathogen infects *Sorghum halepense* (aleppo grass), *Sorghum bicolor* (sorghum), *Sorghum sudanense* (Sudan grass), *Trifolium repens* (white clover), *Vicia sativa* (common vetch). *Bougainvillea* sp; *Ceratonia siliqua* (locust bean), *Cicer arietinum* (chickpea), *Dianthus caryophyllus* (carnation), *Gypsophila paniculata* (babysbreath), *Limonium sinuatum* (sea pink), *Trifolium pratense* (purple clover), *Trifolium subterraneum* (subterranean clover), *Tulipa* (tulip), *Vaccinium* (blueberries) and *Zea mays* (maize).

Symptoms

The disease is characterized by its vascular nature. Where in the lesions appears to be narrow within 5-10 mm long which appears on the leaves all along the vascular region, giving reddish brown color to the sheath. Often several small lesions coalesce to form large spots. A brownish exudate containing masses of bacteria collects on the spotted region.

Causal Agent

The disease is caused by *Burkholderia andropogonis* (Smith, 1911; Gillis *et al.*, 1995). (Syn.: *Pseudomonas andropogonis* (Smith, 1911) Stapp 1928. The colonies of these bacteria are mostly smooth on the medium. The bacteria are rod shaped, slightly curved with rounded ends measuring 1-2 X 0.7 - 1.04 µm, usually motile due to bearing polar single flagellum. The flagellum is sheathed. As these two bacterial diseases are of limited importance only in certain parts so no control measures has been worked out.

Disease Cycle

Plant debris is considered to be the primary over wintering source of *B. andropogonis* for bacterial stripe infection of sorghum (Tarr, 1962).

4. Maize (*Zea mays*)

Bacterial Stalk Rot

Five bacterial diseases of maize have been reported in India of these, *Pseudomonas syringae* pv. *lapsa, Erwinia carotovara* subsp. *dissolvens* and *E. carotovara* subsp. *zeae* are important causing severe damage to the crop. These bacteria infect the basal portions of the stem, causing soft rot and sudden wilting of the plant. *Erwinia carotovara* subsp. *zeae* also causes rotting of the ear head and shoot. However, *Xanthomonas campestris* pv. *maydis* and *X. campestris* pv. *rubrilineans* infect the leaves and are less common, though the latter is reported to cause severe damage on a few cultivars in some parts of North India.

This disease was first reported in 1922 in the USA by Rosen, who named it *Pseudomonas dissolvens.* Since then it has been known to occur in South Rhodesia, Israel, Greece and South Africa. In India, Prasad (1930) reported the occurrence of the disease in Pusa, Bihar. The disease is more prevalent in Himachal Pradesh, Punjab, Delhi, Uttar Pradesh, Bihar, West Bengal, Rajasthan, Andhra Pradesh and Telangana. In 1959, Hingorani *et al.* identified the pathogen as *Erwinia carotovora* f. sp. *zeae* Sabet. Bacterial pathogen attacks tubers of potato and sweet potato, onion bulbs, bean pods, roots of carrot, turnip, radish and sugar beet, fruits of tomato, brinjal, chillies and papaya and plants of rice, pearl millet, sorghum, brinjal, potato, tomato, tobacco and cabbage *etc.*

Symptoms

The basal portion of the leaves rots. The basal internodes become soft and discoloured, emitting an unpleasant odor. The presence of an unpleasant odor is one of the things that distinguishes the symptoms of stalk rot disease caused by bacteria and fungi. The stalk becomes soft and pliable (flexible) and the affected plants topple down rapidly. Ear shoots and cobs occasionally get infected which then droop down and hang simply.

Causal Agent

The disease is caused by *Dickeya zeae* Samson *et al.,* 2005 (Syn.: *Erwinia carotovora* var. *zeae* Sabet. Maize stalk rot organism is a typical soft rot bacterium. However, only maize isolates can attack maize and onion. *E. carotovora* isolates do not infect maize. Therefore, the variety *zeae* has been retained to indicate its specificity. Specific identification of the presence of this bacterium is its ability to produce indigoidine, a blue pigment that is insoluble in water. *D. zeae* is the only species in the bacterial genus *Erwinia* that is able to give a blue color to bacterial colonies as a chemotaxonomic property for rapid identification. The bacterium is rod shaped and gram-negative; size varies from 0.8-3.2 x 0.5-0.8 μm and motile containing 3-14 peritrichous flagella. The bacterium is facultative and non-fluorescent anaerobic, pectinolytic in potato tubers, causing hypersensitivity reactions in tobacco leaves, producing catalase and lecithinase, not producing oxidase or arginine dehydrolase, capable of reducing nitrate and can grow at a temperature of 37°C. These bacteria were producing white, slimy and shiny colonies on King's B medium, while on Nutrient Agar medium the bacterial colonies are gray and slightly prominen. Phages specific to this pathogen do occur in nature.

Disease Cycle and Epidemiology

The organism is soil borne. The soil represents a favorable habitat for microorganisms and is inhabited by a wide range of microorganisms, including bacteria, fungi and protozoa. *D. zeae* survives in plant debris but the survival period varies from different environmental conditions. The survival period was longest (9 months) in soil which contained naturally and artificially infected maize plants as debris. The maize borer (*Chilo partellus*) was shown to act as a carrier of this bacterium. It spreads the pathogen from diseased to healthy plants (Thind and Singh, 1976). Injuries and wounds serve as avenues to enter the host. The isolates of stalk rot pathogen considerably affect the seed germination. All the varieties tested were found to be susceptible to stalk rot pathogen.

D. zeae is preferred high temperatures and high relative humidity for infection and disease development. It can be a problem with areas of heavy rainfall or where overhead irrigation is used and the water is pumped from a lake, pond, or slow-moving stream. A temperature of 35°C, 70 per cent RH (relative humidity) and inoculum level of 2 x 10^8 cells favour the disease development.

Management

- ✰ The pathogen infection can be suppress via organic manure amendment which stimulates the population of beneficial microflora and avoid flooding and excessive irrigation.
- ✰ Ridge sowing method also helps to the farmer to manage that disease.

- ✰ CM 600, CM 104 and CM 105 maize lines and their crosses show partial resistance against *E. chrysanthemi* pv. *zeae* in the field. Two inbred lines CM101, CM- 110 and two open pollinated varieties CM600, Basi are found to be tolerant against *E. chrysanthemi* pv. *zeae*.
- ✰ Water- logging and dense planting should be avoided to decrease the disease pressure under field conditions.
- ✰ Agricultural antibiotic combinations, such as streptocycline and agrimycin-100, chlorinated water (100 ppm) as well as bleaching powder were found to be effective in controlling the disease.
- ✰ Bio-agent (*Pseudomonas fluorescens, Bacillus subtilis* and VAM fungi (*Glomus fasciculatum*) are found effective control the disease.

5. Bajra/(Pearl millet) (*Pennisetum glaucum* and *P. americanum*)

i. Bacterial Leaf Streak

The disease occurs in severe form in parts of Southern India affecting the leaves. The organisms are confined to the leaves, causing leaf spot symptoms. Leaf spots are characteristically reddish brown, almost rectangular, with 5-10 X 2-4 μm in size and mostly bound by vascular margins. The disease is caused by *Xanthomonas campestris* pv. *pennamericanum* (Young *et al.*, 1996). Bacterium is rod shaped posing a single polar flagellum.

ii. Leaf Blotch

The disease is characterized by narrow, reddish brown linear streaks with the length of 5-10 mm, coalescing to form large blotches and is distributed throughout the leaf blade, except mid rib. The under surface of the affected portions is dull brown in colour. The affected portions dry off and cause withering of the leaf. Control measure is not yet identified. The disease is caused by *Xanthomonas campestris* pv. *annamalaiensis.* Bacterium is rod shaped with a single polar flagellum.

iii. Bacterial Brown Spot

The diseases caused by this bacterium are very important throughout the world. For instance, bacterial brown spot occurs wherever beans are grown and canker diseases of fruit trees caused by *Pseudomonas syringae* pv. *syringae* are widespread and may be devastating, causing economical losses. The symptoms first appear on the lower sides of the leaves as small, water-soaked spots. The spots enlarge, coalesce and form larger areas that later become necrotic. The bacteria also enter the vascular tissues of the leaf and spread into the stem. The infected area, which is surrounded by a narrow zone of bright, lemon yellow tissue, turns brown, becomes rapidly necrotic and through coalescence of several small spots, may produce large dead areas of various shapes. The disease produces identical symptoms on the stems, pods and seeds. In addition, light-cream or silver bacterial exudates are often produced on the lesions under moist conditions. The pathogen has multilateral hosts like gramineae and leguminosae.

P. syringae pv. *syringae* is a rod shaped, gram-negative with polar flagella. *P. syringae* tests negative for arginine dihydrolase and oxidase activity and forms the polymer levan on sucrose nutrient agar. It is known to secrete the lipodepsinona peptide plant toxin syringomycin and it owes its yellow fluorescent appearance when cultured *in vitro* on King's B medium to product the siderophore pyoverdin. Pathogen is seed borne and it spreads through the infected seed from season to season (Gaudet and Kokko, 1986). Diseases by *P. syringae* are favoured by wet, cool conditions- optimum temperature for disease development is around 12 - 25°C, although this can vary according to the pathovar involved. The pathogen is seedborne and is dispersed between plants via rain splash (Hirano and Upper, 1990).

6. Finger millet (*Eleusine coracana*)

i. Bacterial Leaf Spot

It is present in parts of Tamil Nadu, Karnataka and Maharashtra. The spots appear on both surfaces of the leaf, they are linear, spreading along the veins extending up to 2.5 cm or more. In the beginning they appears to be light yellow brown, soon they turn dark brown in color. As the disease proceeds, the leaf blade gets shredded by splitting along the affected region. Often the bacterium affects the tender shoots causing severe damage. The disease is of minor importance at present and no control measures are known. The disease is caused by *Xanthomonas euvesicatoria* pv. *caracanae* (Syn.: *Xanthomonas axonopodis* pv. *caracanae* Vauterin *et al.*, 1995; *Xanthomonas campestris* pv. *eleusineae* Dye). The bacterium is a rod shaped bearing a single polar flagellum.

ii. Bacterial Blight

The disease has been reported from Uganda and it is caused by *Acidovorax avenae* subsp. *avenae*. Seedlings of finger millet grown in soil under controlled conditions (30°C, 12 h light/dark cycles) showed severe blight and necrotic stripe symptoms (Mudingotto *et al.*, 2002). Translucent, whitish-grey bacterial colonies measuring 2–3 mm and having fringed margins appear after 24 h on nutrient agar or King's B media agar plates. The strains were Gram-negative, non-fluorescent on King's B medium, arginine dihydrolase negative, nitrate positive and starch-negative.

7. Foxtail Millet (*Setaria italica*)

Bacterial Leaf Spot

This disease is reported from some parts of Tamil Nadu. This disease is characterized by the formation of minute, water soaked lesions on the leaves, later turning into light brown spots, 1-2 mm long and about 1 mm broad. These spots coalesce to form large patches which later become necrotic. The disease is caused by *Xanthomonas campestris* pv. *indica.* Bacterium is rod shape with single polar flagellum.

8. Cotton (*Gossypium* spp.)

Angular Leaf Spots and Black Arm

The disease occurs in all the cotton growing regions of the world, including USA, South America, Egypt, Sudan, Russia, Pakistan, Sri Lanka, China, Australia, *etc.* In India, the disease was first observed during 1918 in Tamil Nadu. Since then it has been found to occur in various degrees of intensities in all the major cotton growing areas of the country where the exotic American cotton is grown. The disease symptom and intensity depend largely on the climate conditions prevailing in each tract. In Egypt, Sudan and Uganda the black arm infection is most serious, whereas, in USA, seedling blight is more common and in India, angular leaf spot and boll rot symptoms are severe.

Symptoms

The bacterium is capable of infecting all the portions of the plant which are above the ground. There are four major types of symptoms.

1. **Seedling infection:** The early symptom of the disease is noticed on the cotyledons of the germinating seed. Water-soaked, circular lesions appear first on the lower surface of the cotyledons. The infection soon spreads and forms angular patches. The diseased tissues turn brown, distort the shape of the cotyledons and finally lead to the drying and withering of the seedlings. The whole seedling may collapse and wilt in severe infections.
2. **Leaf spots:** On the leaves, the spots are usually angular. These first appear on the under surface and then on the upper surface. The disease advances along the main veinlets and veins to cause finger - like lesions on the leaf blade and then the disease is identified as 'vein blight'. The leaves affected by such vein blights become crinkled and twist inwards and dry. The infection also spread through the leaf blade to the petiole causing severe defoliation. On the surface of the affected tissues there may be profuse exudation of the bacterial slime which may dry and form a thin film.
3. **Black arm:** The pathogen causes dark brown to black lesions on the stem and fruiting branches. The lesions are linear and sunken from the surface after covering the whole circumference of the stem or branch. Under favourable climatic conditions most of the branches and the main stem of the plant are affected by the disease. As the disease advances, the branches dry and soon the plant withers.
4. **Boll lesions:** The bacterium affects the boll or fruit also. Initially symptoms appear as water-soaked lesions on the surface. Such lesions turn dark brown and finally black. They are invariably sunken on the surface. When young bolls are attacked with several such lesions drop off from the plant. In severe conditions the bacterial infection spreads inside the boll or the fruit and causes severe reduction in size quantity and

quality of the lint formed inside. As the disease advances, the seeds invaded by the bacterium suffer reduction in their size and losses the viability.

Causal Agent

The disease is caused by *Xanthomonas citri* pv. *malvacearum* (Smith, 1901) Constantin *et al.,* 2016 [Syn.: *Xanthomonas citri* subsp. *malvacearum* (Smith, 1901) Schaad *et al.,* 2007]. It is an aerobic, Gram negative rod shaped bacterium with 0.3-0.6μm X 1.3-2.7 μm. The bacterium is motile by a single polar flagellum. On a 2 per cent sucrose peptone agar, a yellow, copiously mucoid, wet, shining growth is noticed. Gelatin, casein and soluble starch are strongly hydrolysed by it. Nitrite is not produced from nitrate by the bacterium. Aesculin is hydrolysed by it. A total of 14 races of *X. axonopodis* pv. *malvacearum* in Delhi. Cellulases, proteases of Indian isolates of *axonopodis* pv. *malvacearum* were worked out by Verma and Singh (1975).

Disease Cycle and Epidemiology

The pathogen is both externally as well as internally seed-borne in nature (Verma and Singh, 1974). The primary infection starts mainly in the seed where the bacterium may be present as a slimy mass on the fuzz or inside the seed. They found that the pathogen remained in a viable state for six and three months in the trash applied on the surface of the soil and buried 15 cm deep respectively. In the cases of flooding or sowing of wheat following trash application; the pathogen survives for less than one month. Self sown seeds germinated in March - April upto July to produce diseased seedlings which serve as a focus of disease spread. Standing diseased plants remains infective till the next season. The bacterium can enter the plant through openings such as stomates, lenticels and hydathodes and wounds when plants are damaged by wind-blown sand. Infected cotton bolls, leaves and twigs which fall on the soil surface also form an important source for the carry-over of the disease. The secondary spread of the pathogen occurs mainly through the agency of wind-driven rain.

The disease will be more severe in a field if the disease develops early in the season, especially if plants at the seedling stage become infected. Once the bacteria are established in a field, rain-fall, especially shortly after planting, can lead to rapid increase and spread. After the canopy develops, periods of heavy rainfall followed by warm and humid conditions when relative humidity is greater than 85 per cent further increases the development and spread of the disease. Throughout the season, conditions favorable for the development and spread of the disease occur when daytime temperatures are 32.2 – 37.7°C and nighttime temperatures are at least 16.6-20°C. Under favorable conditions, it has been reported that one seed infected with *Xcm* out of 6,000 is sufficient to start a bacterial blight epidemic in a given field.

X. citri subsp. *malvacearum* shows a broad range of virulence (Bayles and Verhalen, 2007). Hunter *et al.* (1968) identified 19 races and However, Verma and Singh (1975) classified the Xcm isolates into 32 races by testing them against 7 cotton differentials. Race 18 is the most virulent and occurs throughout the world (Thaxton *et al.* 2001; Verma and Singh 1975). Other races including 20, 21 and 22 were reported from Africa and were classified as highly virulent strains (HVS). HVS of Xcm were first reported in Africa (Follin *et al.*, 1988). Meshram *et al.* (2002) reported 14 different races such as 1–8, 10, 11, 12, 13, 15 and 18 from major cotton-growing regions of India. Gholve *et al.* (2005) found that among the 27 isolates obtained from the Marathwada region of Maharashtra (India), 89 per cent were race 18 and 11 per cent were race 7. (Kumar *et al.*, 2018) Thirty-four *X. citri* subsp. *malvacearum* (Xcm) isolates collected from three cotton-growing zones of India were subjected for virulence and race documentation and further correlated with genetic diversity as revealed by repetitive elements [repetitive extragenic palindromic (REP), enterobacterial repetitive intergenic consensus (ERIC) and BOX elements] and intersimple sequence repeat (ISSR)—PCR analyses. Among the 34 isolates tested for virulence on susceptible cultivar LRA 5166, 7 were recorded as highly virulent (HV), 16 were moderately virulent (MV) and 11 were less virulent (LV). Eight different races were recorded by using ten cotton host differentials. Twenty-two isolates (65 per cent) belonged to race 18. Twelve isolates (35 per cent) pertained to races 3, 5, 6, 7, 8, 11 and 13.

Management

- ☆ Seed treatment for the control of bacterial blight is an effective means of reducing the crop loss.
- ☆ Delinting with sulphuric acid followed by dusting with copper compounds, the former being more effective but more harmful to seed germination than the latter.
- ☆ Delinting by flaming is claimed to be both effective and non-dangerous to seed.
- ☆ Treating the seeds with antibiotic substances, such as Streptomycin sulphate, results in complete eradication of both the internal as well as external infection of the seed.
- ☆ Hot water treatment of seeds at 56°C for 10 minutes eliminates both external and internal infection of bacterial pathogen, without affecting seed germination.
- ☆ *Gossypium herbaceum* and *G. arboreum* are mostly immune to the disease, whereas *G. barbadense* and *G. herbaceum* var. *typicum* and *G. hirsutum* are susceptible. Five main genes governing the resistance of *Gossypium* spp. have been isolated and identified, B_1, B_2, B_3, B_4 and B_5 and subsequently so have several other genes.
- ☆ Neem (margosa) products (extracts of neem seed) are reported to possess extreme potential for control of the disease (Verma, 1995). Hulloli *et al.* (1998) have reported that certain neem based formulations such as Plantolyte and Agricate were much more effective than the best antibiotics

such as aminoglycosides, streptomycin and kanamycin. Combination of half dose of the neem based formulation with antibiotics not only increased efficiency of the treatment but also excluded the possibility of resistance development in the bacterium against the antibiotics.

- ☆ Seed inoculation with *P. fluorescens*, *P. putida* and *P. alcaligenes* suppressed the bacterium on seedlings. Bhowmik *et al.* (2002) isolated some endophytic strains of *Pseudomonas fluorescens* from cotton seedlings and demonstrated that following seed bacterization with the endophytes their population kept on increasing up to 15 days after seedling emergence which is considered the active phase when infection of cotyledons occurs. This reduced infection of cotyledons. Certain strains of *Pseudomonas* in the phylloplane of cotton that could reduce populations of *X. axonopodis* pv. *malvacearum,* when co-inoculated with the pathogen by multiplying and heavily colonizing the acive multiplication sites on cotton leaves.

9. Pigeon Pea (Red gram) (*Cajanus cajan*)

Leaf Spot and Stem Canker

In India the disease usually appears in the rainy season during July and August.

Symptoms

Small necrotic spots surrounded by bright yellow color halo appear on the lower leaves of about one month old plants. Later rough raised cankerous lesions appear on the stem. Cankers can cause stem to break, but the broken part usually, remains attached to the plant. In cases of severe infection the affected branches dry.

Causal Agent

The disease is caused by *Xanthomonas citri* pv. *cajani* (Kulkarni *et al.*, 1950) Ah-You *et al.* [Syn.: *Xanthomonas axonopodis* pv. *cajani* (Kulkarni *et al.*, 1950) Vauterin *et al.*, 1995]. The bacterium is a rod shape with single polar flagellum. It is an aerobic, Gram negative rod shaped bacterium with 0.3-0.6µm X 1.3-2.7 µm. The bacterium is motile by a single polar flagellum. On a 2 per cent sucrose peptone agar, a yellow, copiously mucoid, wet, shining growth is noticed. Gelatin, casein and soluble starch are strongly hydrolysed by it. Nitrite is not produced from nitrate by the bacterium.

Disease Cycle and Epidemiology

Infected plant materials left over in the field surve as a primary sourceof inoculum. Warm (25-30°C) and humid weather favors disease development. Disease incidence is generally higher in low-lying waterlogged areas of the field than in well-drained areas. African germplasm is more susceptible to the disease than Indian germplasm. Pigeonpeas with purple stems are less susceptible to stem canker than those with green stems. While leaf infection can occur at all stages of plant growth, stem infection usually occurs in younger plants.

Management

- ☆ It may be possible to eradicate the pathogen through seed treatment with appropriate antibiotics like agrimycin 250ppm
- ☆ Spraying of antibiotics like streptocycline @100 ppm at 10 day intervals effectively controls the disease.
- ☆ Spraying copper oxychloride @0.3 per cent alternately with antibiotics is recommended
- ☆ Genotypic differences in susceptibility have been reported.

Chapter 12

Bacterial Diseases of Spices, Medicinal and Aromatic Crops

A. SPICES CRPOS

1. Chilli (*Capsicum annum*)

i. Bacterial Leaf Spot

It was first observed in USA during 1912, at present it is prevalent in many countries. It is the most common of the bacterial diseases of chilli in India. In Rajasthan, the disease caused the yield loss from 8 to 16 per cent.

Symptoms

The leaves exhibit small, circular or irregular, dark brown or black greasy spots. As the spots enlarge in size, the centre becomes lighter surrounded by a dark band of tissue. The spot coalesce to form irregular lesions. Severely affected leaves become chlorotic and fall off. Petioles and stems are also affected. Stem infection leads to formation of cankerous growth and wilting of branches. On the fruits, round, raised water soaked spots with a pale yellow border are produced. The spots turn brown developing a depression in the centre where in shining droplets of bacterial ooze may be observed.

Causal Agent

The disease is caused by *Xanthomonas* spp. like *X. vesicatoria, X. euvesicatoria* pv. *euvesicatoria* (Jones *et al.*, 2004) Constantin *et al.*, 2016 and *X. gardneri* Jones *et al.*, 2004 (Syn.: *Xanthomonas campestris* pv. *vesicatoria* (*Xanthomonas euvesicatoria).* The bacterium is Gram negative, rod shaped, 0.7 to 1.0 x 2.0 to 2.4 µm in size, aerobic, single polar flagellum. rod-shaped bacterium that causes leaf and fruit spots on peppers and tomatoes. *X. vesicatoria* is an aerobic, mobile, Gram-negative rod, occurring singly or in pairs, 0.6 x 1.0-1.5 μm, with a single polar flagellum. On yeast dextrose chalk agar and nutrient dextrose agar, colonies are large, smooth-domed, mucoid-fluidal and yellow with entire edges. Unlike *P. syringae* pv. *tomato*, they are non-fluorescent on King's medium B. *X. vesicatoria* is sensitive to triphenyltetrazolium chloride and is oxidative. This bacterium has been separated into three pathotypes that differ in physiological traits. *X. campestris* pv. *vesicatoria (Xcv)* is composed of a 5.17-Mb circular chromosome, four plasmids and an essential type III protein secretion system for pathogenicity. In the early 1990s, it was shown that two distinct genetic groups (possibly species) existed within strains of pv. *vesicatoria*. In 1995, Vauterin *et al.* restructured the classification of the genus *Xanthomonas* by proposing species status for these groups: *X. vesicatoria* and *X. axonopodis* (syn. *campestris*) pv. *vesicatoria*. More recently, it has been shown that bacteria belonging to four distinct groups (previously designated A, B, C and D) cause bacterial spot. Jones *et al.* (2004) proposed that these should have species status: *X. euvesicatoria* = *X. campestris* (*axonopodis*) pv. *vesicatoria* (group A), *X. vesicatoria* = *X. vesicatoria* (group B), *X. perforans* = group C strains and *X. gardneri* = group D strains. Strains of groups A and B are most widely distributed. The vast majority of strains that infect pepper are in group A and possibly some in groups B and D. No pepper strains have been found in group C; however, strains from all four groups have been isolated from tomato. Some strains infect only pepper (designated pepper strains), some infect only tomato (designated tomato strains) and some can infect both pepper and tomato (designated pepper/tomato strains).

Disease Cycle and Epidemiology

Infected plants displaying symptoms act as a rapid means of dispersal of the leaf spot bacteria. The spread of the pathogen is favored by rainfall splash and by farm operations in fields when foliage is wet. Upon impacting on lesions, rain droplets disperse bacterial cells through many micro-droplets from infected plants to neighboring healthy plants or over distances as far as a few meters, especially under windy conditions. When foliage is wet, farm operations allow bacterial cells from infected plants to be carried on tools and clothing and transported to healthy plants in other field areas. Bacteria can become airborne as aerosols and be carried afar.

The bacterium needs a temperature of 22 to 34° C with high relative humidity for maximum infection. Infection takes place at a wide range of temperatures from 15 to 35°C but the optimum is 24 to 30°C. Forty to fifty days old plants are most susceptible to this disease.

Management

- Hot water treatment of seeds at 50° C for 25 min is effective.
- Seeds also must be obtained from disease free plants.
- Nursery should be raised on a soil where chillies were not grown for several years.

- ✫ The soil may be disinfected with formaldehyde.
- ✫ Seedlings may be sprayed with Bordeaux mixture 1.0 per cent or copper oxychloride 0.25 per cent.
- ✫ Spraying with streptocycline 200 ppm thrice at on interval of 15 days between July and August gives good control of the disease. Spraying with streptocycline should not be done after fruits begin to form.
- ✫ Some varieties of *Capsicum annuum, C. chinense* and *C. pendulum* are resistant. The variety Pusa Jwala and a line 23-1-7 BK are field tolerant. No 23, IC 9110, G2 and X3 are moderately resistant.
- ✫ Seed treatment with *Bacillus subtilis* and *B. amyloliquefaciens* and *Brevibacillus brevis* enhances seed quality parameters like seed germination and seedling vigour and protect the plant from disease.

ii. Soft Rot

Symptom

Starts as small soft spot near the stem. They are usually sunken. The rot spreads rapidly until the entire fruit collapse into a soft slimy mass. Later the entire fruit rots and the fruit resembles a sack of liquid retained by the skin. When the skin breaks and the contents flow out and the skin may be dry up. The bacterium can over winter in soil and they spread by insects, machinery, workers and by rain splashes.

Casual Agent

The disease is caused by *Erwinia carotovora* subsp. *carotovora.* The bacterium is facultative anaerobic, rod-shaped. It is mostly found mostly in tropical, warm regions of the world. Because the organism is spread in so many ways, there is speculation that it was introduced to water through aerosols and runoff into water bodies.

Disease Cycle and Epidemiology

There are many ways in which a plant can become infected by a bacterial soft rot. They can be host to the bacteria either by being infected as seed, or from direct inoculation into wounds or natural openings (stomata or lenticels) in mature plants, which is most common. But, when a plant is infected and the conditions are favorable, the bacteria immediately begin feeding on liquids released from injured cells and start replicating. As they replicate, they release more and more pectolytic enzymes that degrade and break down cell walls. Because of the high turgor pressure within the cells, this maceration effectively causes the cells to explode and die providing more food for the bacteria. The bacteria continue to multiply in the release free water and move into the intercellular spaces, with their cell-wall-degrading enzymes ahead of them preparing the plant tissues for digestion. Often the epidermis is left unscathed, keeping the rotten flesh contained within until a crack allows the ooze to leak out and infect others around it.

When the plant organs are harvested and placed into storage, those that are infected will automatically infect the others placed with it. When certain insects are present, the eggs laid over the stored vegetables will be invaded by the bacteria, becoming host and transporter, able to infect others as they grow. The bacteria then overwinter within the plant tissues, insect hosts, or in the soil and lay dormant until the conditions are right again to reproduce. If the infected storage organs are being used to propagate the plant, or if infected seed was produced, then when spring comes the bacteria will begin to grow just as its host does.

Management

- Field sanitation and removal crop residue is required to control the disease.
- Injuries by insects should be prevented.
- Harvest may be made during dry weather.
- Fruit washing with 25 ppm of free chlorine eliminates the soft rot bacteria.

2. Ginger (*Zingiber officinale*)

Bacterial Wilt

It was first reported in India from Tamil Nadu by Thomas in 1941. It occurs during south west monsoon. It occurs in India, the Philippines, Queensland and Western hemisphere. In India, it was reported from Bihar, Himachal Pradesh, Kerala, Tamil Nadu and Karnataka. The disease is increasingly serious in Wynad district of Kerala (India) with a 50 to 60 per cent incidence thus causing resulting in the serious loss of affected clumps in the garden.

Symptoms

The earliest symptoms appear as water soaked, linear streaks/patches on the collar regions of the pseudo stems followed by yellow to bronze colouration on the margin of the lower most leaf with intense yellowish bronze coloured margins and droop exhibiting typical wilt symptoms. The ligules and leaf sheaths in the infected plants appear yellow to dull green. Finally the leaves roll and whole plant dries up. Pseudo stem comes off easily when it is pulled. The affected pseudo stems and rhizomes in advanced stages of infection are slimy to touch and give off milky bacterial exudate when pressed gently. The affected pseudo stems when cut open longitudinally show dark streaks. Rhizomes show translucent water soaked lesions enlarging gradually. Later, these lesions spread deep in to the rhizomes and show varying degrees of coalescence thus involving major part of the rhizome.

Causal Agent

The disease is caused by *Ralstonia solanacearum* (Smith) Yabuuchi *et al.* (1995). The bacterium is aerobic, non spore forming, Gram negative, rod shaped with polar flagella.

The *R. solanacearum* isolated from ginger were identified as biovar 3 and race 4 (Kumar *et al.*, 2004) whereas severe and frequent epidemics of ginger bacterial wilt in Hawaii and surrounding islands and in China were reported as biovar 4 (Xu *et al.*, 2009). Host range studies undertaken in countries *viz*,. Thailand and China revealed that ginger is infected only by the virulent *R. solanacearum* strains isolated from ginger (Tsuchiya *et al.*, 2005).

Disease Cycle and Epidemiology

The bacterial pathogen infects *Ageratum conzoides, Capsicum* sp., eggplant, *Eupatorium odorotum*, groundnut, potato, *Solanum* sp., tobacco, tomato and turmeric. The bacterium survives even after 20 months period of drought in the soil debris. The bacterium survives in the rhizome. In the field, it spreads through soil water.

The disease is noticed during June- July and its incidence is reported maximum during August- September coinciding with the south west monsoon. The disease is favoured by soil moisture and relative humidity and low temperature. Nematode (*Meloidogyne incognita*), infestation predisposes the plants to infection by the pathogenic bacterium.

Management

- ✰ Avoiding growing low land area or ill drained area.
- ✰ Avoid monocropping.
- ✰ Crop rotation and planting of healthy rhizomes are important.
- ✰ Avoid growing of ginger in field grown with solanaceous crops or inter cropping with solanaceaous crops.
- ✰ Select healthy rhizome for planting.
- ✰ Deep summer ploughing of the field.
- ✰ Provision of good drainage, removal and destruction of infected plants help in reducing the disease.
- ✰ Plantomycin sprays after 30, 45 and 60 days of planting is effective in controlling bacterial wilt of ginger.
- ✰ Copper oxychloride + plantomycin combination is also effective. Soil fumigation with methyl bromide at 1.44 kg/30.5 m^2 gives good control.
- ✰ Add bleaching powder before planting @ 8-10Kgs per acre.
- ✰ Planting should be done in Feb-March in endemic area.
- ✰ Planting is done on 15-20cm raised beds.
- ✰ Pre-sowing soil application with FYM enriched *Trichoderma harzianum* (@ 2 kg/ton FYM) + FYM enriched *Pseudomonas fluorescens* (@5 kg/ton FYM)+ Neem cake (1qt/ac.)+ Pre-sowing rhizome treatment with 0.05 per cent streptocycline + copper oxychloride @3.0 g/lit for 15 minutes followed by drenching with 0.2 per cent bleaching powder and 0.1 per

cent metalaxyl MZ three times at 20 days intervals starting from onset of disease in the management of Rhizome rot and wilt disease complex.

3. Onion (*Allium cepa*)

i. Bacterial Soft Rot

Symptoms

The bacterium enters through injuries or wounds at the collar region or through the dead or senescent leaves and progress downwards into the bulbs. The outer scales become completely rotten. The infection can be easily identified by pressing the bulbs gently in which watery fluid comes out. The affected bulbs may show slimy ooze on the outer scales.

Causal Agent

The disease is caused by *Erwinia carotovara* sub sp. *carotovara.* The bacterium is Gram negative, rod shaped, capsulated, non-spore forming and non-acid fast, motile with peritrichous flagella. It is positive for pectate degradation on CVP medium, growth at 36°C and 39°C, catalase production and negative for indole production. The bacterium is positive in production of acid from various sugars such as lactose, trehalose, maltose and cellobiose.

Disease Cycle and Epidemiology

Sometimes soft rot of the bulbs may also occur during transit and in storage in warm, humid atmosphere. The bacterium remains as a saprophyte in the soil debris. The primary inoculum comes from soil where the bacteria exist in crop debris. Bacteria enter only through wounds. Onion maggots (*Hylemya antigua*) knight may carry the bacteria and introduce while feeding. Onions with mechanical injuries, bruises or sunscald are particularly susceptible to bacterial soft rot. This bacterium also infects vegetables like carrot, celery and parsnip and potato.

Moisture is very important for bacterial reproduction and infection. Optimum temperature for growth of bacteria is 30 to 35°C and maximum is 41°C.

Management

- ☆ Affected bulbs should be separated from healthy lots.
- ☆ A rigid sorting at the time of harvest and packing is necessary.
- ☆ Handling should be done in such a way that a minimum injury is caused to the bulbs during harvest, transit and storage.
- ☆ Control onion maggots. Onion maggots contribute to bacterial infection by creating wounds.
- ☆ Allow the tops and necks of the onions to dry out as much as possible before topping.

- ✩ When using mechanical toppers, aim to prevent as much bruising as possible.
- ✩ Cure bulbs thoroughly to allow outer scales and neck tissue to completely dry.

ii. Sour Skin Rot

Symptoms

Sour skin rot is a decay that affects only some of the outer scales and is characterized by slimy yellow appearance of affected scales which give off a vinegar like odour. The scales outside the rotting ones slip off readily in handling. Infection probably takes place when the tops are cut at harvest.

Causal Agent

The disease is caused by *Burkholderia cepacia* (ex *Barkholder)* Yabuuchi *et al.*, 1992 [Syn.: *Pseudomonas cepacia* (ex. Burkholder) Palleroni and Holmes. Cells of the bacterium are rod shaped, 1.6 to 3.2 X 0.8 to 1.0 µm, single or in pairs, Gram negative and motile having 1 to 3 polar flagella. Non fluorescent pigments are produced in culture media. Some strains produce slime on media containing 2 to 4 per cent sucrose.

Disease Cycle and Epidemiology

The bacterium attacks garlic bulb also. Contaminated surface irrigation water from drainage channels may carry the bacteria in some areas. The disease is always associated with rain storms and warm weather. The pathogen enters onion leaves only through wounds. Once inside, the bacteria spread rapidly in water soaked tissues than in tissues that are not congested.

Management

- ✩ Proper maturing of the crop and quick drying after topping and harvest is required to control the disease.
- ✩ The disease can be controlled successfully by substituting furrow irrigation for overhead sprinkler irrigation (Teviotdale *et al.,* 1989).
- ✩ Since contaminated irrigation water has been implicated in the spread of the pathogen, the use of recycled or irrigation runoff water should be avoided.

iii. Slippery Skin Disease

Bacterial soft rot of onion was reported for the first time at Khanpur in Uttar Pradesh in 1973. It was also found in Bulgaria and Hungary. The incidence ranges from 12.8 to 35.6 per cent in the seed crop (raised from bulb and 8.5 to 30.0 per cent bulb crop raised from seed).

Symptoms

The disease is characterized by internal browning and dry rotting in the basal portion of the flower stalk, seed stalk or scape resulting into partial to complete withering of the inflorescence prior to seed formation. Such rotting may extend internally upto the seed bulb. The plants appear small till either inflorescence gets dried up or the stalks topple down from the rotten portion. In the bulb crop yellowing of outer leaves from tip downwards followed by their drying is seen. During storage the inner scales of affected bulbs become water soaked followed by their browing and drying resulting into hollowing of the bulbs which lose weight considerably.

Causal Agent

The disease is caused by *Burkholderia gladioli* pv. *alliicola* Burkholder [Syn. *Pseudomonas gladioli* pv. *alliicola* (Burkholder) Young *et al., Pseudomonas alliicola* (Burkholder) Starr and Burkholder].The cells are Gram negative, rod shaped with rounded ends, occurring singly or in pairs, 2.0 X 0.8µm size, strictly aerobic, non-sporulating, straight and motile by 1-3 polar flagella. It grows well on yeast-glucose-chalk, peptone -glycerol, sabourands, glucose – yeast extract and nutrient media. Gelatin is slowly liquefied. Good growth is made in tryptone broth, but neither indole nor hydrogen sulfide is produced. Nitrates are reduced to nitrites. Ammonium salts, nitrates and peptone are utilized as sources of nitrogen and asparagine as a combined source of carbon and nitrogen. In Koser's medium without the citrate the organism utilizes arabinose, xylose, dextrose, laevulose, lactose, maltose, sucrose, glycerol, mannitol and salicin, as well as the sodium salts of citric, hippuric, malonic and tartaric acids. Starch is not hydrolysed. Optimum temperature of 30°C and pH of 8°C is necessary for the growth. Ammonium oxalate, peptone and 1-tyrosine, peptone and p-tyrosine are the best nitrogen sources for maximum growth.

Disease Cycle and Epidemiology

The bacterium infects *Amaryllis belladona* L., *Asphadelus teniufolius* Cav., *Cooperia drummondii* Herb. *Narcissus* sp., *Polianthus tubereosus* L., *Tradescantia bicolor* Kunth, *Zephyranthus atamasco* L., *Z. longifolia* Hemst., bean, cowpea, maize, peapods, pearl millets, rice, sorghum and tomato fruits. The bacterium enters the leaves, flower stalks or bulbs only through injuries. During storage of bulbs at room temperature (25 to 32°C), the bacterium from infected bulbs can infect neighbouring healthy bulbs. Particularly rotten bulbs can produce diseased plants in the next season. The bacterium survives upto 12 months at 6 to 8°C and 6 months at 25 to 39°C in diseased plant debris. Seed, soil and irrigation water do not seem to play any important role in the disease cycle.

Management

- ☆ The variety Kalyanpur Red is highly susceptible to soft rot.
- ☆ Onion culture 5001-12 and 6308 (Raised from seed) and 5223-2, 5404-3, 6106, 6621 and VL-14 (Raised from bulb) remain free from infection.

- ☆ Soaking infected bulbs in agrimycin or streptocycline 100 ppm solution for 2 h reduces severity of the disease in the field.

4. Betelvine (*Piper betle* L.)

Bacterial Leaf Spot

Symptoms

The disease appears as small water soaked lesions on the leaf lamina or at the margins. The spots on the leaf lamina are seen scattered and are surrounded by yellowish haloes. As the spots enlarge, they become dark brown in colour. In some cases, the infection is noticed all around the leaf margin. At times, a dark brown pigmentation originating from the infected region diffuse into the rest of the leaf lamina. During the rainy periods and under highly humid atmospheric conditions, the infection spreads rapidly involving a major portion of the leaf lamina. Under such conditions, the infected portions are involved in a sort of wet rot. Defoliation of infected vines are often noticed in severe cases.

Causal Agent

The disease is caused *Xanthomonas axonopodis* pv. *betlicola* (Patel *et al.*) Vauterin *et al.*, 1995 [Syn. *Xanthomonas campestris* pv. *betlicola* (Patel *et al.)* Dye]. The bacterium is Gram negative rod shaped, single polar flagellum.

Disease Cycle and Epidemiology

The bacteria which are viable in the infected vines and leaves serve as a primary source of inoculum. Rain splashes and splash irrigation water help in the secondary spread. Cloudy weather with intermittent rains and high relative humidity favour the disease development. Two to three years old vines are highly susceptible to the disease.

Management

- ☆ Remove and burn the infected vines and stubbles in the field.
- ☆ Regulate irrigation during cold weather season.
- ☆ Spray streptocycline 400 ppm+Bordeaux mixture 0.25 per cent at 20 days intervals, after plucking the leaves.

5. Cardamom (*Elettaria cardomomum*)

Bacterial Blight

Symptoms

It causes canker symptoms on capsules. First water soaked lesions of 0.5 to 1.0 mm appear on the pericarp of the fruit. They later on assume the appearance of spongy eruptions of tissue on the rind, which on fresh green capsules usually show glazed rather than oily and at times a pale yellow halo. Beyond this halo, there is normal green colour of the pericarp. This typical symptom is called as canker. The disease is caused by *Xanthomonas* sp. The bacterium is Gram-negative, rod shaped with 1.5-2.0 x 0.5-0.75 µm in size. Bacterium forms chains and capsules and is motile by one polar flagellum. No spores are formed and they aerobic. Colonies of bacteria on nutrient agar are circular, straw yellow, slightly raised and glistening.

B. PLANTATION CROPS

1. Areca Nut (*Areca catechu*)

Bacterial Leaf Stripe

Rao and Mohan (1970) reported its occurrence from Tumkuru areas of Karnataka state in an endemic form.

Symptoms

The initial symptoms include 1 to 4 mm wide, dark green water soaked, translucent, linear lesions or stripes alongside and parallel to them the midrib of the leaf let of its other main veins. The lesions may develop at any point on the lamina, but more commonly from the base or towards the tip of the leaf let. The margin of the lesions is usually straight and well defind, but occasionally it may appear wavy. The lesions are covered with abundant bacterial exudates on the lower surface. The exudate is creamy white and slimy. On drying, it forms a waxy film or creamy white to yellowish flakes or fine granules or irregular yellowish masses (Figure 1). In the advanced stages, the lesions may measure 1 cm or more wide and several centimeters long involving the midrib also. The affected midrib and veins of the leaf let get discoloured and turn black. All the leaf lets of a leaf may be affected resulting in complete or partial blighting of the leaf and in severe cases the entire crown may be killed particularly in seedlings.

Leaf Spot **Leaf Strife** **Budrot**

Figure 1. Disease Symptom on Leaf as Spot, Stripe and Rotting on Bud.

Causal Agent

The disease is caused by *Xanthomonas vasicola* pv. *arecae* (Rao and Mohan

1970) Studholme *et al.*, 2020 (Syn.: *Xanthomonas campestris* pv. *arecae* (Rao and Mohan 1970) Dye. The bacterium is Gram negative rod shaped, single polar flagellum, strictly aerobic. It is positive for catalase reaction, liquefaction of gelatin, KOH test, protein digestion, methyl red test and produced hydrogen sulphide and negative in ammonia production, starch hydrolysis and fluorescent test.

Disease Cycle and Epidemiology

The bacterium infects areca nut and other ornamental palms. The disease remains aggressive during and after the rainy season and it is of little significance during the hot dry summer months. The incidence is high during the months of July- October when the average monthly rainfall is 130 mm or more with more than 10 rainy days per month. Three to five year old palms are highly susceptible to the disease than the older palms.

The disease remains aggressive during the active monsoon season (July-October) when the average monthly rainfall is 130 mm or above with more than 10 rainy days per month. Afterwards, there is a decline in disease incidence. Temperature above 30°C and below 17°C was found to slacken the disease spread (Kumar, 1981). The organism does not survive in soil for long, indicating that the soil may not be a primary source of inoculum. Palms of younger age group (3-5 years) are highly susceptible to the disease than older palms. As age advances, susceptibility character also varied both with age of the palm as also with the age of the leaf (Kumar, 1981).

Close spacing's were found to be congenial for disease development and spread. Frequent irrigation at 5-10 days intervals aggravates the disease incidence. Higher levels of nitrogen and green leaf manure were found to favour the disease development. While phosphorus and potash did not have any influence. Intercropping with banana in locations with bacterial leaf stripe disease was found to aggravate the disease intensity and was detrimental to areca nut plantations (Kumar, 1981).

Management

- Field sanitation practices include removal and destruction of the diseased plant parts that will helps in reduction of the inoculum.
- Avoidance of closer spacing, frequent irrigation and intercropping with banana plants will reduces the disease.
- Prophylactic spraying of copper hydroxide (2g/L) can reduce the disease effectively.
- Remove the affected plant parts and spraying with streptocycline 500 ppm + copper oxychloride 3000 ppm twice at 20 days intervals.

C. MEDICINAL AND AROMATIC PLANTS

1. Poppy (*Papaver somniferum* L.)

Stem and Capsule Rot

The disease has been reported from India and other poppy growing countries. The occurrence of *E. carotovora* subsp. *carotovora* on *Papaver somniferum,* causing extensive damage in the poppy growing regions of Uttar Pradesh (Sattar *et al.,* 1988).

Symptoms

The disease is characterized by external blackening and internal disintegration, accompanied by the discolouration of the midrib of the leaves. The disease commences at the apex and extends down wards and the whole plant is turned into a slimy mass.

Causal Agent

The disease is caused by *Erwinia carotovora* subsp. *carotovora* (Syn.: *Erwinia papaveris* (Ayyar) Magrou. The bacterium is Gram negative, rod shaped 0.5 to 2.5 X 0.5 µm in size, facultatively anaerobic, oxidase negative, catalase positive, grow at 37°C and do not produce gas from D-glucose. Acid is produced from D (+) – arabinose, lactose and D(+)-trehalose, but not from α- D- methylglucoside. Optimum temperature for growth is 30°C and thermal death point lies between 50°C and 60°C. The bacterium can retain it's viability on poppy seeds for over 20 months.

Disease Cycle and Epidemiology

The disease is transmitted mainly through infected seed, which is reached by the organism both from the stem and through the pod wall. The bacterial pathogen make entry through the stomata and extension is chiefly through the intercellular spaces around the phloem and xylem parenchyma cells and the bast fibres.

Management

The control measures recommended are the use of healthy seed, removal of diseased plants from the fields as soon as they appear, as well as of the stubble remaining from the preceding crop and crop rotation where practicable.

2. Geranium (*Pelargonium* spp.)

Bacterial Leaf Spot

The disease was first described by Brown in 1923 from USA, Later it was reported from many countries including India. It was reported from Shimla, India. The disease is also called blight, stem rot or wilt of Pelargonium. The main host of the pathogen is Pelargonium and their hybrids. The pathogen also infects *P. graveolens* (scented geranium), *P. peltatum* (ivy geranium and its hybrids), *Geranium* spp. (cranesbill geranium including *G. sanguineum*), *Pelargonium* x *domesticum* (regal geranium). *P. acerfolium, P. tomentosum, P. torento* and *P. scarboroviae* are hosts, but they show some tolerance.

Symptoms

Irregular to circular brown spots appear on the under surface of the leaves as water- soaked dots. They become sunken as they enlarge and with collapse of the tissues. When more number of spots appear on the leaves, the diseased leaves turn yellow, brown, shrivel and drop down. Leaves some times wilt and droop but hang on the plant for a week. Diseased stem changes into grey colour and dull. The pith and cortex become black and disintegrate into a dry rot. Roots become black but they do not decay. Cutting fails to put forth roots and rot from the base upward. Bacteria can live upto three months in moist soil. The disease spreads by handling, splashing water, cutting knives and white flies.

Causal Agent

The disease is caused by *Xanthomonas hortorum* pv. *pelargonii* (Brown) Vauterin *et al.*, 1995 (Syn.: *Xanthomonas campestris* pv. *pelargonii* (Brown) Dye.

Disease Cycle and Epidemiology

Infected stock plants are the main source of infection and cuttings from these plants have systemic infection. The bacterial pathogen may survive fro 3- 7 moths in plant debris and it survives less well in soil. The bactyerium is not known to be seed transmitted. The bacterium can be spread in the greenhouse by tools, hands, rain splash, plant contacts and soil occasionally by whitefly and aphids.

Management

- ✰ Remove and burning of infected plant debris.
- ✰ Select healthy mother plants.
- ✰ Certification and indexing strict hygiene and avaoidance of orverhead irrigation and good ventilation reduce the disease incidence.
- ✰ Copper or streptomycin and insecticide sprays may reduce spreading of the bacterium.

D. ORNAMENTAL CROPS

1. Jasmine (*Jasminum* spp.)

Bacterial Leaf Spot

The disease was reported from Annamalainagar, Tamil Nadu during 1958-60 on the Jasmine plant. The same disease has also been reported from Salem and Coimbatore during this period. Several varieties of *J.sambac* were found affected and disease symptoms also varied on the some of the varieties.

Symptoms

The disease symptoms appear on the upper surface of leaf as minute water soaked lesions. These spots soon become yellowish green; several spots coalesce to form linear or irregular spots, which give a characteristic mosaic like appearance to the leaf. On each leaf several spots may appear in a crowded form but in some varieties there may be only a few localized spots. On other parts of the plant is affected by the bacterium. In case of severely affected plants, there may be defoliation and the affected plants could be seen chlorotic even from a distance (Rangaswami and Easwaran, 1961).

Causal Agent

The disease is caused by *Xanthomonas campestris* pv. *jasmini* (Syn.: *Xanthomonas jasminii* Rangaswami and Easwaran. The bacterium is short rods, 2.1 – 4.2 x 0.7 – 1.4 µm in size, single, rarely in chains of 2 or 3, Monotrichus, with single polar flagellum, aerobic, nonacid fast, gram negative, non- capsulated and non- spore forming, forms flat, circular, entire opaque and non- glistening yellow colonies on nutrient agar and yellow growth with sedimentation in nutrient broth, no soluble pigment is produced in the liquid or agar media. Gelatin is liquefied rather slowly but starch is hydrolyzed rapidly and litmus milk is coagulated with acid production, but litmus in the milk not reduced. H2 S is not produced by the bacterium, but ammonia produced, gives positive M. R. and lypolytic activity tests and negative indol and V. P. tests (Rangaswami, 1962). Bacterium utilizes glucose, sucrose, maltose, lactose raffinose, galactose, xylose and laevulose as carbon source with acid production but no gas formation. It also utilizes potassium nitrate, ammonium hydrogen phosphate. Urea, glutamic acid, aspartic acid and tryptophan are utilized as nitrogen sources.

Management

Removal of affected leaves spraying antibacterial antibiotic

2. Chrysanthemum (*Dendranthema grandiflora*)

i. Bacterial Blight and Wilt

The host range is very large and comprises many ornamentals, food crops such as *Allium* spp., *Chicorium intybus, Dacus carota, Oryza sativa, Solanum tuberosum, Ipomea batatus, Zea mays*, tomato, cabbage, lettuce, celery, carrot and radish *etc.* The disease is distributed worldwide.

Symptoms

Cuttings of the plant turn dark brown and later collapse. The surviving cuttings may be infected but have no symptoms. Established plants wilt during the day when infected and recover at night. Water-soaked lesions appear on stems, darkening and death of buds and stems. Stem may break or split. Wilting of one or more branches of the plant appears which later become blighted at or near the succulent growing tip. Red to red-brown lesions produces on the stems and sometimes on the roots. Generally the pathogen induces a canker on the stem, starting from the crown. Pith area becomes reddish to reddish brown and jelly like rot which later dries leaving completely hollow pith and pith maceration.

Causal Agent

The disease is caused by *Dickeya chrysanthemi* (Burkholder *et al.,* 1953) Samson *et al.,* 2005 (Syn.: *Erwinia chrysanthemi* Burkholder *et al.* 1953). The bacterium is a Gram negative, plant pathogenic, pectolytic bacteria belonging to Enterobacteriaceae family. In a 24-hour-old culture on beef-extract-peptone agar the average dimensions of the straight rod with rounded ends, occurring singly or in pairs, were 2.1 by 0.7 µm. Colonies on King's B medium are white and non-fluorescent and grow at 26°C and induces a hypersensitive response on tobacco. The bacterium is positive for β-galactosidase and citrate utilization, acetoin and indole production, gelatinase and utilization of glucose, mannitol, saccharose, melibiose and arabinose.

Disease Cycle and Epidemiology

It can also survive in plant debris. Succulent fast growing tissues are severely attacked. Pathogen produces several extracellular enzymes that can attack components of the plant cell wall: pectinases, cellulases, hemicellulases and proteases. Pathogen spread through wounds during pinching and pruning operations, by fingernails or tools. High temperature (26-32°C), high humidity and absence of free water favours survival of the pathogen. Transmission by insects and other invertebrates-not reported.

Management

- ✰ Disinfect the propagation beds between crops.
- ✰ Disease free cuttings should be used.
- ✰ Destroy infected cuttings.
- ✰ Follow good sanitation.
- ✰ Reduce the humidity in growing areas.
- ✰ Antibiotics like agrimycin 500 can be used.
- ✰ Dip the cuttings in Streptomycin.

ii. Bacterial Leaf Spot

It is described by Jones *et al. (*1983) USA. Pathogen is fluorescent *Pseudomonas* originally described from Chicory (*Chicorium intybus*). reported that *Pseudomonas cichorii* have a wide host range affecting high economic importance plants from different family including tomato, lettuce and watermelon. The pathogen was isolated in Florida on chrysanthemum, celery, geranium and hibiscus recently on *Stevia rubidian* and *Duranta erecta* (Gumtow *et al.,* 2013).

Symptoms

Small dark brown to black spots on lower leaves enlarge and become irregular in shape. When infected leaves dry, the spots become brittle and crack. Discoloration may be prominent along leaf veins or lesions which become angular as bacteria growth is limited by major veins. Dark cankers may also appear on stems. Leaf wilting and death often appear. The disease often spreads up plants in one side of the pot, eventually to the flowers. On stem, elongated black necrotic streaks appear.

Causal Agent

The disease is caused by *Pseudomonas cichorii* (Swingle) Stapp. The bacterium is Gram negative, motile straight rods bearing 1-6 polar flagella and measures around 1.1-3.5 x 0.7 µm. On Nutrient Agar, it produces irregular raised glistering opaque colonies. The bacterium is oxidase and catalase positive. It produces green to193 brownish-green fluorescent diffusible pigment on King's B agar when viewed under UV light, the bacteria produces H_2S. Bacteria show negative in starch hydrolysis and gelatin hydrolysis.

Disease Cycle and Epidemiology

The bacteria is found in the soil and survives in the debris for months. Wounds caused by feeding and oviposition of leaf minor *Liriomyza trifoli* facilitate bacterial infection. High temperature (up to 28°C), high humidity, close planting favours the disease development. High moisture is extremely important for disease development. The disease is transmitted by insects and other invertebrates.

Management

- ☆ Do not plant infected cuttings.
- ☆ Avoid overhead irrigation.
- ☆ Irrigate in such a manner that leaf surface remain dry at all the time.
- ☆ Protect plants grown outdoor from splashing.
- ☆ Host plant resistance-no information on host rage is available.
- ☆ Application of silicon to soil in the form of potassium silicate (more than 200 ppm) reduces the incidence of leaf miners significantly.
- ☆ Weekly sprays of tribasic $CuS0_4$ at 4 lb./100 gal and agrimycin 500 at 3 lb./100 gal without injury to leaves or flowers.
- ☆ Agristrep and agrimycin 100 can also be used to control the disease.

3. Carnation (*Dianthus barbatus*)

i. Bacterial Wilt

The disease was first described from the USA by Jones in 1941. Carnation wilt, a vascular disease may be induced by mechanical plugging, a toxin or a combination of factors. *Burkholderia caryophylli* is an EPPO A2 quarantine pest, in view of the limited number of EPPO countries in which it has been reported and the fact that it is readily carried on cuttings in international trade. Carnations are the main host; however, *Dianthus barbatus* and *D. allwoodii* can be infected through artificial inoculation.

Symptoms

The upper parts of established plants turn pale in color and wilt. The stem develops elongated discoloured stripes and split open which is characteristic of the disease. The roots are rotted partially and the cortical tissues become sticky and show discolouration a tendency to straighten out instead of remaining curled. The leaf lamina is twisted. The roots are generally absent on one side and remain discoloured. The base of the cutting is discoloured with an elongated brown color area extending upwards. The bacteria invade xylem vessels and spread to infect the young shoots. The vessels are distrupted and the host plant produces meristematic tissues. This new tissue causes an uneven development of cortex which splits open longitudinally. In the final stages, the vascular tissue in stem and roots is destroyed and showing a yellow exudates and plants also die. In the naturally infected carnation plants the xylem vessels are found partly plugged lysis of xylem vessels is also observed. Infection is carried to upper portions of the branch into the leaves. Bacteria are not found in parenchyma tissues.

Causal Agent

The disease is caused by *Burkholderia caryophylli* (Barkholder) Yabuuchi *et al.* 1993 [Syn.:*Pseudomonas caryophylli* (Burkholder) Starr and Burkholder]. In culture, colonies of pathogenic bacteria are round, smooth and shining with regular margins: while cream-coloured at first, colonies darken with age in PDA culture. On nutrient agar, growth is slow and cells die rapidly; subculturing is not possible after about a week. *B. caryophylli* is a straight or slightly curved rod with rounded ends, occurring singly or in pairs, aerobic, non-sporing, motile with one or several polar flagella, Gram-negative, sudanophilic, 0.35-0.95 x 1.05-3.18 μm in size.

Disease Cycle and Epidemiology

The disease is spread by use of infected cuttings. The bacteria are very quickly sucked up the vessels exposed at the base of the cuttings. The bacteria can be dispersed by infected planting material and through infected soil or substrate and root to root contact. The bacterium can only enter plants through wounds and subsequently colonizes the vascular system of the stem and roots. The primary infection source is infected cuttings taken from mother plants with a latent infection. Bacteria can pass from one cutting to another in the water of the propagating bed or, if the cuttings are held in water, before planting out. The observed slow, scattered spread of the disease indicates that spread occurs only from one root system to another. Bacterial slime is exposed when stems crack and this inoculum may be transferred from one plant to another.

The disease development is strongly enhanced upto 35°C temperatures. However, over 20°C temperatures accelerate bacterial growth and therefore symptom expression, while at low temperatures infected plants may show no symptoms.

Management

- ✰ Use of cuttings taken from upper parts of the healthy stock plants are less liable for infection and hence recommended as a control measure.
- ✰ Diseased plant debris should be collected and burnt.
- ✰ Overhead watering and splash watering should be avoided.
- ✰ Disease free planting materials are to be used.
- ✰ Foliar application of bacterial antibiotics like streptomycin sulphate @ 250 ppm along with copper fungicides @ 0.25 per cent controls the diseases.

ii. Bacterial Leaf Spot

Burkholderia andropogonis has a broad host range including 52 species of 15 families of unrelated monocot and dicot plants such as white clover, carnation, bougainvillea and other ornamental plants.

Symptoms

This bacterium infects leaves and produces elongate brown spots on them. When infection is severe the entire plant is killed. The disease is severe during May-June and September - November.

Causal Agent

The disease is caused by *Burkholderia andropogonis* (Smith) Gillis *et al.*, 1995 [Syn. *Pseudomonas androponis* (Smith) Stapp = *P. woodsii* (Smith) Stevens]. *B. andropogonis* has a single, polar sheathed flagellum, a unique feature of this bacterium.

Disease Cycle and Epidemiology

The bacterial pathogen is soil borne. It spreads through farm equipments during field operations like weeding, pruning and other intercultural operation. Besides, it also spread through irrigation water and infected plant debris.

Management

- ✰ Destruct diseased plants and use of disease free planting materials.
- ✰ Foliar application of bacterial antibiotics like streptomycin sulphate @ 250 ppm along with copper fungicides @ 0.25 per cent controls the diseases.

4. Gladiolus (*Gladiolus* spp.)

i. Bacterial Neck Rot/Bacterial Scab

In India the disease has been reported from Shimla during 1976. It causes substantial loss in field as well as during storage.

Symptoms

It appears as reddish brown specks on leaves which become dark brown in the late stages. On the leaves the rough spots run together destroying parenchymotous tissues and leaves start drying from tip when the attack is severe and the tissues at the base decay in wet weather and the plant topples over. In sever cases, the leaves dry up from the tip. Symptoms are also observed on the corms. On corm husks, scab appears as elongated lesions with black rough margins, which some times penetrate entirely through husks. On corms these lesions are brown, circular and sunken with black raised margins. This depression is then filled with a gummy exudation which glues husks to the corms. The corms having latent infection of previous year develop weak shoots which die afterwards. These lesions on corms become hard, brittle and Shiny in dry weather but soft rot during wet weather.

Causal Agent

The disease is caused by *Pseudomonas marginata* (MeCull.) stapp.

Disease Cycle and Epidemiology

Warm and humid weather are favourable for this disease. Presence of *Meloidogyne javanica* increases the severity.

Management

- ✰ It is controlled by dipping corms in mercuric chloride 1: 1000 solution for 12 h before planting.
- ✰ Varieties such as American express, George Mazure, Inez, Pfitzers sensation, Salmonia, Syliva, Thunder bird and vinks Glory are highly susceptible to this disease.

ii. Bacterial Blight and Spot

The disease was first time reported by McCulloch from the USA in 1924. It occurs during rainy periods. It was reported from Egypt, Finland and India.

Symptoms

The first symptoms of the disease as dark green, water- soaked spots on the leaf which are more or less rectangular in shape. Later these spots turn brown, necrotic and coalesce. The plant and corm growth is retarded. Bacterial ooze is present on the lesions under wet conditions.

Causal Agent

The disease is caused by *Xanthomonas campestris* pv. *gummisudans* (McCulloch, 1924) Dye, 1978.(Syn.: *Xanthomonas gummisudans*).

Diease Cycle and Epidemiology

The bacteria enter the plant through stomata. Free water (irrigation and wet weather) favour the disease development.

Management

- ☆ Keep foliage dry and avoid working on wet plants.
- ☆ Dipping corms in mercuric chloride 0.1 per cent at the time of planting and storage gives better crop.

5. Indian Shot (*Canna orientalis*)

i. Bacterial Blight

The disease has been reported from Chennai (Tamil Nadu state) in 1980 during rainy months (July- December).

Symptoms

Small yellow spots about one mm dia, appear on the young leaves. With age, they become necrotic, dark brown, sunken and are surrounded by a yellow halo. More extension damage occurs when several lesions coalesce to form large irregular areas. Infection spreading to young unfolding leaves becomes aggressive and often leads to leaf distortion.

Causal Agent

The disease is caused by *Xanthomonas campestris* pv. *cannae* Easwaramurthy *et al.* The bacterium is Gram negative, rod shaped, non-sporulating and motile with single polar flagellum and 1.0 to 1.9 X 0.5 to 0.7 μm.

ii. Bud Rot

Symptoms

The bacterium enters through stomata of young leaves and flower buds before they open. As the leaf lamina unfolds, numerous spots are visible which enlarge and run together along the veins giving the leaf a striped appearance. The spots are white, at first but soon turn black in color. Affected flower buds turn black and die before they open. The disease may proceed down the leaf stalk and kill the young stem and buds. If older leaves are affected, the diseased area spreads slowly, forming irregular, yellowish spots with water soaked margins. The bacteria survive through the winter on the root stocks. The disease occurs early in the season on young plants.

Causal Agent

The disease is caused by *Xanthomonas campestris* pv. *cannae.*

Management

- ☆ Healthy root stocks should be used for propagation.
- ☆ Spraying streptomycin on buds and leaves helps to control the disease.
- ☆ Over watering, over crowding of plants and poor ventilation should be avoided.

6. Petunia (*Petunia axillaries* lamk)

Bacterial Wilt

The bacterium attacks the plant at any stage of the crop and kills. Diseased plants rapidly wilt and die.

Causal Agent

The disease is caused by *Ralstonia solanacearum* (Smith) Yabuuchi *et al.* on *P. axillaries.*

Management

- ☆ In new plantings, the soil should be treated with bleaching powder (neutral) before taking up sowing.
- ☆ Sick soil should be avoided.
- ☆ Crop rotation with non-host crops like cereals is recommended to reduce the incidence of the disease.

7. Orchids (*Cattleya* spp.)

i. Bacterial Soft Rot

Symptoms

The disease initiates at the upper and of the leaf as a small, water soaked and some what darker than normal green spot. Pseudo bulbs of affected plants turn soft and pulpy and become yellow. A foul smelling liquid ooze from the bulbs.

Causal Agent

The disease is caused by *Erwinia carotovara* subsp. *carotovora* (Jones) Bergey *et al.*

Management

- ☆ Spraying with agrimycin is beneficial.
- ☆ The use of 8-quinolinol benzoate or sulphate or the sodium salt of 0-hydroxy diphenyl is also recommended for its control.

ii. Bacterial Brown Spot

Symptoms

This is a disease of both seedlings and full grown plants. The disease appears as small, water soaked spot on the leaves. The lesion increases and soon the leaf becomes soft. The disease may proceed into the growing point and kills the plant.

Causal Agent

The disease is caused by *Xanthomonas cattleyae.*

8. *Paphiopelium* spp.

Bacterial Brown Rot

Symptoms

The bacterium attacks the plants through cracks and wounds. Yellowing and browning of leaves indicate the presence of the disease. The infection spreads into the crown and causes death of the plant.

Causal Agent

The disease is caused by *Erwinia cypripedii.*

Management

The disease could be reduced by merging infected plants in 1: 2000 solution of 8 -quinolinol benzoate or natriphene (Na-o-hydroxydiphenyl) for 1 to 2.5 h.

9. Marigold (*Tagetes* spp.)

i. Bacterial Blight

Symptoms

Minute, light green, water-soaked lesions appear on the leaf and the size of the lesion increases along with the advancement of the disease. The spots are found on older leaves as they are highly susceptible. In advanced stages, the lesions coalesce and produce an yellow discolouration of the leaf followed by shrivelling and drying up of the affected portions.

Causal Agent

The disease is caused by *Xanthomonas tagetis.*

ii. Leaf Spot Disease

Symptoms

Symtpoms as cirular and necroticspots appear on the leaf. Lesions appear on petioles also. The lesions have purple margin.

Causal Agent

The disease is caused by *Pseudomonas syringae* pv. *tagetis* (Hellmers) Young *et al.* (1996)

10. Tuberose (*Polianthes tuberosa*)

Flower Bud Rot

Symptoms

Mainly occurs on young flower buds and results in dry rot of buds and brown, scorched and necrotic discolouration of peduncles. Later, affected buds shrivel and become dry.

Causal Agent

The disease is caused by *Erwinia* sp.

Management

The disease is controlled by segregation of plants and applications of corrosive sublimate (1:1000).

11. Zinnia (*Zinnia angustifolia*)

Bacterial Leaf Spot

Symptoms

First, minute water-soaked dots develop on the leaves and they gradually increase in size and turn into translucent brownish to black spots. They are irregular in shape and mostly found in the interveinal space even though the larger veins are free. Frequently they are surrounded by yellowish green haloes. They are 1 to 8 mm in size and several spots are found on each leaf. The older leaves are more commonly affected than the younger one. The disease confines itself only to the leaf.

Causal Agent

The disease is caused by *Xanthomonas nigromaculans* f.sp. *zinniae.*

12. Dahlia (*Dahlia australis*)

i. Crown Gall

Symptoms

Larger tumours of abnormal tissue develop at the base of the plant and on the roots. Infected plants are stunted and the shoots are spindly.

Causal Agent

The disease is caused by *Rhizobium radiobacter* (Sy.: *Agrobacterium tumefaciens* (Smith and Townsend) Conn.

Management

- ☆ Dipping the roots in streptomycin sulphate solution and planting prevents the disease.
- ☆ Roots and crowns of plants with tumours should be collected and destroyed.

ii. Bacteriosis

Symptoms

The disease appears as browning and softening of the stem. The layers of pith that line the hollow stem are moist and blackish with the rot, which extends into the bark region. The rot gives off a foul odour. The disease also affects the tubers.

Causal Agent

The disease is caused by *Erwinia carotovara* subsp. *carotovara* (Jones) Bergey *et al.*

iii. Bacterial Wilt

Symptoms

The affected plants usually droop and wilt rather suddenly. The conducting system is found plugged with yellowish color masses of bacteria, which ooze out when the stems are cut. The wet soft rot of the stems near the soil is characteristic feature and distinguishes this wilt from that caused by *Fusarium*. The bacterium over winters in the soil.

Causal Agent

The disease is caused by *Ralstonia solanacearum* (Smith) Yabuuchi *et al.* 1995, (Syn.: *Pseudomonas solanacearum* (Smith) Smith].

Management

- All the diseased plant should be removed and destroyed to reduce the incidence.
- Infected plant parts should not be thrown into compost pits but should be destroyed.
- Heavily infested soil should be steam pasteurized before use.
- Adoption of crop rotation is the best method.

References

Abbasai, P.A., Cuppies, D. A. and Lazarovits, G. (2003). Effect of foliar applications of neem oil and fish emulsion on bacterial spot and yield of tomatoes and peppers. *Can. J. Plant Pathol.*25 (1): 41.

Abd El-Rahman, A. F., Shaheen, H. A., Abd El-Aziz, R. M. and Ibrahim, D. S. (2019). Influence of hydrogen cyanide-producing rhizobacteria in controlling the crown gall and root-knot nematode, *Meloidogyne incognita*. *Egyptian J. Biol. Pest Control*, 29(1): 1-11.

Adekanmbi, E. O. and Srivastava, S. K. (2016). Dielectrophoretic applications for disease diagnostics using lab-on-a-chip platforms. *Lab on a Chip* 16: 2148– 2167.

Adhikari, R. S., Bore, S. S. and Singh, S. B. (1988). *Xanthomonas campestris* pv. *jugalendis*. New report from India. *Cur. Sci.* 57: 725.

Agarwala, R. K. (1964). Diseases of peaches. Part II, State Regional Fruit Research Station, Mashobra, Agr. Res. 4: 139.

Agarwala, R. K. and Sharma, V. C. (1970). Natural varietal infectibility of stone fruits with bacterial canker. *Indian J. Agri. Sci.* 40: 559-561.

Agarwala, R. K. and Sharma, V. C. (1971). Further studies on Mashobra paste for the control of bacterial canker. *Indian Phytopath.* 24: 31-33.

Agrios, G. N. (2005). Plant Pathology fifth edition. *Ed Elsevier Academia Press. San Diego Calf. USA.*

Aley, E.F.L.G.P. and Elphinstone, J. (1995). Culture media for *Ralstonia solanacearum* isolation, identification and maintenance. *Fitopatologia*, 30: 126-130.

Al-Fallooji, S. A., Al-Hasnawi, S. L. and Al-Yasseen, A. K. (2021). Evaluate the efficiency of some biological and chemical factors to control of *Pseudomonas syringae* pv. *lachrymans* parasitizing on cucumber. *J. Al-Muthanna for Agril. Sci. 8*(2): DOI 10.52113/mjas04/8.2/16.

Anzai, H., Yoneyama, K. and Yamaguchi, I. (1989). Transgenic tobacco resistant to a bacterial disease by the detoxification of a pathogenic toxin. *Mol. Gen. Genet.* 219: 492–494.

Arasu, M. V., Al-Dhabi, N. A., Choi, K. C., Bensy, A. D. V. and Rajaselvam, J. (2022). Bioactive potential of *Albizia lebbeck* extract against phytopathogens and protective properties on tomato plant against speck disease in greenhouse. *Physiol. Mol. Plant Pathol.* 117: 101750.

Arwiyanto, T, Aryani, W.T., Gusti, E. and Restina, D. (2019). Control of tomato bacterial wilt (*Ralstonia solanacearum*) by grafting and bacteriophage application. 4th international symposium on biological control of bacterial plant diseases. J Pl Pathol Biocontrol, July 9-11, Viterbo, Italy.

Asrul, T. A. (2004). The influence of tomato seeds treatment with *Pseudomonas putida* Pf-20 against bacterial wilt disease. *Agrosains,* 17: 419–430

Ayana, G. and Fininsa, C. (2017). Effect of crop rotation on tomato bacterial wilt (*Ralstonia solanacearum*) and survival of the pathogen in the rhizospheres and roots of different crops in Ethiopia. *Int. J. Phytopathol.* 5(3): 81-88.

Azegami, K. (1983). Bacterial palea browning, a new disease of rice caused by *Erwinia herbicola. Bull. Natl. Inst. Agric. Ser C,* 37: 1-12.

Azegami, K., Nishiyama, K. and Kato, I. I. (1987a). Growth of *Pseudomonas plantarii* and production of tropolone and protein under limited iron deficiency. *Ann. Phytopathal. Soc. Jpn,* 411-412.

Azegami, K., Nishiyama, K., Watanabe, Y., Kadota, I., Ohuchi, A. and Fukazawa, C. (1987b). *Pseudomonas plantarii* sp. nov., the causal organism of rice seedling blight, *Int. J. Syst. Bacteriol.* 37: 144-152.

Babadoost, M. and Zhang, X. (2018). Eradication of *Xanthomonas cucurbitae* in pumpkin seed by hot-water treatment." In *XXX Int. Horticultural Congress IHC2018: II Int. Symp. Innov. Plant Protec. in Horti.* 1269, 9-16.

Bagri, R.K., Goyal, S.K., Singh J. and Shekhawat, S.S. (2019). Management of bacterial leaf blight disease of cluster bean in semi-arid tropic of Rajasthan. *Plant Dis. Res.* 34 (1): 58-60.

Balaraman, K. and Purusotman, R. (1981). Control of citrus canker on acid lime. *South Indian Hort.* 29: 175-177.

Barik, S., Ponnam, N., GC, A., TH, S., Kumari, M. and Gouri Shankar, S. (2021). Genetic analysis of bacterial wilt resistance in eggplant (*Solanum melongena* L.). *Euro. J. Plant Pathol.* 160(2): 349-364.

Barik, S., Reddy, A. C., Ponnam, N., Kumari, M., DC, L. R., Petikam, S. and Gouri Shankar, S. (2020). Breeding for bacterial wilt resistance in eggplant (*Solanum melongena* L.): Progress and prospects. *Crop Protection*, 137: 105270.

Bayles, M.B. and Verhalen, L.M. (2007). Bacterial blight reactions of sixty-one upland cotton cultivars. *The J. Cotton Sci.* 11: 40-51.

Behlau, F., Canteros, B.I., Minsavage, G.V., Jones, J.B. and Graham, J.H. (2011). Molecular characterization of copper resistance genes from *Xanthomonas citri* subsp. *citri* and *Xanthomonas alfalfae* subsp. *citrumelonis*. *Appl. Environ. Microbiol.* 77: 4089–4096.

Belete, T., Bastas, K. K., Francesconi, S. and Balestra, G. M. (2021). Biological effectiveness of *Bacillus subtilis* on common bean bacterial blight. *J. Plant Pathol.* 103(1): 249-258.

Bhowmik, B., Singh, R. P., Jayaraman, J. and Verma, J. P. (2002). Population dynamics of cotton endophytic *Pseudomonas*, their antogonis and protective action against the major pathogens of cotton. *Indian Phytopath.* 55 (2): 124.

Bondar, G. (1912). Una nova molestia bacteriana das hastes da mandioca. *Characas e Quintais Sao Paulo.* 5 (4): 15- 18.

Bophela, K. N., Petersen, Y., Bull, C. T. and Coutinho, T. A. (2020). Identification of *Pseudomonas* isolates associated with bacterial canker of stone fruit trees in the Western Cape, South Africa. *Plant Dis.*, 104(3): 882-892.

Bose, P. C. and Singh, C. (1980). Chemical control of bacterial canker of mango. *Pesticides*, 14: 30-31.

Brenner, D. J., Krieg, N. R. and Staley, J. T. (2005). *Bergey's Manual of Systematic Bacteriology* Second Edition. Springer. 2: 340–345.

Brown, N. A. (1923). Bacterial leaf spot of geranium in the eastern United States. *Ibid.* 23: 361-72.

Bryaskova, R., Pencheva, D., Nikolov, S. and Kantardjiev, T. (2011). Synthesis and comparative study on the antimicrobial activity of hybrid materials based on silver nanoparticles (AgNps) stabilized by polyvinylpyrrolidone (PVP). *J. Chem. Biol.*, 4: 185.

Buddenhagen, I.W. and Elsasser, T. A. (1962). An insect-spread bacterial wilt epiphytotic of Bluggoe banana. *Nature*, 194 (4824): 164-165.

Burkholder, W. H. (1926): A new bacterial disease of the bean. *Phytopathology*, 16: 915-28.

Burkholder, W. H., McFadden, L. A. and Dimock, J. W. (1953): A bacterial blight of chrysanthemums. *Ibid.*, 43: 522-526.

Burrill, T. J. (1882). The bacteria: an account of their nature and effects, together with a systematic description of the species. 111. *Indus. Univ. Ann. Rpt.*, 11: 93-157.

Cairns, J. (1963). The bacterial chromosome and its manner of replication as seen by autoradiography. *J. Mol. Biol.* 6(3): 208-IN5.

Carlson, R. R. and Vidaver, A. K. (1982). Taxonomy of *Corynebacterium* plant pathogens, including a new pathogen of wheat, based on polyacrylamide gel electrophoresis of cellular proteins. *Int. J. Syst. Bacteriol.* 32: 315-326.

Cavara, F. (1897). Tubercolosi della vite. Intorno alla eziologia de alcune malattie di piante coltivate. *Stazoni Sperimentali Agrarie Italiane* 30:483-487.

Cement, A., Saygili, H., Horuz, S. and Aysan, Y. (2018). Potential of bacteriophages to control bacterial speck of tomato (*Pseudomonas syringae* pv. *tomato*). *Fresenius Environ. Bull.* 27(12A): 9366-9373.

Chahardoli, M., Fazeli, A. and Ghabooli, M. (2017). Antimicrobial activity of LF chimera synthetic peptide against plant pathogenic bacteria. *Arch. Phytopathol. Plant. Protec.* 50: 1008–1018.

Chand, J. N. and Pal, V. (1982). Citrus canker in India and its management. In: *Problems of Citrus Diseases in South* (S. P. Ray Chaudhuri and Y. S. Ahlawat. Eds.) Sarabhai printers and publishers. New Delhi. pp. 21-36.

Chand, R. and Kishun, R. (1987). Studies on bacterial diseases of fruit crops. *Annual Report*, IIHR, Bangalore. pp. 58.

Chand, R. and Kishun, R. (1988 a). Outbreak of grape vine bacterial canker in India. *Vitis* 29: 183–188.

Chand, R. and Kishun, R. (1988b). April pruning and bacterial blight in grape vine. *Drakashavritta Suvenir*. pp. 49-51.

Charlermroj, R., Himananto, O., Seepiban, C., Kumpoosiri, M., Warin, N., Oplatowska, M. and Elliott, C. T. (2013). Multiplex detection of plant pathogens using a microsphere immunoassay technology. *PloS one*, *8*(4): e62344.

Chien, C.C., Chang, Y. C., Liao, Y.M. and Ou, S.H. (1983). Bacterial grain rot of rice-a new disease in Taiwan. *J. Agri. Res.f China*, 32(4): 360-366.

Choi, O. and Hu, Z. (2008). Size dependent and reactive oxygen species related nanosilver toxicity to nitrifying bacteria. *Environ. Sci. Technol.* 42: 4583–4588.

Chung, Y.C., Su, Y.P., Chen, C.C., Jia, G., Wang, H.L., Wu, J.G. and Lin, J.G. (2004). Relationship between antibacterial activity of chitosan and surface characteristics of cell wall. *Acta Pharma,* 25: 932–936

Constantin, E.C., Cleenwerck, I., Maes, M., Baeyen, S., Van Malderghem, C., De Vos, P. and Cottyn, B. (2016). Genetic characterization of strains named as *Xanthomonas axonopodis* pv. *dieffenbachiae* leads to a taxonomic revision of the *X. axonopodis* species complex. *Plant Pathol.* 65: 792-806.

Cordon, G., Andrade, C., Barbara, L. and Romero, A. M. (2021). Early detection of tomato bacterial canker by reflectance indices. *Information Processing in Agri.* 9 (2): 184-194.

Coteance, O. (1971). Diseases of mulberry, their prevention and control. *Sericultura* 7: 22- 29.

Daughtrey, M. (2003). Southern bacterial wilt, caused by Ralstoniasolanacearum. In: *Society of American Florists' 19th Annual Conference on Insect and Disease Management on Ornamentals.*

Davis, M. J., Gillaspie, A. G. Jr., Vidaver, A. K. and Harris, R. W. (1984). *Clavibacter*: A new genus containing some phytopathogenic Coryneform bacteria, including *Clavibacter xyli* subsp. *xyli* sp. nov., subsp. nov. and *Clavibacter xyli* subsp. *cynodontis* subsp. nov., pathogens that cause ratoon stunting disease of sugarcane and bermudagrass stunting disease. *Int. J. Syst. Bacteriol.* 34: 107–117.

De Boer, S. H., Li, X. and Ward, L. J. (2012). Pectobacterium spp. associated with bacterial stem rot syndrome of potato in Canada. *Phytopathology,* 102: 937-947.

den Dooren de *long* L.E. (1926). *Bijdrage tot de kennis van het mineralisatieprocess,* Thesis. Technische Hogeshool Delft. Nijgh and Van Ditmar Rotterdam. pp. 1–199.

Dong, Y.H., Wang, L.H., Xu, J.L., Zhang, H.B., Zhang, X.F. and Zhang, L.H. (2001). Quenching quorum-sensing-dependent bacterial infection by an N-acyl homoserine lactonase. *Nature,* 411: 813–817.

Dong, Y.H., Xu, J.L., Li, X.Z., and Zhang, L.H. (2000). AiiA, an enzyme that inactivates the acylhomoserine lactone quorum-sensing signal and attenuates the virulence of *Erwinia carotovora. Proc Natl Acad Sci USA* 97: 3526– 3531.

Du, W.L., Niu, S.S., Xu, Y.L., Xu, Z.R. and Fan, C.L. (2009). Antibacterial activity of chitosan tripolyphosphate nanoparticles loaded with various metal ions. *Carbohydr. Polym.* 75: 385–389

Dutta, B., Gitaitis,R. D., Lewis,K. J. and Langston, D. B. (2013). A new report of *Xanthomonas cucurbitae* causing bacterial leaf spot of watermelon in Georgia, USA. *Plant Dis.* 97(4): 556.

Dye, D. W. and Kemp, W. J. (1977). A taxonomic studv of plant pathogenic Corynebacterium species. *N.Z. J. Agri. Res.* 20: 563-582.

Elphinstone, J.G. (2005). The current bacterial wilt situation: a global overview. In: *Bacterial wilt disease and the Ralstonia solanacearum species complex*, pp.9-28.

Escobar, M.A., Civerolo, E.L., Summerfelt, K.R. and Dandekar, A.M. (2001). RNAi-mediated oncogene silencing confers resistance to crown gall tumorigenesis. *Proc. Natl. Acad. Sci.* 98: 13437–13442.

Ewbank E. (1992). Etude du catabolisme d'acides aminés conduisant à la formation de phytotoxines par *Xanthomonas campestris* pv. *manihotis* (agent pathogène de la bactériose du manioc).[PhD Thesis]. Louvain-la-Neuve, Belgium: Université Catholique de Louvain; 1992. 191 p

Fang, C. T., Ren, H. C., Chen, T. K., Chu, Y. K., Faan, H. C. and Wu, S. C. (1957). A comparison of rice bacterial leaf blight organism with the bacterial leaf streak organism of rice and *Leersia hexandra* Swartz. *Acta Phytopathol. Sin.* 3: 99-124.

Fayette, J., Roberts, P. D., Pernezny, K. L. and Jones, J. B. (2012). The role of cymoxanil and famoxadone in the management of bacterial spot on tomato and pepper and bacterial leaf spot on lettuce. *Crop Protection*, 31(1): 107-112.

Fegan, M. and Prior, P. (2005). How complex is the *Ralstonia solanacearum* species complex. In: *Bacterial wilt disease and the Ralstonia solanacearum species complex, 1*, 449-461.

Feng, Y., Zhang, Y., Ying, C., Wang, D. and Du, C. (2015). Nanopore-based fourth-generation DNA sequencing technology. *Genomics, Proteomics and Bioinformatics* 13: 4– 16.

Fessehaie, A., Hopkins, D., Gitaitis, R., Langston, D. and Walcott, R. (2005). Role of honey bees in watermelon seed infestation by *Acidovorax avenae* subsp. *citrulli. Phytopathology,* 95: S29–S29.

Fillippone, M. P., J. C. Diaz-Ricei *et al.* (2001). Effect of fragarin on the cytoplasmic membrane of the phytopathogen *clavibacter michiganensis. Mol. Plant-Microbe. Interactions.* 14 (7): 925.

Floyd, J. (2007). New pest response guidelines: *Ralstonia solanacearum* race 3 biovar 2. USDAAPHIS-PPQ, emergency domestic programs,45 pp. Riverdale, MD, Online, www.usdapest

Follin, J.C., Girardot, B., Mangano, V. and Benitez, R. (1988). New results on inheritance of immunity to bacterial blight (*Xanthomonas campestris* pv. *malvacearum* (Smith) Dye, strains 18 and 20) in the cotton plant (*Gossypium hirsutum* L.). *Cotonet Fibres Tropicales (France)*, 43 (3): 167-175.

Francis, D. M., E. Kabelka, J. Bell *et al.* (2001). Resistance to bacterial canker in tomato (*Lycopersicon hirsutum* LA 407) and its progeny from crosses to *Lycopersicum escutentum. Plant Dis.* 85: 383.

Gadewar, A. V., Trivedi, T. P. and Sekhawat, G. S. (1991). Potato in Karnataka. *Tech. Bull.* 17: 33

Gagnevin and Pruvast, O. (2001). Epidemiology and control of mango bacterial black spots. *Plant Dis.* 85 : 928.

Gagnon, Z. R. (2011). Cellular dielectrophoresis: applications to the characterization, manipulation, separation and patterning of cells. *Electrophoresis* 32: 2466 – 2487.

Gagnon, Z., Senapati, S., Gordon, J. and Chang, H. C. (2008). Dielectrophoretic detection and quantification of hybridized DNA molecules on nano-genetic particles. *Electrophoresis*, 29(24): 4808-4812.

Gama, M. A. S., Mariano, R. L. R., Silva, W. J. Jr., Farias, A. R. G., Barbosa, M. A. G., Ferreira, M. A. S. V., *et al. (*2018). Taxonomic repositioning of *Xanthomonas campestris* pv. *viticola* (Nayudu, 1972) Dye 1978 as *Xanthomonas citri* pv. *viticola* (Nayudu, 1972) Dye 1978 comb. nov. and emendation of the description of *Xanthomonas citri* pv. *anacardii* to include pigmented isolates pathogenic to cashew plant. *Phytopathology,* 108: 1143–1153.

García, R. O., Kerns, J. P. and Thiessen, L. (2019). *Ralstonia solanacearum* species complex: a quick diagnostic guide. *Plant Health Progress*, 20(1): 7-13.

Gaudet, D. A. and Kokko, E. G. (1986). Seedling disease of sorghum grown in southern Alberta caused by seedborne *Pseudomonas syringae* pv. *syringae*. *Can. J. Plant Pathol.* 8(2): 208-217.

Ghimire, B. (2017). Use of Crispr/Cas9 for development of disease resistant cultivars in plant breeding. *Int. J. Appl. Sci. Biotechnol.* 5: 403–409.

Gholve, V.M., Kurundkar, B.P., Meshram, M.K. and Wasule, D.L. (2005). Prevalence of races of *Xanthomonas axonopodis* pv. *malvacearum* in Marathwada region of Maharashtra State. *J. Cotton Res. Develo.* 19(2): 254-256.

Gilbertson, R. L. (2002). Bacterial leaf blight of carrot. In: *Compendium of Umbelliferous Crop Diseases*. American Phytopathological Society, St. Paul, MN. Pages 11-12

Gillaspie, A. G. and Davis, M. J. (1992). Ratoon starting of sugar cane. *In: Plant Diseases of International Importance : Diseases of Sugar, Forest and Plantation Crops"* edited by A. N. Mukhopadhyay, J. Kumar, H. S. Chaube and U. S. Singh. Vol. 4.prentice – Hall, Englewood cliffs, New Jersey.

Gillis, M., Van, T.V., Bardin, R., Goor, M., Hebbar, P., Willems, A., Segers, P., Kersters, K., Heulin, T. and Fernandez, M.P. (1995). Polyphasic taxonomy in the genus *Burkholderia* leading to an emended description of the genus and proposition of *Burkholderia vietnamiensis* sp. nov. for N2-fixing isolates from rice in Vietnam." *Int. J. Syst. Bacteriol.* 45: 274-289.

Goto, M. (1979a). Bacterial foot rot of rice caused by a strain of *Erwinia chrysanthemi*. *Phytopathology*, 69(3): 213-216.

Goto, M. (1979b). Dissemination of *Erwinia chrysanthemi*, the causal organism of bacterial foot rot of rice. *Plant Dis. Repor.* 63(2): 100-103.

Griffith, P. (1928). Significance of pneumococcal types. *J. Hyg., Camb.* 27: 113.

Gumtow, R. L., Khan, A. A., Bocsanczy, A. M., Yuen, J. M. F., Palmateer, A. J. and Norman, D. J. (2013). First report of a leaf spot disease of golden dewdrop (*Duranta erecta*) caused by *Pseudomonas cichorii* and a *Xanthomonas* species in Florida. *Plant Dis.* 97(6): 836-836.

Gutierrez-Barranquero, J.A., de Vicente, A., Carrion, V.J., Sundin, G.W. and Cazorla, F.M. (2013). Recruitment and rearrangement of three different genetic determinants into a conjugative plasmid increase copper resistance in *Pseudomonas syringae*. *Appl. Environ. Microbiol.* 79: 1028–1033.

Hailu, N. and Tontosha, H. (2021). Management of common bacterial blight (*Xanthomonas campestris* pv. *phaseoli* (Smith) Dye.) of common bean. *J. Plant Pathol. Microbiol.* 12: 533.

Hao, J. J., Meng, Q. X., Yin, J. F. and Kirk, W. W. (2009). Characterization of a new *Streptomyces* strain, DS3024, that causes potato common scab. *Plant Dis.* 93: 1329–1334.

Harveson, R. M. (2019). Improving yields and managing dry bean bacterial diseases in Nebraska with new copper-alternative chemicals. *Plant Health Progress,* 20(1): 14-19.

Hauben, L., Moore, E. R., Vauterin, L., Steenackers, M., Mergaert, J., Verdonck, L. and Swings, J. (1998). Phylogenetic position phytopathogens within Enterobacteriaceae. *Syst. Appl. Microbiol.* 21: 384–397.

Hayward, A. C. (1994). The hosts of *Pseudomonas solanacearum*. *Bacterial wilt: the disease and its causative agent, Pseudomonas solanacearum.*, 9-24.

Helman, Y. and Chernin, L. (2015). Silencing the mob: disrupting quorum sensing as a means to fight plant disease. *Mol. Plant Pathol.* 16: 316–329.

Higgins, M. L. and Shockman G.D. (1971). Procaryotic cell division with respect to wall and membranes., *CRC Crit. Rev. Microbiol.* 1: 29–72.

Hingorani, M. K. and Mehta, P. P. (1952). Bacterial leaf spot of pomegranate. *Indian Phytopath.* 5: 55-56.

Hingorani, M. K. and Singh, N. J. (1959). *Xanthomonas punicae* sp. nov. on *Punica granatum* L. *Indian J. Agric. Sci.* 29 : 45-48.

Hirano, S. S. and Upper, C. D. (1990). Population biology and epidemiology of *Pseudomonas syringae*. *Annu. Rev. Phytopathol.* 28: 155-177.

Honda, M., Kondo, T., Tanaka, K. *et al.* (2022). Bacterial rot of Chinese yam caused by *Pseudomonas allii*. *J. Gen. Plant Pathol.* https: //doi.org/10.1007/s10327-022-01073-6

Huang, H. C., Hsieh, T.F. and Erickson, R.S. (2003). Biology and epidemiology of *Erwinia rhapontici*, causal agent of pink seed and crown rot of plants. *Plant Pathology Bull.* 12: 69–76.

Huang, Y., Nordeen, R. O., Di, M., Owens, L.D. and McBeath, J. H. (1997). Expression of an engineered cecropin gene cassette in transgenic tobacco plants confers disease resistance to *Pseudomonas syringae* pv. *tabaci*. *Phytopathology*, 87: 494–499.

Hulloli, S. S., Singh, R. P. and Verma, J. P. (1998). Management of bacterial blight of cotton induced by *Xanthomonas axonopodis* pv. *malvacearum* with the use of neem based formulations. *Indian Phytopath.* 51: 21.

Hunter, R.E., Brinkerhoff, L.A. and Bird, L.S. (1968). Development of a set of upland cotton lines for differentiating races of *Xanthomonas malvacearum*. *Phytopathology*, 58(6): 830- 832.

Hutchinson, C. M. (1917). A bacterial disease of wheat in the Punjab. Memoirs of the Department of Agriculture in India. *Bacteriol. Ser.* 1: 169-175

Hyman, L.J., Toth, I.K. and Pérombelon, M.C.M. (1998). Isolation and identification In: Pérombelon, M.C.M. and Van Der Wolf, J.M. (Eds.). Methods for the detection and quantification of *Erwinia carotovora* subsp. *atroseptica* on potatoes. *Laboratory Manual. Invergowrie. Scottish Crop Research Institute.* 60-65.

Ilagan, Y.A., Lavina, W.A., Natural, M.P. and Raymundo, A.K. (2003). Genetic homogeneity of the banana-infecting strains of *Ralstonia solanacearum* (Smith) Yabuuchi *et al.,* in the Philippines. *Philippine Agricultural Scientist,* 86(4): 394-402.

Inoue, Y. and Nakaho, K. (2019). Case studies of various soil disinfection methods on *Ralstonia solanacearum* population reduction in soil (in Japanese with English summary). *Bull NARO Agric Res Cent Reg* 7: 1–10.

Jain, A. C. (1951). A bacterial disease (canker) of tomatoes new to India. *Sci. and Cul.* 17: 46-47.

Janse, J. D. (2005). *Phytobacteriology: Principles and practice*. CABI Publishing, UK. 360p.

Jaynes, J. M., Nagpala, P., Destéfano-Beltrán, L., Huang, J.H., Kim, J., Denny, T. and Cetiner, S. (1993). Expression of a cecropin B lytic peptide analog in transgenic tobacco confers enhanced resistance to bacterial wilt caused by *Pseudomonas solanacearum*. *Plant Sci.* 89: 43–53.

Jia, H., Orbovic, V., Jones, J. B., Wang, N. (2016). Modification of the PthA4 effector binding elements in Type I Cs LOB 1 promoter using Cas9/sg RNA to produce transgenic Duncan grapefruit alleviating XccΔpthA4: dCs LOB 1.3 infection. *Plant Biotechnol. J.* 14: 1291–1301.

Jindal, J. K., Patel, P. N and Singh, R. (1972). Bacterial leaf spot disease on *Amorphophallus campanulatus*. *Indian Phytopath.* 25(3): 374-377

Jindal, K. K. and Sharma, R. C. (1988), Occurrence of biovar -1 of *Agrobacterium tumefaciens* on almond in India. *Indian Phytopath*. 41: 614-615.

Jones, J. B., Engelhard, A. W. and Raju, B. C. (1983). Outbreak of a stem necrosis on chrysanthemum incited by *Pseudomonas cichorii* in Florida. *Plant Dis.* 67: 431- 433.

Jones, J.B., Lacy, G.H., Bouzar, H., Stall, R.E. and Schaad, N.W. (2004). Reclassification of the Xanthomonads associated with bacterial spot disease of tomato and pepper. *Syst. Applied Microbiol.* 27 (6): 755-762.

Jones, L. K. (1941). Bacterial wilt of carnation. *Phytopathology*, 31: 199.

Kadota, I. and Ohuchi, A. (1983). Symptoms of bacterial brown stripe of rice seedlings in nursery boxes. *Japanese J. Phytopathol.* 49(4): 561-564.

Kannan, V. R. and Bastas, K. K. (2015). *Sustainable Approaches to Controlling Plant Pathogenic Bacteria*. CRC Press.

Kashyap, P.L., Xiang, X. and Heiden, P. (2015). Chitosan nanoparticle based delivery systems for sustainable agriculture. *Int. J. Biol. Macromol.* 77: 36–51.

Kemboi, V. J., Kipkoech, C., Njire, M., Were, S., Lagat, M. K., Ndwiga, F. and Tanga, C. M. (2022). Biocontrol potential of chitin and chitosan extracted from black soldier fly pupal exuviae against bacterial wilt of tomato. *Microorganisms*, 10(1): 165.

Kendrick, J. B. (1934). Bacterial blight of carrot. *J. Agr. Res.* 49: 493- 510.

Kishun, R. (1981 b). Studies on bacterial canker of mango. *Proc.* 3rd *Int. Symp. Plant Pathol.*, New Delhi. 151-152.

Kishun, R. (1981a). Bacterial canker of mango. *Annual Report*. IIHR, Bangalore. pp.70.

Kishun, R. (1987). Control of bacterial wilt in India. *Bacterial Wilt Newsl.* 2: 7–8.

Kishun, R. (1988b). Stem injection of chemicals to control of bacterial canker of mango. *Acta Horticulturae* 231: 518-521.

Kishun, R. (1989). Appraisal of loss in yield of cowpea due to *Xanthomonas campestris* pv. *vignicola*. *Indian Phytopath.* 42: 241-246

Kishun, R. and Chand R. (1987). Studies on germplasm resistance and chemical control of citrus canker. *Indian J. Hort.*, 44: 126-132.

Kishun, R. and Chand, R. (1988). Survival of *Xanthomonas campestri* spv. *mangiferae indicae* on weed hosts. *Indian Phytopath.* 41: 269.

Kishun, R. and Chand, R. (1989). Effect of sowing time and plant population on bacteria blight and yield of cowpea. *Indian J. Hort.* 46(2): 2144- 2149.

Kishun, R. and Sohi, H. S. (1983). Bacterial canker in mangoes. *Indian Farmer's Digest* 14: 21-23.

Kishun, R. and Sohi, H. S. (1984). Control of bacterial canker of mango by chemicals. *Pesticides*, 18: 32-33.

Kloepper, J.W. (1978). Plant growth-promoting rhizobacteria on radishes. In: *Proc. of the 4th Internet. Conf. on Plant Pathogenic Bacter, Station de Pathologie Vegetaleet Phytobacteriologie, INRA, Angers, France*, 1978. 2: 879–882.

Kulkarni, Y. S., Patel, M. K. and Abhankar, S. G. (1950). A new bacterial leaf spot and stem canker of pigeon pea. *Cur. Sci.* 19 (12): 384.

Kumar, A., Sarma, Y.R. and Anandaraj, M. (2004). Evaluation of genetic diversity of *Ralstonia solanacearum* causing bacterial wilt of ginger using REP–PCR and PCR–RFLP. *Cur. Sci.* 11(10): 1555-1561.

Kumar, A.S., Aiyanathan, K.E.A., Nakkeeran, S. and Manickam, S. (2018). Documentation of virulence and races of *Xanthomonas citri* pv. *malvacearum* in India and its correlation with genetic diversity revealed by repetitive elements (REP, ERIC and BOX) and ISSR markers. *3 Biotech*, 8(11): 479.

Kumar, M. R. and Khan, A.N.A. (1996), Prevailence of races and survival of *Xanthomonas campestris* pv. *vesicatoria* causing bacterial spot of tomato. *Plant Pathogenic Bacteria. Proc. 9th International conference, CASE, Madras*, 384-391.

Kumar, S. N. S. (1981). Bacterial leaf stripe disease of Arecanut (*Areca Catechu* L.) caused by *Xanthomonas campestris* pv. *arecae*. *Ph. D. Thesis*, Indian institute of science, Bangalore, India.Pp. 190.

Kumud, J. and Shyam, K. R. (2003). Integrated management of black rot of cauliflower caused by *Xanthomonas campestris* pv. *campestris*. *Plant Dis. Res. (Ludhiana)* 18(2): 152-158.

Kuwata, H. (1985). *Pseudomonas syringae* pv. oryzae. Nov., causal agent of bacterial halo blight of rice. Japanese *J. Phytopathology*, 51 (2): 212-218.

Lambani, K. and Jahagirdar, S. (2017). Morphological, cultural, physiological and biochemical characteristics of bacterial pustule of soybean caused by *Xanthomonas axonopodis* pv. *glycines*. *J. Pure and Appl. Microbiol.* 11(2): 1155-1159.

Lambani, K., Jahagirdar, S. and Nirmalnath P. Jones. (2017). *In vitro* evaluation of antibiotics, chemicals and bioagents against *Xanthomons axonopodis* pv. *glycines*. *J. Pure and Appl. Microbiol.* 11 (1): 423-426.

Lang, M.D. and Evans, K.J. (2010). Epidemiology and status of wal-nut blight in Australia. *J. Plant Pathol.* 92: S1.49- S1.55.

Lei, L., Zhang Z., Zhao, Y., Hua, X., Shi, Y., Chai, A., Xie, X. and Li, B. (2022). Isolation and identification of the pathogen of bacterial stem soft rot of muskmelon in some areas of Shandong Province [J]. *Acta Phytopathologica Sinica*, 52(3): 508-510.

Leu, L. S. and Su, C. C. (1993). Isolation, cultivation and pathogenicity of *Xylella fastidiosa*, the causal bacterium of pear leaf scorch disease in Taiwan. *Plant Dis.* 77: 642-646.

Li, B., Wang, X., Chen, R., Huangfu, W. and Xie, G. (2008). Antibacterial activity of chitosan solution against *Xanthomonas* pathogenic bacteria isolated from Euphorbia pulcherrima. *Carbohydr. Polym.* 72: 287–292.

Li, J.F., Norville, J.E., Aach, J., McCormack, M., Zhang, D., Bush, J., Church, G.M. and Sheen, J. (2013). Multiplex and homologous recombination-mediated genome editing in Arabidopsis and *Nicotiana benthamiana* using guide RNA and Cas9. *Nat. Biotechnol.* 31: 688–691.

Li, T., Liu, B., Spalding, M.H., Weeks, D.P. and Yang, B. (2012). High-efficiency TALEN-based gene editing produces disease-resistant rice. *Nat. Biotechnol.* 30: 390.

Li, Y., Hutchins, W., Wu, X., Liang, C., Zhang, C., Yuan, X., Khokhani, D., Chen, X., Che, Y., Wang, Q. and Yang, C.H. (2015). Derivative of plant phenolic compound inhibits the type III secretion system of *Dickeya dadantii* via HrpX/HrpY two-component signal transduction and Rsm systems. *Mol. Plant Pathol.* 16: 150–163.

Li, Y., Peng, Q., Selimi, D., Wang, Q., Charkowski, A.O., Chen, X. and Yang, C.H. (2009). The plant phenolic compound p-coumaric acid represses gene expression in the *Dickeya dadantii* type III secretion system. *Appl. Environ. Microbiol.* 75: 1223–1228.

Lia, R. S., Ali, M. R., Jahan, M. S., Akter, A., Sumi, M. S. E., Hasan, M. F., — and Sikdar, B. (2018). Detection of *Xanthomonas campestris* pv. *cucurbitae* from bacterial leaf spot disease of cucumber and evaluation of its biological control. *Adv. Biores*, 9: 41-46.

Lindow, S. E. and Brandl, M. T. (2003). Microbiology of the phyllosphere. *Appl. Environ. Microbiol.* 69: 1875–1883.

Lindow, S., Newman, K., Chatterjee, S., Baccarai, C., Iavarone, A.T. and Ionescu, M. (2014). Production of *Xylella fastidiosa* diffusible signal factor in transgenic grape causes pathogen confusion and reduction in severity of Pierce's disease. *Mol. Plant–Microbe Interact.* 27: 244–254.

Lok, C.N., Ho, C.M., Chen, R., He, Q.Y., Yu, W.Y., Sun, H., Tam, P.K., Chiu, J.F. and Che, C.M. (2007). Silver nanoparticles: partial oxidation and antibacterial activities. *J. Biol. Inorg. Chem.* 12: 527–534.

Loria, R., Coombs, J., Yoshida, M., Kers, J. and Bukhalid, R. (2003). A paucity of bacterial root diseases *Streptomyces* succeeds where other fail. *Physiol. Mol. Plant Pathol.* 62 (2): 65.

Lozano, J.C. and Sequeira, L. (1970). Differentiation of races of *Pseudomonas solanacearum* by a leaf infiltration techniques. *Phytopathology,* 60: 833-838.

Luthra, J. C. and Sattar, A. (1940) A note on Citrus canker (*Phytomonas citri*) in the Punjab. *Proc. 27th Ind. Sci. Congr. Part.III*, 228-229.

Ma, X. (2015). *Characterization and management of bacterial leaf spot of processing tomato in Ohio.* The Ohio State University.

Madhiazhagan, K., Ramadoss, N. and Anuradha, R. (2002). Effect of botanicals on bacterial blight of rice. *J. Mycol. Plant. Pathol.* 32 (1): 68.

Maji, M. D. and Qadri, S. M. H. (1999). Bacterial diseases of mulberry (*Morus* spp.). *Sericologia* 39(1): 1- 7.

Maji, M. D., Banerjee, R., Das, N. K. and Ghosh, A. (2012). Forecasting model of bacterial leaf spot disease of mulberry caused by *Xanthomonas campestris* pv. *mori*. *Arch Phytopathol. Plant Protec.* 45 (13): 1570-1574.

Makovitzki, A., Viterbo, A., Brotman, Y., Chet, I. and Shai, Y. (2007). Inhibition of fungal and bacterial plant pathogens *in vitro* and in planta with ultrashort cationic lipopeptides. *Appl. Environ. Microbiol.* 73: 6629–6636.

Manickam, R., Jaw-Rong, Chen, Paola Sotelo-Cardona, Lawrence Kenyon and Ramasamy Srinivasan. (2021). Evaluation of different bacterial wilt resistant eggplant rootstocks for grafting tomato. *Plants* 10: 75. https://doi.org/10.3390/plants10010075.

Manjula, C. P., Khan A.N.A., Jalikop, S. A. and Ravikumar, M. R. (2003). Chemical and biological control of bacterial blight of pomegranate caused by *Xanthomonas campestris* pv. *punicae*. In: National seminar on plant disease management and sustainable Agriculture. Organised by UGC/Annamalai nagar held at Annamalai University, 20-21st March, 2003.

Mansilla, A.Y., Albertengo, L., Rodríguez, M.S., Debbaudt, A., Zúñiga, A. and Casalongué, C.A. (2013). Evidence on antimicrobial properties and mode of action of a chitosan obtained from crustacean exoskeletons on *Pseudomonas syringae* pv. *tomato* DC3000. *Appl Microbiol Biotechnol* 97: 6957–6966

Mbega, E. R., Mortensen, C. N., Mabagala, R. B. and Wulff, E. G. (2012). The effect of plant extracts as seed treatments to control bacterial leaf spot of tomato in Tanzania. *J. Gen. Plant Pathol.* 78(4): 277-286.

McCulloch, L. (1924). A bacterial blight of gladioli. *J. Agri. Res.* 27: 225-230.

McKeague, M., Bradley, C. R., De Girolamo, A., Visconti, A., Miller, J. D. and Derosa, M. C. (2010). Screening and initial binding assessment of fumonisin B_1 aptamers. *Int. J. Mol. Sci.* 11: 4864– 4881.

Mellano, M.A. and Cooksey, D.A. (1988). Nucleotide sequence and organization of copper resistance genes from *Pseudomonas syringae* pv. *tomato*. *J. Bacteriol.* 170: 2879–2883.

Meng, X., Chai, A., Shi, Y., Xie, X., Ma, Z. and Li, B. (2017). Emergence of bacterial soft rot in cucumber caused by *Pectobacterium carotovorum* sub sp. *brasiliense* in China. *Plant Dis.* 101: 279-287.

Merchan-Vargas, V.M. (2003). Situación actual delMoko (*Ralstonia solanacearum* race 2) enmusáceas. *Manejo Convencionaly Alternativo de Sigatokanegra, Nematodos y OtrasPlagasAsociadas al Cultivo de Musáceas, eds G. Rivas and F. Rosales (Guayaquil: MUSALAC/INIBAP)*, 181-182.

Meshram, M.K., Sheo, R. and Wasule, D.L. (2002). Bacterial blight. In: Sheo R (ed) Twenty five years of achievements in cotton pathology at CICR (1976–2001). Central Institute for Cotton Research, Nagpur

Mitra, A. and Zhang, Z. (1994). Expression of a human lactoferrin cDNA in tobacco cells produces antibacterial protein(s). *Plant Physiol.* 106: 977–981.

Mohammadi, A. S., Foster, R.E., Maynard, E.T. and Egel, D.S. (2015). Timing early season cucumber beetle control to manage bacterial wilt in muskmelon. *Hort. Sci.* 50:

Molina, A.B. (2006). Managing bacterial wilt/fruit rot disease of banana in Southeast Asia. In: *Developing a regional strategy to address the outbreak of banana Xanthomonas wilt in East and Central Africa. Proceedings of the banana Xanthomonas wilt regional preparedness and strategy development workshop held in Kampala, Uganda, 14-18 February, 2005* (pp. 26-31). INIBAP Regional Office for Asia and the Pacific.

Molina, L., Constantinescu, F., Michel, L., Reinmann, C., Duffy, B. and Defago, G. (2003). Degradation of pathogen quorum-sensing molecules by soil bacteria: A preventive and curative biological control mechanism. *FEMS Microbiol. Ecol.* 45: 71–81.

Moller, W.J., Schroth, M.N. and Thomson, S.V. (1981). The scenario of fire blight and streptomycin resistance. *Plant Dis.* 65: 563–568.

Moniz, L., Sabley, J. E. and More, W. D. (1964). A new bacterial canker of *Carissa congesta* in Maharashtra. *Indian Phytopath.* 17: 256.

Mudingotto, P. J., Veena, M. S. and Mortensen, C. N. (2002). First report of bacterial blight caused by *Acidovorax avenae* ssp. *avenae* associated with finger millet seeds from Uganda. *Plant Pathol.* 51: 396.

Mura, Y. and Goto, M. (1996). Bacterial Leaf Spot of Amaranthus cruentus L. Caused by *Burkholderia andyopogonis* (Stapp) Gills *et al. Ann. Phytopathol. Soc. Jpn.* 62: 181-183.

Nakaho, K. (2021). Mechanisms of resistance to *Ralstonia solanacearum* in tomato rootstocks and integrated management of bacterial wilt using high grafting. *J. Gen. Plant Pathol.*, 87(6): 398-402.

Ocsoy, I., Paret, M.L., Ocsoy, M.A., Kunwar, S., Chen, T., You, M.X. and Tan, W.H. (2013). Nanotechnology in plant disease management: DNA-directed silver nanoparticles on graphene oxide as an antibacterial against *Xanthomonas perforans. ACS Nano,* 7: 8972–8980.

Opio, F. and Male- Kayiwa, S. (1995). The status of bean breeding in Uganda. Breeding for disease resistance with emphasis on durability. In: Proc. Regional Workshop for Eastern, Central and Southern Afgrica, held at Njoro, Kenya, Octobober 2-6, 1994, 110- 113.

Orzali, L., Corsi, B., Forni, C. and Riccioni, L. (2017). Chitosan in agriculture: A new challenge for managing plant disease. *Biol. Activ. Appl. Mar. Polysacch*: 17–36. 10.5772/66840.

Ou, S. H. (1984). *Rice diseases,* 2nd edition. Slough: Commonwealth Agriculture Bureaux.

Paret, M. L., Vallad, G.E., Averett, D.R., Jones, J.B. and Olson, S.M. (2013). Photocatalysis: effect of light activated nanoscale formulations of TiO2 on *Xanthomonas perforans* and control of bacterial spot of tomato. *Phytopathology,* 103: 228–236.

Patel, M. K. and Diwan, N. D. (1950). Bacterial blight of cowpea. *Indian Phytopath.* 3: 75- 80.

Patel, M. K., Kulkarni, Y. S. and Dhande, G. W. (1950). Bacterial leaf-spot of chillies. *Indian Phytopath.* 3(1): 95–97.

Patel. M. K., Monize, L. and Kulkarni, Y. S. (1948). A new bacterial disease of *Mangifera indica*. Cur Sci. 17: 180-190.

Pazarlar, S. E. R. C. A. N., Sanver, U. T. K. U. and Cetinkaya, N. E. D. Ý. M. (2021). Exogenous pipecolic acid modulates plant defence responses against *Podosphaera xanthii* and *Pseudomonas syringae* pv. *lachrymans* in cucumber (*Cucumis sativus* L.). *Plant Biol.* 23(3): 473-484.

Peng, A., Chen, S., Lei, T., Xu, L., He, Y., Wu, L. and Zou, X. (2017). Engineering canker-resistant plants through CRISPR/Cas9-targeted editing of the susceptibility gene Cs LOB 1 promoter in citrus. *Plant Biotechnol. J.* 15: 1509–1519.

Perez, S. M., Biondi, E., Laurita, R., Proto, M., Sarti, F., Gherardi, M. and Colombo, V. (2019). Plasma activated water as resistance inducer against bacterial leaf spot of tomato. *PloS one*, 14(5): e0217788.

Perreaux, D., Maraite, H and Meyer, J. A. (1986). Detection of 3 (methylthio) propionic acid in cassava leaves infected by *Xanthomnas campestris* pv. *manihotis*. *Physiol. Mol. Plant Pathol.* 28: 323- 328.

Perry, J.A. and Wright, G.D. (2013). The antibiotic resistance "mobilome": searching for the link between environment and clinic. *Front. Microbiol.*4: 138.

Piepenburg, O., Williams, C.H., Stempl,e D.L. and Armes, N.A. 2006). DNA detection using recombination proteins. *PloS Biol.* 4 doi: 10.1371/journal.pbio.0040204.

Poplawsky, A. R., Robles, L., Chun, W., Derie, M. L., du Toit, L. J., Meng, X. Q. and Gilbertson, R. L. (2004). Identification of a *Xanthomonas* pathogen of coriander from Oregon USA.(Abstr.) *Phytopathology* 94: S85.

Prabhu, Y.T., Rao, K.V., Kumari, B.S., Kumar, V.S.S. and Pavani, T. (2015). Synthesis of Fe3O4 nanoparticles and its antibacterial application. *Int. Nano Lett.*, 5: 85–92.

Prasad K. K. S. and Siddaramaiah, A. L. (1978). Studies on the bacterial leaf blight of mulberry in Karnataka. *India.J. Seric.*17: 61- 63.

Prasad, H. H. (1930). A bacterial stalk rot of maize. *Agric. J. India, 25*(1): 72.

Prasad, H. H. (1931). A note on bacterial leaf spot of khira. (*Cucumis sativus*). *Indian J. Agri. Sci.* 1: 289- 290.

Prasad, N. (1959). Citrus canker. *Proc. Seminar on Disease of Horticultural Plants, Simla,* 87-88.

Pruvast, O., B. Boler, C. Brocherieux *et al.* (2002). Survival of *Xanthomonas axonopodis* pv. *citri* in leaf lesions under tropical environmental conditions and simulated splash dispersal of inoculum. *Phytopathology*. 92 (4): 336.

Pysher, M. D. and Hayes, M. A. (2007). Electrophoretic and dielectrophoretic field gradient technique for separating bioparticles. *Analytical Chem.* 79: 4552–4557.

Rajamanickam, S., Karthikeyan, G., Kavino, M. and Manoranjitham, S. K. (2018). Biohardening of micropropagated banana using endophytic bacteria to induce plant growth promotion and restrain rhizome rot disease caused by *Pectobacterium carotovorum* subsp. *carotovorum*. *Sci. Horti.* 231: 179-187.

Ramakrishnan, T.S. (1954). Common diseases of citrus in Madras state. Govt. of Madras publication.

Rangarajan, S., Saleena, L. M., Vasuderan, P. and Nair, S. (2003). Biological suppression of rice diseases by *Pseudomonas* spp. under saline soil conditions. *Plant Soil*. 251 (3): 73.

Rangaswami, G. and Eswaran, K. S. S. (1961). A bacterial leaf spot disease of jasmine. *Cur. Sci* 30(9): 352-352.

Rangaswami, G. (1962). Bacterial Plant Diseases in India. Asia Publishing House, Bombay, 163 pages.

Rangaswami, G. and Rangarajan, M. (1965). A bacterial leaf-spot disease of banana. *Phytopathology*, 55: 1035–1036.

Rao, K. S. P. and Hiremath, R. V. (1985). Influence of host nutrition on the severity of bacterial blight of cowpea. *Indian Phytopath*. 38(1): 81- 84.

Rao, Y. P. and Mohan, S. K. (1976). Bacterial leaf stripe of areca nut caused by *Xanthomonas arecae* sp. *Indian Phytopath.* 29: 251-255.

Rao, Y.P. and Mohan, S.K. (1970). A new bacterial leaf stripe disease of arecanut (*Areca catechu*) in Mysore State. *Indian Phytopath*. 23: 702–704.

Ravikumar, M. R. and Khan, A.N.A. (2000), *In vitro* evaluation of tomato cultivar for their resistance to bacterial leaf spot of tomato. *Karnataka J. Agril. Sci.*, 13 (4): 1016- 1018.

Ravikumar, M. R. and Khan, A.N.A. (2000). Chemical control of bacterial spot of tomato. *Karnataka J. Agril. Sci.* 13 (4): 1011- 1015.

Ravikumar, M. R. and Khan, A.N.A. (2000). Host range studies of *Xanthomonas campestris* pv. *vesicatoria* causing bacterial leaf spot of tomato. *Karnataka J. Agril. Sci.*, 13 (4): 1009-1010.

Ravikumar, M. R. and Khan, A.N.A. (2001). Detection of *Xanthomonas campestris* pv. *vesicatoria* from insect vectors. *Karnataka J. Agril. Sci.* 14 (1): 160-161.

Ravikumar, M. R. and Khan, A.N.A. (2001). Influence of weather parameters on the incidence of bacterial leaf spot of tomato. *Indian J. Agril. Res.* 35 (1): 60-62.

Ravikumar, M. R. and Khan, A.N.A. (2002). Detection of *Xanthomonas campestris* pv. *vesicatoria* in tomato seeds by enzyme linked immuno sorbent assay (ELISA). *Indian J. Agril. Res.* 36(1): 39-43.

Ravikumar, M. R. and Khan, ANA. (2001). Field Evaluation of antibiotics and chemicals against bacterial spot of tomato in Karnataka, *Pestology*, XXV (6): 11-14.

Ravikumar, M. R., Shamrao Jahagirdar and Khan A.N.A. (2003). Management of bacterial leaf spot of grape through chemicals and antibiotics in northern Karnataka. *Indian Phytopath.* 56 (3): 341.

Ravikumar. M. R. and Khan, A.N.A. (2000). Studies on development and standardization of inoculation techniques in management of Bacterial leaf spot of tomato. *Karnataka J. Agril. Sci.* 13 (4): 1022-1024.

Raymundo, A.K., Orlina, M.E., Lavina, W.A. and Opina, N.L. (2005). Comparative genome plasticity of tomato and banana strains of *Ralstonia solanacearum* in the Philippines. *Bacterial wilt disease and the Ralstonia solanacearum species complex*, pp. 387-393.

Reddy, G. S. (1968). Citrus diseases in India and their control. *Tech. Bull.* no. 9. *Indian Coun. Agr. Research, New Delhi.* 70 p.

Rettcher, S., Jungk, F., Kühn, C., Krause, H. J., Nölke, G., Commandeur, U. and Schröper, F. (2015). Simple and portable magnetic immunoassay for rapid detection and sensitive quantification of plant viruses. *Appl. Environ. Microbiol.* 81(9): 3039-3048.

Riker, A. I., Banfield, W. M., Wright, W. H., Keitt, G. W. and Sagan, H. E. (1930). Studies on infectious hairy root of nursery apple trees. *J. Agric. Res.* 41: 507- 540.

Rodriguez, C., Lang, L., Wang, A., Altendorf, K., Garcia, F. and Lipski, A. (2006). Lettuce for human consumption in Costa Rica contains complex communities of culturable oxytetracycline- and gentamicin-resistant bacteria. *Appl. Environ. Microbiol.* 72: 5870–5876.

Rorer, J.B. (1911). A bacterial disease of bananas and plantains. *Phytopathology*, 1(2): 45- 49.

Rutherford, S.T. and Bassler, B.L. (2012) Bacterial quorum sensing: its role in virulence and possibilities for its control. Cold Spring Harb. Perspect. Med. 2, a012427.

Sackett, W. G. (1916). A bacterial stem blight of field and garden peas. *Bull. Colo. St. Univ. agric. Exp. Stn.* 218.

Safenkova, I. V., Pankratova, G. K., Zaitsev, I. A., Varitsev, Y. A., Vengerov, Y. Y., Zherdev, A. V. and Dzantiev, B. B. (2016). Multiarray on a test strip (MATS): Rapid multiplex immunodetection of priority potato pathogens. *Analytical and Bioanalytical Chemistry*, 408(22): 6009-6017.

Sakthivel, K., Manigundan, K., Gautam, R. K., Singh, P. K., Nakkeeran, S. and Sharma, S. K. (2019). *Bacillus* spp. for suppression of eggplant bacterial wilt pathogen in Andaman Islands: Isolation and characterization. *Indian J. Expt. Biol.* 57 (2): 138- 144.

Samson, R., Legendre, J. B., Christen, R., Fischer-Le Saux, M., Achouak, W. and Gardan, L. 2005. Transfer of *Pectobacterium chrysanthemi* (Burkholder *et al.*, 1953) Brenner *et al.*, 1973 and *Brenneria paradisiaca* to the genus Dickeya gen. nov. known as *Dickeya chrysanthemi* comb. nov and *Dickeya paradisiaca* combi. nov. and delineation of four novel species, *Dickeya dianthi* sp nov., *Dickeya dianthicola* sp. nov., *Dickeya dieffenbachiae* sp. nov. and *Dickeya zeae* sp. nov. *Int. J. Syst. and Evol Microbiol.* 55: 1415–1427.

Sanogo, S. Etarock, B.F. and Clary, M. (2011). First report of bacterial wilt caused by *Erwinia tracheiphila* on pumpkin and watermelon in New Mexico. *Plant Dis.* 5(12): 1583.

Sato, M. and Takahashi, K. (1973). Ecological studies on the bacterial blight of mulberry 2. Seasonal differences of the population of *Pseudomonas mori* phage in the soil of mulberry field. *J. Seric. Sci. Japan*, 42 (3): 207-212.

Sattar, A., Janardhanan, K. K. and Husain, A. (1988). Out breaks and new records. India. Bacterial stem rot of opium poppy caused by *Erwinia carotovora* subsp. *carotovora* in India. FAO Plant Protection Bulletin 36(3): 163- 164.

Schaad, N.W., Berthier-Schaad, Y. and Knorr, D. (2007). A high throughput membrane BIO-PCR technique for ultra-sensitive detection of *Pseudomonas syringae* pv. *phaseolicola. Plant Pathology* 56: 1-8.

Schaad, N.W., Jones, J.B. and Chun, W. (2001). *Laboratory Guide for Identification of Plant Pathogenic Bacteria*, 3rd ed. American Phytopathological Society Press, 373p

Schaad, N.W., Sowell, G., Goth, R.W., Colwell, R.R. and Webb, R.E. (1978). *Pseudomonas pseudoalcaligenes* subsp. *citrulli* subsp. nov. *Int. J. Syst. Bacteriol.* 28: 117–125.

Sefah, K., Shangguan, D., Xiong, X., O'Donoghue, M. B. and Tan, W. (2010). Development of DNA aptamers using Cell-SELEX. *Nat.Protoc* 5: 1169–1185.

Sequeira, L. (1998). Bacterial wilt: the missing element in international banana improvement programs. In *Bacterial Wilt Disease*.Springer, Berlin, Heidelberg. 6-14.

Sequeira, L. and Averre, C.W. (1961). Distribution and pathogenicity of strains of *Pseudomonas solanacearum* from virgin soils in Costa Rica. *Plant Dis. Repor.* 45(6): 435.

Shakya, D.D., Vinther, F. and Mathur, S.B. (1985). World wide distribution of a bacterial stripe pathogen of rice identified as *Pseudomonas avenae*. *J. Phytopathol.* 114(3): 256-259.

Sharma, J. P., Kumar S. and Singh D. (2015). *Bacterial wilt of solanaceous crops.* I. K. International Publishing House Pvt. Ltd. New Delhi, 286 pages

Shekhawat, G. S. and Patel, P.N. (1975). Studies on bacterial canker of mango. *Z. Pflkrankh Pflschutz.* 82: 129-138

Shekhawat, G. S., Srivastava, D. N. and Rao, Y. P. (1969). Seed infection and transmission of bacterial leaf streak of rice. *Plant Dis. Rep.* 53: 115- 116.

Sherbakoff, C. D. (1918). Report of the associate plant pathologist. *Fla. Agric. Exp. Stn. Rep.* 1916-1917: 66R-86R.

Shtienberg, D., Zilberstaine, M., Oppenheim, D., Herzog, Z., Manulis, S., Shwartz, H. and Kritzman, G. (2001). Efficacy of oxolinic acid and other bactericides in suppression of *Erwinia amylovora* in pear orchards in Israel. *Phytoparasitica,* 29: 143–154.

Singh D., Yadav, D.K. and Sinha S. (2017). Improvement of bio-efficacy of bacterial antagonists by using bleaching powder and resistant cultivars to control bacterial wilt of tomato. *Indian J. Hort.* 74(4): 552-556.

Singh, D., Kashyap, A. S. and Kesharwani, A. K. (2020). Biointensive management for soil borne diseases in cucurbitaceous crops. In: *Biointensive Approaches: Application and Effectivenesss in Plant Diseases Management* edited by Khan M. R., A. N. Mukhopadhyay, R. N. Pandy, M. P. Thakur, Dinesh Singh, M. A. Siddiqui, Mohd. Akram, F. A. Mohiddin and Ziaul Haque, Indian Phytopathological society, New Delhi-110 012, pp 357- 380.

Singh D., Shri Dhar and Yadav, D. K. (2010b). Effect of endophytic bacterial antagonists against black rot disease of cauliflower caused by *Xanthomonas campestris* pv. *campestris. Indian Phytopath.* 63 (2): 122 – 126.

Singh, D., Sinha, S., Yadav, D.K., Sharma, J.P., Srivastava, D.K., Lal, H.C., Mondal, K.K. and Jaiswal, R.K. (2010a). Characterization of biovar/races of *Ralstonia solanacearum* the incitant of bacterial wilt in solanaceous crops. Indian Phytopath. 63 (3): 261-265.

Singh, U. B. (1943). Control of fruit diseases in Kumaon. *Indian Farm* 4: 411–412.

Sinha, P. P. (1989). Preliminary studies on bacterial leafspot of cucumber. *Indian Phytopath.* 42: 146- 149.

Sinha, S. K. and Saxena, S. F. (1966). First record of bacterial blight of mulberry in India caused by *Pseudomonas mori* (Boyer et Lambert) Stevens. *Indian Phytopath.*19: 498- 504.

Skodowska, M., Mielczarz, K., Chojak-KoŸniewska, J., Naliwajski, M., ¯yŸniewska, M. and Góralczyk-Biñkowska, A. (2021). Antioxidant response of cucumber leaf tissues treated with prohexadione-Ca to infection with *Pseudomonas syringae* pv. *lachrymans. Scientia Horti.* 289: 110452.

Smith, E. F. (1903). Observations on a hitherto uneejported bacterial disease, the cause of which enters the plant through ordinary stomata. *Science* (n. s.) 17: 456-457.

Smith, E. F. and Townsend, C. O. (1907). A plant tumor of bacterial origin. *Science* 25: 671-673.

Smith, E.F. (1911). Bacteria in relation to plant diseases. *Carnegie Inst. Wash. Publ.* 2: 1-368.

Soylu, S., Baysal, Ö. and Soylu, E. M. (2003). Induction of disease resistance by the plant activator, acibenzolar-S-methyl (ASM), against bacterial canker (*Clavibacter michiganensis* subsp. *michiganensis*) in tomato seedlings. *Plant Sci.* 165(5): 1069-1075.

Spieckermann, A. and Kotthoff, P. (1914): Untersuchungen tiber die Kartoffelpflanze und ihre Krankheiten. *Landwirtschajtliche Jahrbucher, Berlin* 46: 659-732.

Srinivasan, M. C., Thirumalachar, M. J. and Patel, M. K. (1959). Bacterial blight disease of rice. *Cur. Sci.* 28: 469-470.

Srinvaschary Khan A. N. A. and Ravikumar M. R. (1996). Incidence of bacterial blight of mulberry and its control. In: *Annual meeting and symposium on Epidemiology and management of plant diseases*. Nov. 21-22. 1996 at UAS, Dharwad.

Stackebrandt, E., Murray, R.G.E. and Truper, H.G. (1988). Proteobacteria classis nov., a name for the phylogenetic taxon that includes the "purple bacteria and their relatives". *Int. J. Syst. Bacteriol.* 38:321-325.

Stall, R. E. and Giverolo, E. L. (1991). Research relating to the recent out break of citrus canker in Florida. *Annu. Rev. Phytopathol.* 29: 399-420.

Stall, R. E., Jones, J. B. and Minsavage, G. V. (2009). Durability of resistance in tomato and pepper to xanthomonads causing bacterial spot. *Annu. Rev. Phytopathol.* 47: 265-284.

Stapp, C. (1928). Schizomycetes (Spaltpilze oder Bakterien). In: "*Sorauer, Handbuch der Pflanzenkrankheiten*", 5th edition. 2: 1-295. Pau: Parey, Berlin. 758 pp.

Stavrinides, J., McCann, H.C. and Guttman, D.S. (2008). Host–pathogen interplay and the evolution of bacterial effectors. *Cell Microbiol.* 10: 285–292.

Stover, R.H. (1972). Banana, plantain and abaca diseases. Commonwealth Mycological Institute, United Kingdom. 316.

Studholme, D.J., Wicker, E., Abrare, S.M., Aspin, A., Bogdanove, A., Broders, K., Dubrow, Z., Grant, M., Jones, J.B., Karamura, G., Lang, J., Leach, J., Mahuku, G., Nakato, G.V., Coutinho, T., Smith, J. and Bull, C.T. (2020). Transfer of *Xanthomonas campestris* pv. *arecae* and *X. campestris* pv. *musacearum* to *X. vasicola* (Vauterin) as *X. vasicola* pv. *arecae* comb. nov. and *X. vasicola* pv. *musacearum* comb. nov. and description of *X. vasicola* pv. *vasculorum* pv. nov. *Phytopathology,* 110(6): 1153-1160.

Su, L., Qiu, P., Fang, Z., Sun, J., Mo, X., Liu, Y.,. and Shen, Q. (2022). Potassium phosphite enhances the antagonistic capability of *Bacillus amyloliquefaciens* to manage tomato bacterial wilt. *Plant Dis.* 106(2): 654-660.

Sulley, S., Huang, Y., Hind, S. R. and Babadoost, M. (2021). Screening and identification of Cucurbita germplasm resistant to *Xanthomonas cucurbitae*, incitant of cucurbit bacterial spot. *Plant Pathol.* 70(9): 2188-2196.

Tanil, A., Miyajima, K. and Akita, T. (1976). The sheath brown rot disease of rice plant and its causal bacterium, *Pseudomonas fuscovaginae* A. Tanii, K. Miyajima et T. Akita sp. nov. *Japanese J. Phytopathol.* 42(5): 540-548.

Tarr, S.A.J. (1962). Diseases of sorghum, sudan grass and broom corn. The Commonwealth Mycological Institute, Kew, Surrey, 298 pp.

Taylor, J.D., Teverson, D. M., Allen, D. J. and Pastor-Corrales, M. A. (1996). Identification and origin of races of *Pseudomonas syringae* pv. *phaseolicola* from Africa and other bean growing areas. *Plant Pathol.* 45:469–478.

Teviotdale, B. L., Davis, R. M., Guerard, J. P. and Harper, D. H. (1989). Effect of irrigation management on sour skin of onion. *Plant Dis.* 73(10): 819-822.

Thangavelu, R. and Gopi, M. (2015). Field suppression of Fusarium wilt disease in banana by the combined application of native endophytic and rhizospheric bacterial isolate possessing multiple function. *Phytopathol. Medit.* 54 (2): 241-252.

Thaxton, P.M., Brooks, T.D. and El-Zik, K.M. (2001). Race identification and severity of bacterial blight from natural infestations across the cotton belt. In: *Proc. Beltwide Cotton Conf., Anaheim, CA* (pp. 9-13).

Thind, B.S. and Singh, N. (1976). Maize borer [*Chilo partellus* (Swinhoe)] as carrier of *Erwinia carotovora* var. *zeae*, the causal agent of bacterial stalk rot of maize. *Curr. Sci.* 45: 117–18.

Thomson, S. V., Schroth, M. N., Hills, J. and Whitney, E. D. (1973). Bacterial vascular necrosis and rot of sugarbeet. Abstract 754 in Abstracts of Papers, *2nd Int. Cong. Plant Pathol., 5-12 September, Minneapolis, Minnesota.*

Thwaites, R., Eden-Green, S. J. and Black, R. (2000). Diseases caused by bacteria. In: J. D. R. (Ed.), *Diseases of Banana*, Abacá and CABI Publishing, Wallingford, Oxon, UK. Enset. 213–239.

Toribio, A. J., Jurado, M. M., Suárez-Estrella, F., López-González, J. A., Martínez-Gallardo, M. R. and López, M. J. (2021). Application of sonicated extracts of cyanobacteria and microalgae for the mitigation of bacterial canker in tomato seedlings. *J. Appl. Phycol.* 33(6): 3817-3829.

Toth, I. K., van der Wolf, J. M., Saddler, G., Lojkowska, E., Helias, V., Pirhonen, M., Tsror, L. and Elphinstone, J. G. (2011). *Dickeya* species: an emerging problem for potato production. *Euro. J. Plant Pathol.* 60: 385–399.

Trivedi, B. M., Patel, P. N. and Verma, J. P. (1978). Zonate spot of cabbage in India. *Sci. and Cult.* 44 (4): 182-183.

Trueman, C. L. and Goodwin, P. H. (2021). Effects of para-aminobenzoic acid on bacterial speck symptom development and *Pseudomonas syringae* pv. *tomato* populations in tomato leaves. *Euro. J. Plant Pathol.* 160(3): 717-730.

Tsuchiya, K. (2005). Occurrence and epidemic adaptation of new strains of *Ralstonia solanacearum* associated with Zingiberaceae plants under agro-ecosystem in Japan. *Bacterial wilt disease and the Ralstonia solanacearum species complex*, pp.463-469.

Vasudevan P, Kavitha S, Priyadarisini VB, Babujee L, Gnanamanickam SS (2002) Biological control of rice diseases. In: Gnanamanickam (ed) *Biological control of crop diseases*. Marcel Dekker Inc, New York, 11–32.

Vauterin, L., Hoste, B., Kersters, K. and Swings, J. (1995). The relationships within the genus *Xanthomonas* and a proposal for a new classification. *Int. J. Syst. Bacteriol.* 45: 472-489.

Verma, J.P. and Singh, R.P. (1974). Studies on the distribution of races of *Xanthomonas malvacearum* in India. *Indian Phytopath. 28*: 459–63.

Verma, J.P. and Singh, R.P. (1975). Studies on the distribution of races of *Xanthomonas malvacearum* in India. *Indian Phytopath*. 28: 459.

Verma, J P. (1995). Advances in bacterial blight of cotton. *Indian Phytopath.* 48: 1–13.

Vidyasekaran, P., Kamala, N., Ramanathan, A. and Rajapan, K. (2001). Induction of systemic resistance by *Pseudomonas fluorescens* PF1 against *Xanthomonas oryzae* pv. *oryzae* in rice leaves. *Phytoparasitica*. 29 (2): 155–166.

Vinay, J.U. and Nargund, V. B. (2019). Green synthesis of water soluble chitosan based iron nanoparticles and their antibacterial activity. *J. Farm Sci.* 32: 466–470.

Vinay, J.U., Nargund, V. B., Jahagirdar, S., Patil, R. R. and Hegde R.V. (2018). Green synthesis of Zinc nanoparticles using *Pseudomonas fluorescens* extract and their antibacterial activity against *Xanthomonas* spp. *Int. J. Curr. Microbiol. App. Sci.* 7(10): 1280-1291.

Wang, Y., Cheng, X., Shan, Q., Zhang, Y., Liu, J., Gao, C. and Qiu, J.L. (2014). Simultaneous editing of three homoeoalleles in hexaploid bread wheat confers heritable resistance to powdery mildew. *Nat. Biotechnol.* 9: 947–951.

Webb, R.E. and Goth, R.W. (1965). A seedborne bacterium isolated from watermelon. *Plant Dis. Rep.* 49: 818–821.

Williams, R. J., Frederiksen, R. A. and Girard J. C. (1978). Sorghum and pearl millet disease identification handbook. *Information Bulletin No. 2. ICRISAT, Patancheru 502 324, AP, India*. 88pp.

Worthington, R. J., Richards, J. J. and Melander, C. (2012). Small molecule control of bacterial biofilms. *Org. Biomol. Chem.* 10: 7457–7474.

Wulff, E. G., Mguni, M., Giese, K. M., Fels, J., Lubecle, M. and Hockentill, J. (2002). Biochemical and molecular characterization of *Bacillus anyloliquefaciens, B. subtils* and *B. pumilus* isolates with distinct antagonistic potential against *Xanthomonas campestris* pv. *campestris. Plant Pathol.* 51 (5): 574.

Xu, J., Pan, Z.C., Prior, P., Xu, J.S., Zhang, Z., Zhang, H., Zhang, L.Q., He, L.Y. and Feng, J. (2009). Genetic diversity of *Ralstonia solanacearum* strains from China. *Euro. J. Plant Pathol.* 125(4): 641-653.

Yabuuchi, E., Kosako, Y., Yano, I., Hotta, H. and Nishiuchi, Y. (1995). Transfer of two *Burkholderia* and an *Alcaligenes* species to *Ralstonia* gen. nov.: proposal of *Ralstonia pickettii* (Ralston, Palleroni and Doudoroff 1973) comb. nov., *Ralstonia solanacearum* (Smith 1896) comb.nov. and *Ralstonia eutropha* (Davis 1969) comb. nov. *Microbiol. and Immunol.* 39(11): 897-904.

Yamazaki, A., Li, J., Zeng, Q., Khokhani, D., Hutchins, W.C., Yost, A.C., Biddle, E., Toone, E.J., Chen, X. and Yang, C.H. (2012). Derivatives of plant phenolic compound affect the type III secretion system of *Pseudomonas aeruginosa* via a GacS-GacA two-component signal transduction system. *Antimicrob. Agents Chemother.* 56: 36–43.

Yang, C., Powell, C.A., Duan, Y., Shatters, R. and Zhang, M. (2015). Antimicrobial nanoemulsion formulation with improved penetration of foliar spray through citrus leaf cuticles to control citrus huanglongbing. *PLoS One,* 10: e0133826.

Yang, F., Korban, S.S., Pusey, P.L., Elofsson, M., Sundin, G.W. and Zhao, Y. (2014). Small-molecule inhibitors suppress the expression of both type III secretion and amylovoran biosynthesis genes in Erwinia amylovora. *Mol. Plant Pathol.* 15: 44–57.

Yang, S., Peng, Q., San Francisco, M., Wang, Y., Zeng, Q. and Yang, C.H. (2008). Type III secretion system genes of *Dickeya dadantii* 3937 are induced by plant phenolic acids. *PLoS One*, 3: e2973.

Young, J. M., et al. (1978). A proposed nomenclature and classification for plant pathogenic bacteria. *New Zeal. J. Agr. Res.* 21:153–177.

Young, J. M., et al. (1996). Names of plant pathogenic bacteria 1864-1995. *Rev. Plant Pathol.* 75: 721-763.

Young, J. M., Kuykendall, L. D., Martinez-Romero, E., Kerr, A. and Sawada, H. (2001). A revision of *Rhizobium* Frank 1889, with an emended description of the genus and the inclusion of all species of *Agrobacterium* Conn 1942 and *Allorhizobium undicola* de Lajudie *et al.,* 1998 as new combinations: *Rhizobium radiobacter, R. rhizogenes, R. rubi, R. undicola* and *R. vitis. Int. J. Syst. and Evol. Microbiol.* 51(1): 89-103.

Young, J.M., Saddler, G. S., Takikawa, Y., Boer, S.H.de; Vauterin, L., Gardan, L., Gvozdyak, R.I. and Stead, D. E. (1996). Names of plant pathogenic bacteria 1864-1995. *Rev. Plant Pathol.* 75(9): 721-763.

Zeigler, R.S. and Alvarez, E. (1987). Bacterial sheath brown rot of rice caused by *Pseudomonas fuscovaginae* in Latin America. *Plant Dis.* 71(7): 592-597.

Zeigler, R.S. and Alvarez, E. (1990). Characteristics of *Pseudomonas* spp. causing grain discoloration and sheath rot of rice and associated pseudomonad epiphytes. *Plant Dis.* 74(11): 917-922.

Zeriouh, H., Romero, D., García-Gutiérrez, L., Cazorla, F. M., de Vicente, A. and Pérez-García, A. (2011). The iturin-like lipopeptides are essential components in the biological control arsenal of *Bacillus subtilis* against bacterial diseases of cucurbits. *Mol. Plant-Microbe Interactions*, 24(12): 1540-1552.

Zhang, Y., Kong, J., Huang, F., Xie, Y., Guo, Y., Cheng, Y. and Yao, W. (2018). Hexanal as a QS inhibitor of extracellular enzyme activity of *Erwinia carotovora* and *Pseudomonas fluorescens* and its application in vegetables. *Food Chemistry*, 255: 1-7.

Zhou, J., Peng, Z., Long, J., Sosso, D., Liu, B., Eom, J.S. and White, F. F. (2015). Gene targeting by the TAL effector PthXo2 reveals cryptic resistance gene for bacterial blight of rice. *Plant J.* 82: 632–643.

Index

Acidovorax 37
Agrobacterium 37
Alphaproteobacteria 24
Alternating Current Dielectrophoresis Technology 94
Amaranthus 209
Amplified Fragment Length Polymorphism (Aflp) 90
Angular Leaf spots 185
Angular Leaf Spots and Black Arm 225
Animalia Kingdom 22
Annealing step 81
Antimicrobial Peptides 109
Apple 133
Aptamer-based Diagnosis 93
Archaebacteria 20
Areca Nut 240
Arthrobacter 38
Artificial System 27
Bacteria Embedded in Slime 72
Bacterial Blight 117
Bacterial Blight 143, 175, 181, 199, 206, 211, 224, 239, 250, 252
Bacterial Blight and Spot 249
Bacterial Blight and Wilt 244
Bacterial Blight of Rice 66
Bacterial Blight Symptoms 65
Bacterial Brown Rot 252
Bacterial Brown Spot 223, 252
Bacterial Brown Stripe 216
Bacterial Canker 125, 147, 152, 139
Bacterial Fruit Blotch 190
Bacterial Leaf Blight 145, 180
Bacterial Leaf Necrosis 208, 142, 146, 148, 149, 174, 84, 209, 209, 224, 224, 231, 239242, 243, 245, 248
Bacterial Leaf Spot 253
Bacterial Leaf Streak 214
Bacterial Leaf Stripe 240
Bacterial Neck Rot/Bacterial Scab 248
Bacterial Nutrition 51
Bacterial Ooze Test 75

Bacterial Palea Browning 218
Bacterial Panicle Blight and Grain Rot 217
Bacterial Pustule 182
Bacterial Ring Rot 170
Bacterial Rot 208
Bacterial Soft Rot 172, 200, 236, 251
Bacterial Speck 159
Bacterial Spot 115
Bacterial stalk Rot 221
Bacterial Vascular Necrosis and Rot 204
Bacterial Wilt 154, 161, 188, 234, 246, 251, 254
Bacterial Wilt/Brown Rot of Potato 165
Bacteriophage 59, 76, 106
Bacteriosis 253
Bajra/(Pearl millet) 223
Banana 127
Beans/French Beans 176
Beet Root 203
Beta-proteobacteria 25
Betel vine 239
Binary Fission 55
Binomial Nomenclature 16
Bioassay 75
Biological management 103
Bio-PCR 87
Black Rot 193
Blackleg Disease 168
Bud Rot 250
Buffer solution 81
Burkholderia 38
Candidatus Libeobacter 38, 74
Candidatus Phytopasma 74, 39
Canker 120
Cardamom 239
Carnation 246
Carrot 199
Casein Agar Plus Glucose 35
Cassava/Tapioca 206
Category 1 112
Category 2 112
Category 3 112
Cell Division 56
Cell Envelope – Structure 9
Cellular Analysis and Notification of Antigen Risks and Yields (CANARY) 79
Domain 20
Characterization of Phytopathogenic Bacteria 36
Chemical Control 99
Chemoautotrophs 51
Chemoheterotrophs 51
Chemotaxonomy 31
Chilli 231
Chitosan an Organic Elicitor in Management of Plant Bacterial Diseases 109
Chromosome 12
Chrysanthemum 244
Citrus 120
Citrus Canker 71
Citrus Greening 124
Class: Bacilli 26
Class: Clostridia 26
Class: Mollicutes 26
Classification 16
Clavibacter 39
Cluster Bean 180
Common Ancestor 20
Common Bacterial Blight and Fuscous Blight 176

Common Media 34
Common or Fuscous Blight of Beans 67
Concept of Bacterial Species 15
Conjugation 57
Contents of Taxonomy 15
Conventional Taxonomy 28
Co-operational-PCR 87
Copper and Antibiotics 99
Cotton 225
Cowpea 181
CRISPR/Cas9-Novel 108
Crown Gall 69, 135, 203, 253
Cruciferous Vegetables 192
Cucurbits 184
Culm and Root Disease 215
Cultural Practices for Disease Management 113
Culture Media for the Growth of Bacteria 52
Curd Rot/Soft Rot 198
Curtobacterium 39
Cytoplasm 12
Dahlia 253
Death Phase 55
Deltaproteobacteria 25
Denaturation Step 81
Detection of Bacterial Pathogens 75
Diagnosis 65
Diagnosis by Bacteriophage 75
Difference between Gram Positive and Gram Negative Bacteria 10
Disease Escape 113
Disinfection 113
Divalent cations 81
DNA Base Composition 29
DNA polymerase 80
DNA Replication 56
DNA template 80
DNA-DNA Homology 29
DNA-rRNA Homology 29
Domain "Bacteria" 22
Ecology 61
Eggplant/Brinjal 161
Elephant Foot Yam 209
Elongation step 81
Endospores 12
Enterobactor 40
Eradication 113
Erwinia 40
Eubacteria 20
Eukaryotes 22
Eukaryotic Cells 3
Ewingella 40
Excrescences and Galls 68
Exponential (log) Phase 54
Extension 81
External Environment – Cell Responses 55
External Structures 11
Extracellular Polysaccharides 12
Fastidious Vascular Bacteria 48
Field Diagnosis 65
Field hygiene 113
Fimbria 11
Final elongation 82
Final hold 82
Finger millet 224
Fire Blight 133
Flagella 11
Flagellum 11
Fluorescent In situ Hybridization (FISH) 90

Fluorescent Resonance Energy Transfer (FRET) 85
Forms of Bacteria 48
Foxtail Millet 224
Fungi Kingdom 22
Galls 68
Galls Appear on Vine of Grape 69
Gammaproteobacteria 25
Gene Diagnosis 75
Gene Transfer 30
Genetics 56
Genospecies 16
Geranium 242
Gibbsiella 41
Ginger 234
Gladiolus 248
Gluconobacter 41
Grafting Technique 107
Gram-reaction 10
Grapes 115
Growth Rate and Generation Time 55
Hairy Root 137
Halo Blight 66, 178, 216
Herbaspirillum 41
Histoid galls 69
Historical Events 5
History of Classification 16
Indian Shot 250
Infection 59
Infection Process 63
Initialization step 81
Internal Structure 12
Irrigation 113
Isolation 74
Isolation of Plant Pathogenic Bacteria 34
Isozymes 30
Janthinobacterium 42
Jasmine 243
Karonda 147
King's B Medium 35
Lady's Finger/Bhendi/Okra 174
Lag Phase 54
Leaf Blotch 223
Leaf Sheath and Grain Rot of Rice 217
Leifsonia 42
Life Cycles and Dispersal 62
Lonsdalea 42
Lysis of Bacteria 58
Lytic Cycles 59
Maize 221
Mango 125
Marigold 252
Merits and Demerits of Five Kingdom Classification 18
Methods of Classification 28
Micro/Macroarrays 90
Microbial Antagonist 104
Microscopic 74
Microsphere Immunoassays 92
Minor Diseases of Rice 215
Moisture 55
Moko Disease/Bacterial Wilt 127
Molecular Beacons 86
Molecular Species 15
Molecular Taxonomy 29
Morphology of Bacterial Cell 7
Mulberry 145
Multiplex PCR 88
Nanotechnology 101
Natural Forces 55
Natural system 27

Necrosis and Cankers 71
Nested-PCR 86
Nomenclature and Classification 16
Nomenspecies 15
Norcardia 43
Nucleotides 80
Numerical Taxonomy 27
Nutrient Agar 34
Nutrient Broth Yeast Extract Agar 34
Nutrient Glucose Agar 34
Onion 236
Orchids 251
Organoid galls 69
Outer Membrane 9
Pantoea 43
Paphiopelium 252
PCR/Electrospray Ionization-Mass Spectrometry 91
Pea 175
Peach 139
Peppery Leaf Spot of Cauliflower 192
Periplasm 9
Petunia 251
Phenotype MicroArrays 77
Phospholipid Fatty Acid Analysis 79
Photoautotrophs 51
Photoheterotrophs 51
Phylum: Actinobacteria 24, 27
Phylum: Firmicutes 23, 26
Phylum: Proteobacteria 23, 24
Phylum: Tenericutes 24, 26
Pigeon Pea 228
Pigments 12
Pilus 11
Plant density and nutrition 113
Plant Diagnosis 65
Plantae Kingdom 22
Plasmids 12
Plum 141
Polymerase Chain Reaction 80
Pomegranate 117
Poppy 242
Portable Nanopore Sequencing 95
Potato 163
Primers 80
Procedure of PCR 81
Prokaryotic Cells 3
Prokaryotic Domain 22
Protein Profiles 30
Protista Kingdom 22
Pseudomonas 43
Pyrotaxonomy 31
Quarantine 112
Radish 202
Ralstonia 44
Current Infra Subspecific Classification 156
Rathayibacter 44
Real-Time (Quantitative) – PCR 83
Recombinase Polymerase Amplification 92
Release 59
Repetitive Extragenic Palindromic (Rep) – PCR 90
Reproduction 59
Reproduction and Genetics 55
Reserve Materials 12
Resistant Variety 107
Restriction Fragment Length Polymorphisms (RFLP) 30, 89
Rhizobium 37
Rhizome/Pseudostem Soft Rot 132
Rhizomonas 45

Rhodococus 45
Rice 211
Root Rot 202
Rotation 113
Rotting 72
Scab 163
Seedling Blight 215
Serological taxonomy 31
Sheath Rot 217
Silvering Disease 205
Slippery Skin 237
Soft Rot 175
Soft Rot 233
Sorghum/(Jowar) 220
Sour Skin Rot 237
Soybean 182
Spiroplasma 45
Stationary Phase. 54
Stem and Capsule Rot 242
Stem and Soft Rot 187
Streptomyces 46
Survival and Dispersal 63
SX Agar 35
Symptoms and Signs 65
  Bacterial Ring Rot on Potato 70
  Bacterial Wilt of Tomato 70
  Common Scab 68
  Fastidious Bacteria 73
Symptoms of Plant Diseases caused by bacteria 65
System of Classification 27
  Fatty Acid Indicators 80
Tamarind 148
TaqMan Probe 84
Targeting Bacterial Biofilms 105
Targeting Quorum Sensing 110
Targeting the Type Three Secretion System 111
Taro/Coco yam/*Colacasia esculenta* 208
Taxospecies 15
Temperate Phage and Lysogeny 60
Temperature 55
The Archaea 20
The Bacteria 20
The Cell Wall 9
The Eukarya 21
Three Domain System of Life 19
Tip Over and Soft Rot Disease 131
Tomato 149
Transduction 58
Transformation 57
Transgenic Approaches 110
Tuberose 252
Tumours 69
Tundu Disease of Wheat or Yellow Ear 218
Tween Medium 36
Types of Culture Media 52
Types of Culture Media 53
Tyrosine Agar 35
Vascular Disease and Wilting 69
Walnut 143
Wheat 218
Xanthomonas 46
Xylella fastidiosa 74
Xylophilus 47
Yam 208
Yeast Dextrose Calcium Carbonate Agar 34
Zinnia 253
Zonate Spot 197

COLOUR PLATES

Figure 1. Symptoms of Bacterial Blight of Rice (p. 66)

Figure 2. Halo Blight of Beans (p. 67)

(a)

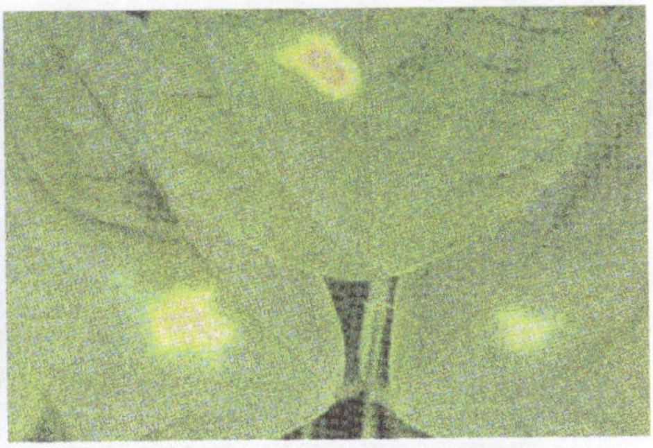

(b)

Figure 3. Symptoms of Common Blight of Beans Caused by X. phaseoli pv. phaseoli (a) and Fuscous Blight of Beans caused by X. phaseoli var. fuscans (b) (p. 67)

(a)

(b)

Figure 4. Bacterial Blight Symptoms on Leaves (a) and Fruit (b). (p. 68)

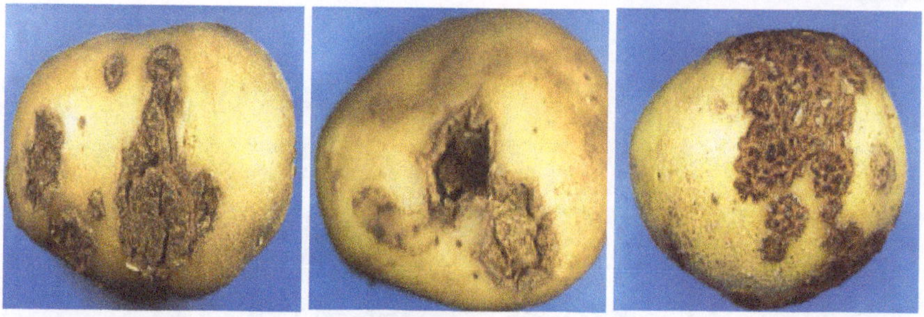

(a) Superficial Scab (b) Deep – Pitted Scab (c) Netted/Russet Scab

Figure 5. Symptoms of Common Scab of Potato Caused by *Streptomyces scabies. (p. 68)*

(a) Crown Gall Symptoms b) Galls Appear on Vine of Grape

Figure 6. Symptoms of Tumour Caused by *Rhizobium radiobacter* (*Agrobacterium tumefaciens)* (a) and *R. vitis* (b) (p. 69)

Figure 7. Symptoms of Bacterial Wilt of Tomato under Field Condition. (p. 70)

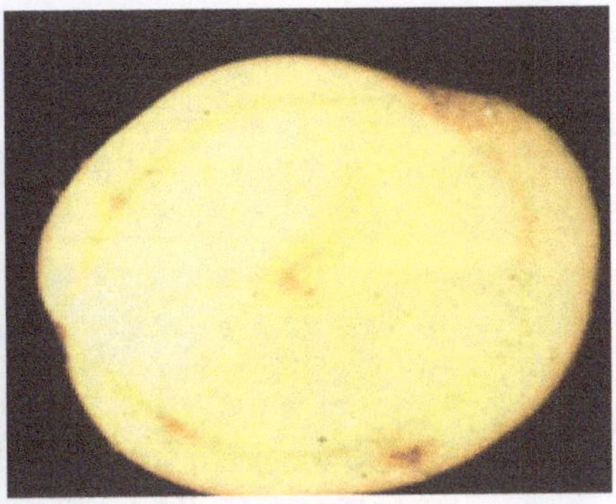

(a) (b)

Figure 8. Symptoms of Bacterial Ring Rot on Potato Plant, (a) advance stages of infection, rolling of leaf margins, mottling and necrotic tissue, (b) Vascular ring appear only as a broken, sporadically, appearing dark line or a continuous yellowing discolouration. (p. 70)

(a) Marginal Browning of Leaves **(b) Fruit Spots Caused by Bacterial Canker**

Figure 9. Bacterial Canker Symptoms Appear on Leaves (a) and Fruit Spot (b) on Tomato. (p. 71)

Figure 10. Symptoms of Black Rot showing V-shaped and Blackening of Veins Symptoms on Cabbage Leaf. (p. 71)

Figure 11. Canker Symptom on Citrus. (p. 72)

Figure 12. Soft Rot of Carrot. (p. 72)

Figure 1. Symptoms of Bacterial Spot on the Leaves and Canker on the crop vine. (p. 116)

a. Foliar Infection **b. Disease Severity on the Branches**

(p. 117)

c. Stages of Infection in the Fruits

d. The Cracking of the Fruits due to Bacterial Infection

e. Devastated orchard due to bacterial blight infection

Figure 2. Bacterial Blight Disease Symptoms on Foliar, Branches, and Fruits. (p. 118)

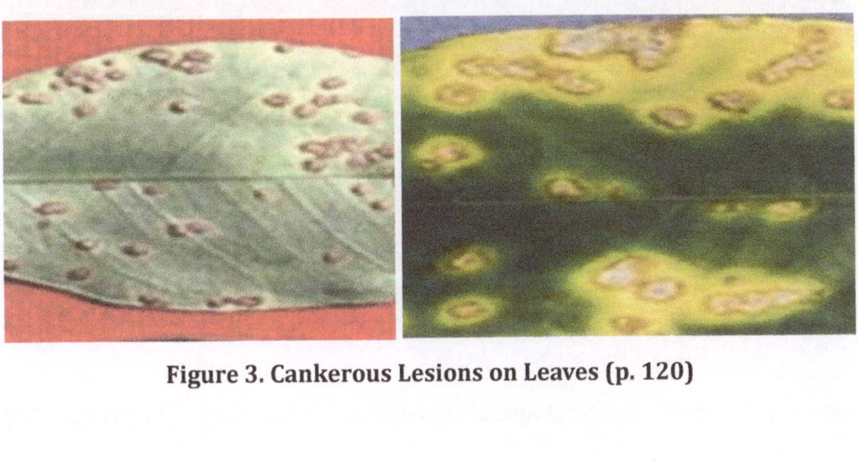

Figure 3. Cankerous Lesions on Leaves (p. 120)

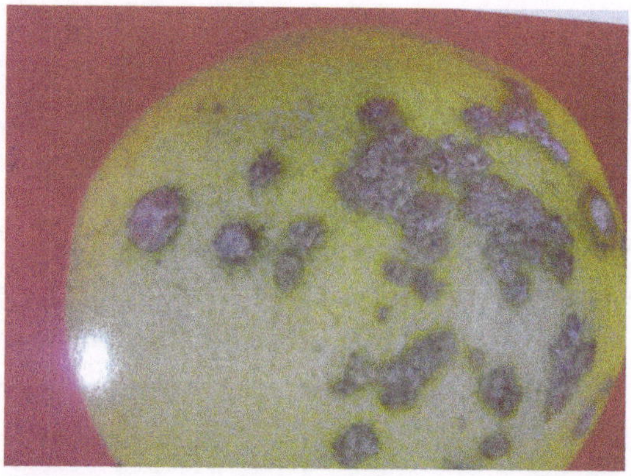

Figure 4. Cankerous Lesions on Fruits (p. 121)

Fig. 5. Bacterial soft rot disease symptoms (p. 132)

(a) **(b)**

Figure 1. Bacterial Leaf Spot Symptoms on Leaves (a) and Fruits (b). (p. 150)

(a) (b)

Figure 2. Black Rot Disease Symptoms on Cabbage (a) and Cauliflower (b). (p. 195)

Leaf Spot **Leaf Strife** **Budrot**

Figure 1. Disease Symptom on Leaf as Spot, Stripe and Rotting on Bud. (p. 240)

9 789359 190945

Printed at Repro India Ltd.